Analyzing Microarray
Gene Expression Data

Analyzing Microarray Gene Expression Data

Geoffrey J. McLachlan
The University of Queensland
Department of Mathematics and
Institute for Molecular Bioscience
St. Lucia, Brisbane
Queensland, Australia

Kim-Anh Do
University of Texas
M. D. Anderson Cancer Center
Department of Biostatistics and
Applied Mathematics
Houston, Texas

Christophe Ambroise
U.M.R. C.N.R.S. Heudiasyc
Université de Technologie
de Compiègne
Compiègne, France

A JOHN WILEY & SONS, INC., PUBLICATION

Published by John Wiley & Sons, Inc., Hoboken, New Jersey.
Published simultaneously in Canada.

For general information on our other products and services please contact our Customer Care Department within the U.S. at 877-762-2974, outside the U.S. at 317-572-3993 or fax 317-572-4002.

Wiley also publishes its books in a variety of electronic formats. Some content that appears in print, however, may not be available in electronic format.

Library of Congress Cataloging-in-Publication Data Is Available

ISBN 0-471-22616-5

Printed in the United States of America.

10 9 8 7 6 5 4 3 2 1

To

Beryl, Jonathan, and Robbie

Brad and Alex

Martine, Manon, Lison, and Liou

Contents

Preface   xv

1 Microarrays in Gene Expression Studies   1
1.1   Introduction   1
1.2   Background Biology   2
  1.2.1   Genome, Genotype, and Gene Expression   2
  1.2.2   Of Wild-Types and Other Alleles   3
  1.2.3   Aspects of Underlying Biology and Physiochemistry   4
1.3   Polymerase Chain Reaction   5
1.4   cDNA   6
  1.4.1   Expressed Sequence Tag   6
1.5   Microarray Technology and Application   7
  1.5.1   History of Microarray Development   8
  1.5.2   Tools of Microarray Technology   10
  1.5.3   Limitations of Microarray Technology   18
  1.5.4   Oligonucleotides versus cDNA Arrays   20
  1.5.5   SAGE: Another Method for Detecting and Measuring
       Gene Expression Levels   23
  1.5.6   Emerging Technologies   24
1.6   Sampling of Relevant Research Entities and Public
    Resources   24

2 Cleaning and Normalization   31
2.1   Introduction   31
2.2   Cleaning Procedures   32
   2.2.1   Image Processing to Extract Information   32
   2.2.2   Missing Value Estimation   36
   2.2.3   Sources of Nonlinearity   38
2.3   Normalization and Plotting Procedures for Oligonucleotide
      Arrays   38
   2.3.1   Global Approaches for Oligonucleotide Array Data   38
   2.3.2   Spiked Standard Approaches   39
   2.3.3   Geometric Mean and Linear Regression Normalization
       for Multiple Arrays   41
   2.3.4   Nonlinear Normalization for Multiple Arrays Using
       Smooth Curves   42
2.4   Normalization Methods for cDNA Microarray Data   44
   2.4.1   Single-Array Normalization   46
   2.4.2   Multiple Slides Normalization   48
   2.4.3   ANOVA and Related Methods for Normalization   49
   2.4.4   Mixed-Model Method for Normalization   50
   2.4.5   SNOMAD   51
2.5   Transformations and Replication   52
   2.5.1   Importance of Replication   52
   2.5.2   Transformations   53
2.6   Analysis of the Alon Data Set   56
2.7   Comparison of Normalization Strategies and Discussion   56

3 Some Cluster Analysis Methods   61
3.1   Introduction   61
3.2   Reduction in the Dimension of the Feature Space   62
3.3   Cluster Analysis   63
3.4   Some Hierarchical Agglomerative Techniques   64
3.5 k-Means Clustering   68
3.6   Cluster Analysis with No *A Priori* Metric   69
3.7   Clustering via Finite Mixture Models   69
   3.7.1   Definition   69
   3.7.2   Advantages of Model-Based Clustering   71
3.8   Fitting Mixture Models Via the EM Algorithm   72
   3.8.1   E-Step   73
   3.8.2   M-Step   74

3.8.3   Choice of Starting Values for the EM Algorithm   75

3.9   Clustering Via Normal Mixtures   75

3.9.1   Heteroscedastic Components   75

3.9.2   Homoscedastic Components   76

3.9.3   Spherical Components   76

3.9.4   Choice of Root   77

3.9.5   Available Software   77

3.10   Mixtures of t Distributions   78

3.11   Mixtures of Factor Analyzers   78

3.12   Choice of Clustering Solution   80

3.13   Classification ML Approach   81

3.14   Mixture Models for Clinical and Microarray Data   82

3.14.1   Unconditional Approach   83

3.14.2   Conditional Approach   84

3.15   Choice of the Number of Components in a Mixture Model   84

3.15.1   Order of a Mixture Model   84

3.15.2   Approaches for Assessing Mixture Order   84

3.15.3   Bayesian Information Criterion   85

3.15.4   Integrated Classification Likelihood Criterion   85

3.16   Resampling Approach   86

3.17   Other Resampling Approaches for Number of Clusters   87

3.17.1   The Gap Statistic   87

3.17.2   The Clest Method for the Number of Clusters   88

3.18   Simulation Results for Two Resampling Approaches   88

3.19   Principal Component Analysis   91

3.19.1   Introduction   91

3.19.2   Singular Value Decomposition   93

3.19.3   Some Other Multivariate Exploratory Methods   94

3.20   Canonical Variate Analysis   94

3.20.1   Linear Projections with Group Structure   94

3.20.2   Canonical Variates   95

3.21   Partial Least Squares   97

4   Clustering of Tissue Samples   99

4.1   Introduction   99

4.2   Notation   100

4.3   Two Clustering Problems   101

4.4   Principal Component Analysis   102

4.5   The EMMIX-GENE Clustering Procedure   103
4.6   Step 1: Screening of Genes   104
4.7   Step 2: Clustering of Genes: Formation of Metagenes   105
4.8   Step 3: Clustering of Tissues   107
4.9   EMMIX-GENE Software   108
4.10   Example: Clustering of Alon Data   108
   4.10.1   Clustering on Basis of 446 Genes   108
   4.10.2   Clustering on Basis of Gene Groups   109
   4.10.3   Clustering on Basis of Metagenes   112
4.11   Example: Clustering of van 't Veer Data   112
   4.11.1   Screening and Clustering of Genes   113
   4.11.2   Usefulness of the Selected Genes   115
   4.11.3   Clustering of Tissues   121
   4.11.4   Use of Underlying Signatures with Clinical Data   123
4.12   Choosing the Number of Clusters in Microarray Data   124
   4.12.1   Some Previous Attempts   124
4.13   Likelihood Ratio Test Applied to Microarray Data   125
   4.13.1   Golub Data   125
   4.13.2   Alizadeh Data   126
   4.13.3   Bittner Data   127
   4.13.4   van 't Veer Data   127
4.14   Effect of Selection Bias on the Number of Clusters   128
4.15   Clustering on Microarray and Clinical Data   128
4.16   Discussion   130

5   Screening and Clustering of Genes   133
5.1   Detection of Differentially Expressed Genes   133
   5.1.1   Introduction   133
   5.1.2   Fold Change   134
   5.1.3   Multiplicity Problem   134
   5.1.4   Overview of Literature   135
5.2   Test of a Single Hypothesis   137
5.3   Gene Statistics   138
   5.3.1   Calculation of Interactions via ANOVA Models   138
   5.3.2   Two-Sample t-Statistics   139
5.4   Multiple Hypothesis Testing   139
   5.4.1   Outcomes with Multiple Hypotheses   140
   5.4.2   Controlling the FWER   140

5.4.3   False Discovery Rate (FDR)   141
5.4.4   Benjamini-Hochberg Procedure   142
5.4.5   False Nondiscovery Rate (FNR)   143
5.4.6   Positive FDR   143
5.4.7   Positive FNR   143
5.4.8   Linking False Rates with Posterior Probabilities   143
5.5   Null Distribution of Test Statistic   144
5.5.1   Permutation Method   144
5.5.2   Null Replications of the Test Statistic   145
5.5.3   The SAM Method   146
5.5.4   Application of SAM Method to Alon Data   146
5.6   Recent Approaches for Strong Control of the FDR   148
5.6.1   The q-Value   148
5.6.2   Technical Definition of q-Value   149
5.6.3   Controlling FDR Strongly   150
5.6.4   Selecting Genes via the q-Value   151
5.6.5   Application to Hedenfalk Data   152
5.7   Two-Component Mixture Model Framework   154
5.7.1   Definition of Model   154
5.7.2   Bayes Rule   155
5.7.3   Estimated FDR   155
5.7.4   Bayes Risk in terms of Estimated FDR and FNR   156
5.8   Nonparametric Empirical Bayes Approach   158
5.8.1   Method of Efron et al. (2001)   158
5.8.2   Mixture Model Method (MMM)   158
5.8.3   Nonparametric Bayesian Approach   159
5.8.4   Application of Empirical Bayes Methods to Alon
         Data   159
5.9   Parametric Mixture Models for Differential Gene
       Expression   160
5.9.1   Parametric Empirical Bayes Methods   160
5.9.2   Finding Clusters of Differentially Expressed Genes   164
5.9.3   Example: Fitting Normal Mixtures to t-Statistic
         Values   165
5.10   Use of the P-Value as a Summary Statistic   166
5.10.1   Beta Mixture for Distribution of P-Values   168
5.10.2   Example: Fitting Beta Mixtures to P-Values   169
5.11   Clustering of Genes   171
5.12   Finding Correlated Genes   173

5.13   Clustering of Genes via Full Expression Profiles   173
5.14   Clustering of Genes via PCA of Expression Profiles   174
5.15   Clustering of Genes with Repeated Measurements   175
   5.15.1   A Mixture Model for Technical Replicates   175
   5.15.2   Application of EM Algorithm   176
   5.15.3   M-Step   176
5.16   Gene Shaving   177
   5.16.1   Introduction   177
   5.16.2   Methodology and implementation   177
   5.16.3   Optimal cluster size via the Gap statistic   178
   5.16.4   Supervised Gene Shaving   179
   5.16.5   Real Data Example   179
   5.16.6   Computer Software   180

6   Discriminant Analysis   185
6.1   Introduction   185
6.2   Basic Notation   185
6.3   Error Rates   187
6.4   Decision-Theoretic Approach   187
6.5   Training Data   189
6.6   Different Types of Error Rates   190
6.7   Sample-Based Discriminant Rules   191
6.8   Parametric Discriminant Rules   192
6.9   Discrimination via Normal Models   193
   6.9.1   Heteroscedastic Normal Model   193
   6.9.2   Plug-in Sample NQDR   194
   6.9.3   Homoscedastic Normal Model   195
   6.9.4   Optimal Error Rates   197
   6.9.5   Plug-in Sample NLDR   197
   6.9.6   Normal Mixture Model   198
6.10   Fisher's Linear Discriminant Function   199
   6.10.1   Separation Approach   199
   6.10.2   Regression Approach   199
6.11   Logistic Discrimination   201
6.12   Nearest-Centroid Rule   202
6.13   Support Vector Machines   203
   6.13.1   Two Classes   203
   6.13.2   Selection of Feature Variables   204

6.13.3 Multiple Classes   205
6.13.4 Computer Software   206
6.14 Variants of Support Vector Machines   207
6.15 Neural Networks   207
6.16 Nearest-Neighbor Rules   208
6.16.1 Introduction   208
6.16.2 Definition of a k-NN Rule   209
6.17 Classification Trees   210
6.18 Error-Rate Estimation   211
6.18.1 Apparent Error Rate   211
6.18.2 Bias Correction of the Apparent Error Rate   213
6.19 Cross-Validation   213
6.19.1 Leave-One-Out(LOO) Estimator   213
6.19.2 q-Fold Cross-Validation   214
6.20 Error-Rate Estimation via the Bootstrap   214
6.20.1 The 0.632 Estimator   214
6.20.2 Mean Squared Error of the Estimated Error Rate   215
6.21 Selection of Feature Variables   216
6.22 Error-Rate Estimation with Selection Bias   218
6.22.1 Selection Bias   218
6.22.2 External Cross-Validation   218
6.22.3 The 0.632+ Estimator   219

7 Supervised Classification of Tissue Samples   221
7.1 Introduction   221
7.2 Reducing the Dimension of the Feature Space of Genes   222
7.2.1 Principal Components   223
7.2.2 Partial Least Squares   223
7.2.3 Ranking of Genes   223
7.2.4 Grouping of Genes   224
7.3 SVM with Recursive Feature Elimination (RFE)   224
7.4 Selection Bias: SVM with RFE   226
7.5 Selection Bias: Fisher's Rule with Forward Selection   228
7.6 Selection Bias: Noninformative Data   230
7.7 Discussion of Selection Bias   232
7.8 Selection of Marker Genes with SVM   233
7.8.1 Description of van de Vijver Breast Cancer Data   233
7.8.2 Application of SVM with RFE   234

7.9	Nearest-Shrunken Centroids	236
7.9.1	Definition	236
7.10	Comparison of Nearest-Shrunken Centroids with SVM	239
7.10.1	Alon Data	239
7.10.2	van de Vijver Data	239
7.11	Selection Bias Working with the Top 70 Genes	245
7.11.1	Bias in Error Rates	245
7.11.2	Bias in Comparative Studies of Error Rates	246
7.11.3	Bias in Plots	248
7.12	Discriminant Rules Via Initial Grouping of Genes	249
7.12.1	Supervised Version of EMMIX-GENE	249
7.12.2	Bayesian Tree Classification	249
7.12.3	Tree Harvesting	249
7.12.4	Block PCA	250
7.12.5	Grouping of Genes via Supervised Procedures	250
8	Linking Microarray Data with Survival Analysis	253
8.1	Introduction	253
8.2	Four Lung Cancer Data Sets	254
8.3	Statistical Analysis of Two Data Sets	255
8.4	Ontario Data set	256
8.4.1	Cluster Analysis	256
8.4.2	Survival Analysis	259
8.4.3	Discriminant Analysis	260
8.5	Stanford Data Set	261
8.5.1	Cluster Analysis of AC Tumors	262
8.5.2	Survival Analysis	263
8.5.3	Discriminant Analysis	266
8.6	Discussion	266
References		267
Author Index		297
Subject Index		313

Preface

In recent times, there has been an explosion in the development of comprehensive, high-throughput methods for molecular biology experimentation. An example is the advent in DNA microarray technologies, such as cDNA arrays and oligonucleotide arrays, that provide means for measuring tens of thousands of genes simultaneously. These technologies benefit biological research greatly and further our understanding of biological processes by drawing together researchers in biology and quantitative fields including statistics, mathematics, computer science, and physics. In addition to the enormous scientific potential of microarrays to help in understanding gene regulation and interactions, microarrays have very important applications in pharmaceutical and clinical research.

This book has been written with two types of readers in mind: biologists who will undertake the statistical analyses of their own experimental microarray data, and biostatisticians entering the field of microarray gene expression data analysis. The primary focus of the book is on data analysis methods for this field; however, the biology and technology behind gene expression microarray experiments, as well as cleaning and normalization of the data, will be briefly covered.

Although biological experiments vary considerably in their design, the data generated by microarray experiments can be viewed as a matrix of expression levels, organized by genes versus tissue samples. In the case where a tissue sample corresponds to a single microarray experiments, we can represent the output from M experiments in the form of a $N \times M$ array (matrix). Each column of the matrix (the expression signature vector) contains the expression levels on the N genes monitored in the microarrays, while each row (the expression profile) contains the expression levels of a gene as it varies over the M tissue samples. Outside this matrix of expression levels, we may have covariate information for samples, genes, or both. The goal of microarray data analysis is to make inferences among samples, genes, and their expression levels and covariates.

The actual measurement of the expression levels raises several statistical issues in experimental design, image processing, outlier detection, transformations, and nonlinear modeling. We consider some of these issues (which are still ongoing as we complete this book) in the first two chapters. The rest of the book then considers the analysis of the microarray data, assuming that they have been appropriately preprocessed.

This analysis is centered on methods for the detection of differential expression, for cluster analysis (unsupervised classification), and for discriminant analysis (supervised classification) of microarray data.

An important and common question in microarray experiments is the detection of genes that are differentially expressed in tissue samples across a number of specified classes. These classes may correspond to tissues (cells) that are at different stages in some process, in distinct pathological states, or under different experimental conditions. A plethora of methods to detect differential gene expression are presented.

Cluster analyses have demonstrated their utility in the elucidation of unknown gene function, the validation of gene discoveries, and the interpretation of biological processes. Discriminant analysis is playing an ever-increasing role in predicting gene function classes and cancer classification.

There are two distinct clustering problems with microarray data. One problem concerns the clustering of the tissues on the basis of the genes. The clusters of tissues can play a useful role in the discovery and understanding of new subclasses of diseases. The second problem concerns the clustering of the genes on the basis of the tissues. The clusters of genes obtained can be used to search for genetic pathways or groups of genes that might be regulated together. Also, in the first problem above, we may wish first to summarize the information in the very large number of genes by clustering them into groups, which can be represented by some metagenes. We can then carry out the clustering of the tissues in terms of these metagenes.

In both the clustering of the tissues and the genes, hierarchical (agglomerative) clustering has been the most widely used method for the analysis of patterns of gene expression. It produces a representation of the data with the shape of a binary tree, in which the most similar patterns are clustered in a hierarchy of nested subsets. Nevertheless, classical hierarchical clustering presents drawbacks when dealing with data containing a non-negligible amount of noise, as is the present case. Also, there is no reason why the clusters of tissues or genes should belong to a hierarchy such as in the evolution of species. In this book, the emphasis is on a model-based approach to clustering. An advantage of model-based clustering is that it provides a sound mathematical framework for clustering. In particular, it provides a principled statistical approach to the practical questions that arise in applying clustering methods, namely, the question of what metric (distance function) to adopt and the question of how many clusters there are in the data.

In recent times, model-based clustering has become very popular in the statistical literature. Unfortunately, as the data to be analyzed from microarray experiments often have gene-to-sample ratios of approximately 100-fold, off-the-shelf parametric methodology does not apply at least to the classification of the tissues on the basis of the genes. This is because the dimension of the feature space (the number of genes) is so much greater than the number of observations (the number of tissues). But even the cluster analysis of the genes on the basis of the tissues is a nonstandard problem, as the genes are not all independently distributed.

An obvious way to handle the very large number of genes is to perform a principal component analysis (PCA) and carry out the cluster analysis on the basis of the leading components. But a potential problem with a PCA is the determination of an appropriate number of principal components (PCs) useful for clustering. A common practice is to choose the first few leading components. But it is not clear where to stop and whether some of these components are caused by some artifact or noises in the data unrelated to the clustering task. Also, there is the difficulty of interpretation of components because each component has loadings generally on all genes.

Hence the focus in the book is on the EMMIX-GENE procedure, which is a normal mixture-based method of clustering that has been especially developed for the clustering of tissue samples or other high-dimensional data. This procedure has an option for an initial selection of the genes where genes that appear to have little clustering capacity are discarded. It then clusters the (standardized) gene profiles into groups, effectively using Euclidean distance as the metric, with the aim that highly correlated genes are put in the same cluster. Each group of genes is then represented by a single metagene (the group-sample mean) and then clustering is performed in terms of the metagenes. This divide-and-conquer approach is becoming popular in the bioinformatics literature for both unsupervised and supervised classification of tissue samples.

The clustering step of EMMIX-GENE makes use of, if needed, mixtures of factor analyzers. That is, it provides a global nonlinear approach to dimension reduction as it postulates a finite mixture of linear submodels (factor models) for the distribution of the full signature vector or a reduced version of metagenes given the (unobservable) factors. Thus, it is a local dimensionality reduction method in contrast with a PCA, which is a global linear method.

A number of discriminant rules are discussed for the supervised classification of the tissue samples. However, the aim of this book is not to provide a comprehensive review of available methods but rather to focus on what we think are useful methods for the analysis of microarray data. To this end, the focus in discriminant analysis of tissue samples is on the support vector machine. It has the advantage that it can be formed from all the genes and its performance is generally not too disadvantaged as a consequence of using all the genes. Its performance can be improved by undertaking feature selection using an easily implemented procedure called recursive feature elimination.

In the statistical analyses, including discriminant and cluster analyses, some form of feature selection will usually be carried out. A consequence of basing the final analysis on a selected "top" subset of the available genes is that there will typically be a selection bias that needs to be corrected for in relating the conclusions to subsequent (new) data. In the case of a discriminant rule, it means that the selection bias has to be allowed for in the estimation of the generalization error. Otherwise, a false overoptimistic impression will be obtained for the discriminatory power of the rule.

This bias has often been overlooked in the bioinformatics literature. Also, this bias arises in an unsupervised context with tests and plots on the number of clusters.

The first two chapters of this book aim to (1) provide a bridge to the biological and technical aspects involved in microarray experiments, and (2) summarize and emphasize the need for basic research in DNA array technologies and statistical thinking through every step of the microarray experiment and analysis to enhance reliability and reproducibility of research results.

Chapter 1 is an introductory chapter and provides a review on DNA microarrays and relevant technology. In particular, we begin with the biological principles behind microarray experiments. Background information on the substrates and technology used in microarray gene expression studies is intended for the biostatistician who is not familiar with the biological experiments. We discuss DNA, cDNA, oligonucleotides, and the development of microarray technology, as well as the steps involved in the manufacture of cDNA microarrays and in generating experimental microarray data. Commercial arrays, primarily the GeneChip®, are also briefly introduced in this chapter.

Chapter 2 discusses cleaning and normalization of gene expression microarray data, as well as the need for designs of experiments with replicated data.

Chapter 3 considers in a general context some methods for the cluster analysis of multivariate data consisting of n independent observations taken on a p-dimensional feature vector associated with the random phenomenon of interest. The focus is on model-based methods of clustering and it covers the use of mixtures of factor analyzers for high-dimensional data such as microarray data. In relation to the problem of how many clusters there are in the data, consideration is given to the problem of assessing the number of components in a mixture model by resampling.

Chapter 4 considers the development of the model-based methodology covered in Chapter 3 for its application to problems in the clustering of tissue samples. The emphasis is on the EMMIX-GENE procedure which has been developed specifically for the clustering of tissue samples. Its application to real microarray data sets is illustrated on two well-known sets in the literature. Also, it is demonstrated on several real data sets how this model-based approach to clustering can be used to consider the question of how many clusters of tissues there are in the data.

Chapter 5 focuses on the selection of differentially expressed genes in known classes of tissue samples. As this problem concerns the selection of significant genes from a large pool of candidate genes, it needs to be carried out within the framework of multiple hypothesis testing. The recent and fruitful literature on the latter topic in the context of microarrays is covered in depth. Distributional problems, including use of the t-distribution and its variants to provide robustness are introduced with a discussion of numerous methods, frequentist and Bayesian, to handle the multiplicity issue. The latter part of this chapter considers the clustering of genes that have been identified as being differentially expressed with a view to finding: (1) groups of genes that are significantly correlated with each other; (2) groups of genes that share similar expressions across the tissues.

Chapter 6 considers methods in discriminant analysis or supervised classification in a general context with a view to their application to microarray data. Discriminant rules covered include the traditional normal-based linear and quadratic discriminant classifiers, more flexible parametric rules based on normal mixtures or mixtures of factor analyzers, support vector machines and their variants, nearest-neighbor and nearest centroid rules, classification trees, and neural networks. The problem of error-rate estimation of a discriminant rule is considered too, along with ways for the provision of standard errors for the estimates of the error rates.

Chapter 7 considers applications of some of the discriminant rules introduced in the previous chapter to the supervised classification of tissue samples. In applications concerned with the diagnosis of cancer, one class may correspond to cancer and the other to benign tumors. In applications concerned with patient survival following treatment for cancer, one class may correspond to the good prognosis group and the other to the poor prognosis group. Also, there is interest in the identification of "marker" genes that characterize the different tissue classes. Attention is focused on applications of the support vector machine and nearest-shrunken centroids, which is a recent version of nearest centroids to handle the very large number of genes. These two approaches are demonstrated on some cancer data sets. Particular attention is paid to the need to correct for the selection bias in estimating the prediction capacity of a discriminant rule formed from a subset of genes selected from a much larger set.

Chapter 8 is concerned with linking results of a model-based clustering of tumor tissues on cancer biology and clinical outcome. Cancer patients with the same stage of disease can have markedly different treatment responses and clinical outcome. Thus there is much interest in whether microarray expression data can be used to provide prognostic information beyond that provided by stage and other traditional clinical criteria. We report some recent results that show that the clustering provides significant prognostic information on the outcome of the disease beyond that available in current systems based on histopathology criteria and extent of disease at presentation.

The authors wish to acknowledge several Houston-based colleagues. LeeAnn Chastain contributed significant efforts to Chapter 1 with respect to the literature search, writing, editing, and proofreading. We are grateful to Wei Zhang, Keith Baggerly, Kevin Coombes, Li Zhang, and Tuyet-Trinh Do for reading and commenting of this chapter. Peter Müller was a great collaborator on the nonparametric Bayesian mixture model for differential gene expression. Sijin Wen assisted the second author in programming the gene shaving method. Bradley Broom contributed significantly with advanced programming using his high-performance computing tool.

Concerning acknowledgments to colleagues in Brisbane, the authors wish to thank Nazim Khan, Abdollah Khodkar, Katrina Monico, and Justin Zhu for their assistance. They wish to acknowledge the significant contributions made by Angus Ng and Liat Ben-Tovim Jones to the research leading to the results reported in Chapter 8. Further thanks are due to Liat for her many very helpful comments and suggestions on drafts of the manuscript. Finally, special thanks are due to Richard Bean who has greatly assisted with all aspects of the book, including proof reading and overseeing the numerous technical issues that arose during the preparation of the manuscript in camera-ready form.

The first author was supported by the Australian Research Council. The second author was partially supported by University of Texas SPORE in Prostate Cancer grant

CA90270, and the Early Detection Research Network grant CA99007. Thanks are due too to the authors and owners of copyright material for permission to reproduce tables and figures, and to Joel Tyndall (Institute for Molecular Bioscience) who used the InsightII Modeling Environment (Accelrys, 2004) to prepare the art work for the cover of the book.

Brisbane, Australia	Geoff McLachlan
Houston, USA	Kim-Anh Do
Compiègne, France	Christophe Ambroise

1

Microarrays in Gene Expression Studies

1.1 INTRODUCTION

Recently, the scientific world has witnessed an explosion in the development of comprehensive, high-throughput methods for molecular biology experimentation. Potentially, these cutting-edge techniques will allow researchers to characterize genetic diseases such as cancer at the molecular level, and will lead to new treatments directed at specific cellular aberrations. The focus in this book is on the output from array technologies, which have made it straightforward to monitor simultaneously the expression pattern of thousands of genes. We are concerned with how to analyze such massive data sets.

In this chapter, we provide background information on the substrates and technology used in microarray gene expression studies. It is intended for biostatisticians who are not familiar with the biological experiments that produce their microarray data. We discuss DNA, cDNA, oligonucleotides, and the development of microarray technology as well as the steps involved in the manufacture of cDNA microarrays and in generating experimental microarray data. Commercial arrays, primarily the GeneChip®, are also introduced briefly in this chapter. In Chapter 2 we discuss cleaning and normalization of gene expression microarray data and their effects on methods for detecting differential expression. Subsequent chapters of the book are devoted to statistical analyses of the data taken to be cleaned and normalized.

1.2 BACKGROUND BIOLOGY

1.2.1 Genome, Genotype, and Gene Expression

The human genome is a representation of our entire gene complement. The human genome map, completed in April 2003, represents the identification and prediction of the base-pair sequences along each of the 23 pairs of chromosomes present in the human cell nucleus. Even as researchers celebrated the completion of the map ahead of its scheduled date, the human genome was not (and is not) a known entity. In addition to chromosomal areas that still prove difficult to map, a multitude of unknown variations in the genome complicate the identification of individual gene complements. In fact, scientists use a different term when speaking about one person's complete gene complement: *genotype*. Each person's genotype may be unique because there are untold numbers of genetic sequence variations in the form of mutations and polymorphisms.

A related research endeavor, the International HapMap project, started in October 2002, will identify and describe the patterns of variations in DNA sequences that are common among humans. This research involves identifying the sites in the human genome where persons differ by a single base (known as a single nucleotide polymorphism, or SNP), and identifying sets of associated SNPs, known as haplotypes. The ultimate goal of this project is to produce a database of the common haplotypes in the human genome and the SNPs that can be used as tags for each of the haplotypes. (See http://hapmap.org for additional information.)

The initial mapping of the human genome has provided a common foundation to which researchers in the field of genetics and in the many overlapping fields of molecular biology, biochemistry and biophysics, biostatistics, pharmacogenetics, bioinformatics, computer science, and many others will contribute from this point forward.

The development of computational models and methods for the investigation of gene expression patterns has already led to important biostatistical research projects, and its importance will continue to grow because of the increasing specialization of biomedicine. The greatest biomedical gains are being realized through knowledge of a specific subtype of a disease or disorder, the specific biochemical pathways affected by the disease and by the therapy prescribed, and the myriad characteristics of the individual patient's genotype and phenotype[1] that result in his or her very unique biological response to the disease or disorder and to the therapy that is prescribed.

What is to be gained from the measurement of gene expression patterns? Experiments are designed to observe the changes in a gene in response to external stimuli and/or to the activation or expression of other genes, allowing the observation and measurement of the relative expression of a gene. Cell samples are exposed experimentally to human hormones, toxins, pharmacologic agents, and so on, and the resulting increase or decrease in the transcription (expression) of a particular DNA

---

[1]A phenotype comprises all the physical, biochemical, and physiological characteristics of a person as determined through genetic and environmental influences. A phenotype is also the manifestation or expression of a gene or gene pair in human characteristics.

segment or gene can be measured. This information will be used to elucidate the potential pathways of genes as well as the interrelations among various genes. It will be applied to the development of pharmacologic agents and genetic therapies with a level of target specificity that is well beyond that of our current ability to analyze disorders, implement preventive measures, or prescribe medical treatments appropriately. Scientists are learning that complex disorders result from the interactions of many genes and are identifying the components of the interactions.

1.2.2 Of Wild-Types and Other Alleles

Researchers in the human genome project have determined that chromosome 20 is made up of approximately 60 million bases and contains 727 genes (Hattori and Taylor, 2001). It is believed that a human being inherits 30,000 to 40,000 genes from each parent. What is a gene? A *gene* is a specific segment of a DNA molecule that contains all the coding information necessary to instruct a cell to synthesize a specific product, such as an RNA molecule or a protein. Contained within the gene are segments that we acknowledge as active in the coding process *(exons)*, as well as segments that are noncoding *(introns)*. Each gene also represents a basic unit of a person's biological inheritance from his or her two parents. Genes can be "mapped" because each occupies a specific location (or locus) on a chromosome, and each chromosome can be specifically identified as well.

Genes are identified according to their apparent general or specific function. It is believed that *housekeeping genes* (for example, GAPDH, B-actin, tubulin) are expressed or functional in all cells because they encode proteins that are needed for basic cellular activity. Additional examples of gene types that have been identified include the immunoglobulin genes, which code to direct the synthesis of specific types of immunoglobulins (antibodies); and a tumor suppressor gene (antioncogene), which functions to limit the formation and growth of malignant cells. By definition, human genes function to promote and regulate biological activity that is considered necessary and productive for the functioning of the organism. It is not correct to state that a gene codes for a disease or predisposes a person to a specific disorder. Rather, it is a deleterious mutation in a gene that may predispose a person to a specific disease or disorder.

A variation or any alternative form of a gene that is found to occupy the same locus on a particular chromosome is known as an *allele*. A *wild-type allele* is the form of a particular gene that is thought to have developed through the evolutionary processes that exist in nature (called "wild" because it is a product of nature itself). A gene that is found to have a mutation will be labeled as a specific allele of that gene, which is different from the wild-type allele and from other alleles that identify other types of mutations occurring at that same chromosomal locus.

1.2.3 Aspects of Underlying Biology and Physiochemistry

Deoxyribonucleic acid (DNA) is contained within chromosomes in the nucleus of each cell. The DNA molecule consists of two anti-parallel strands of sugar–phosphate linkages that are bonded together in a right-handed double helix by the noncovalent hydrogen bonding between pairs of attached amino bases, which lie in a flat plane roughly perpendicular to the long axis of the molecule. The anti-parallel arrangement of the nucleotide chains requires the transcription of a new RNA or DNA chain to run in the opposite direction of the template. Hydrophobic interactions between the stacked bases in the interior of the DNA molecule also stabilize the double helix by packing it tightly to exclude water and other nonpolar molecules. Adenine, thymine, guanine, and cytosine are the amine bases, the sequential order of which contributes to the functioning of a particular segment of the DNA strand (a gene). The bases exhibit a characteristic and specific bonding known as *base pairing*. Base pairing (also known as *Watson–Crick base pairing*) is a chemical bonding process that allows molecular hybridization to occur. Between two strands of DNA, the base known as adenine (A) specifically bonds with thymine (T) through two hydrogen bonds, and guanine (G) specifically bonds to cytosine (C) through two hydrogen bonds, in a manner that creates the double helix. Between a strand of DNA and a strand of ribonucleic acid or RNA (during transcription), adenine from the DNA strand will bond specifically to the base uracil (U) from the RNA strand, and guanine will again bond specifically to cytosine. The amine base that will form a bonding pair with another amine base (A with T or A with U, and G with C) is considered to be its complementary base, and a single strand of DNA or RNA that contains the same sequential order of complementary bases for bonding as a given strand is considered to be its complementary strand. Single DNA or RNA strands will form stable bonds only with a complementary strand. This specificity of bonding allows the "message" of the sequence of base pairing in that segment of DNA to be communicated through the process of transcription.

Transcription is the communication of a genetic code from DNA to RNA through the synthesis of a strand of RNA that has sequences of bases complementary to that of the DNA strand. Genetic transcription is carried out to direct the activity of the cell. The sequence of the bases in a DNA segment comprises the code or genetic instructions that are passed on from the DNA molecule to the RNA molecule because of the specific pairing that occurs between the bases in DNA and RNA. Nucleic acids that guide the production of proteins [2] are transcribed in the nucleus of the cell as messenger RNA (mRNA). Microarray technology utilizes these properties of specific bonding or *hybridization* of a single strand of DNA to a complementary strand of DNA or RNA. The hydrogen bonding between the bases is relatively weak and can be broken by heating the DNA or RNA sample to its melting temperature (approximately 90 °C) through a process referred to as *denaturing*. The single denatured strands of the polynucleotide can then be attached to a solid substrate or used to probe strands

---

[2]The process of synthesizing polypeptide chains from mRNA is known as *translation*, wherein the sequence of bases in the mRNA strand determines the amino acid sequence in the protein that is produced.

of unknown coding order in experiments. Once the denatured DNA is slowly cooled to approximately 60°C, *reassociation* occurs. Reassociation is the process whereby single strands of the polynucleotide associate with complementary strands through random collisions, resulting in the formation of specific amine base pairs through hydrogen bonding. Reassociation is facilitated if the DNA sample is fragmented into short lengths of nucleotides, thus increasing the number of random collisions and increasing the probability that complementary chains will undergo base pairing.

1.3 POLYMERASE CHAIN REACTION

Polymerase chain reaction (PCR) is a technique that "amplifies" or replicates DNA fragments. It is commonly used to create billions of copies of specific fragments of DNA from a single DNA molecule. This technique has numerous applications in medical research, in forensic science, and in many related fields and is used to produce DNA for the manufacture of microarrays. The PCR technique was developed in 1983 through the work of Kary B. Mullis, a biochemist, and his colleagues at Cetus Corporation in Emeryville, California (Mullis, 1990). [F. Hoffman–LaRoche Ltd. and Roche Molecular Systems, Inc., purchased the patent for the PCR technique from Cetus Corporation; however, its recognition as an acceptable patent, since it is based on a naturally occurring enzyme, is currently under dispute in United States appeals courts. European courts upheld the patent in a ruling issued in 2003. See Dalton (2001) and Knight (2003).]

The PCR technique is based on the catalytic action of a DNA polymerase enzyme that is stable at high temperatures, such as those used to denature DNA and RNA molecules. The initial technique utilized a DNA polymerase enzyme isolated from the genetically engineered bacterium *Thermus aquaticus (Taq)*, which was found in thermal springs of Yellowstone National Park in Wyoming. Use of a polymerase enzyme from bacteria with characteristics similar to the *Taq* bacteria enables the DNA replications to be conducted at high temperatures for fast reaction rates and can be rigorously controlled for high fidelity. In human cell division, a primer (a short RNA segment that functions to start the copying of the DNA strands) starts the creation of a template of each single strand of DNA in each chromosome as the base pairing bonds separate. The polymerase then takes over, creating the DNA templates that reproduce the genetic material in the creation of a new cell. For the PCR technique, a *Taq* polymerase from the bacterium is provided, along with the primers and a supply of the four nucleotide bases (adenine, guanine, cytosine, and thymine). The DNA to be duplicated is then added to a vial containing these components. The vial is heated to 90°C for 30 seconds to denature the DNA, separating the strands. The vial is then slowly cooled to 60°C to allow the primers to bind to the DNA strands, and it is again heated to promote the action of the *Taq* polymerase. The entire process, duplicating each piece of DNA in the vial, takes less than 2 minutes. The cycle is then repeated for the same vial approximately 30 times, with each new DNA segment acting as a new template, exponentially reproducing the number of DNA segments in the vial (Mullis, 1990). Recombinant *Taq* polymerase, obtained by the insertion of the gene for the *Taq* polymerase into another type of bacteria (and currently held under a second patent by F. Hoffman–LaRoche Ltd.), is now more commonly used for DNA amplification (Dalton, 2001).

Following the PCR process, the DNA samples are purified to reduce the presence of unwanted components as well as salts and primers used in the PCR process. Purification is done by precipitation, gel-filtration chromatography, or both (Duggan et al., 1999). PCR products representing specific genes are then applied to the array to manufacture DNA microarrays.

1.4 CDNA

Messenger RNA (mRNA) is the form of ribonucleic acid that directs the production of cellular proteins, so it is important in experiments of gene expression. Researchers want to observe what cellular proteins are produced and the function of those proteins in particular types of cells (such as tumor cells) or in response to specific external stimuli, so they are interested in testing the expression patterns of the mRNA. Although protein synthesis and activation are not regulated solely at mRNA levels in a cell, mRNA measurement is used to estimate cellular changes in response to external signals or environmental changes. The mRNA in a biological sample is first chemically bound to a DNA molecule in order to remove it from the other cellular components. The molecule of mRNA is relatively fragile, however, and can easily be broken down by the action of enzymes that are prevalent in biological solutions, so researchers commonly manipulate a form of DNA that possesses the complementary bases of the mRNA while existing in a more stable state. This form of DNA, known as *complementary DNA* (cDNA), is created directly from the sample mRNA through a procedure known as *reverse transcription* (transcribing complementary genetic base sequences from RNA to DNA). cDNA is also called *synthetic DNA*, since it is formed through reverse transcription from RNA rather than through self-replication during cell division. cDNA is generally prepared in strand lengths of 500 to 5,000 bases of known sequence.

1.4.1 Expressed Sequence Tag

Human genes contain base-pairing sequences that are replicated, as well as sequences that are not replicated, during mRNA translation to form specific polypeptide chains in protein synthesis. The sequences that are translated in protein synthesis are coding sequences, known as *exons*, while the noncoding sequences are known as *introns*. Enzymes activated during mRNA transcription recognize the noncoding junctions in the nucleotide sequence and splice together the exons for protein production after removing the introns. *Expressed sequence tag* (EST) is the name given to a short sequential segment from a gene. It is generated to represent the coding portion of a gene; thus, an EST is frequently used as a gene substitute for PCR amplification, microarray production, and experiments. Substituting shorter nucleotide sequences for genomic DNA was proposed in the 1980s and was first undertaken in experiments on cDNA clones derived from human brain tissue by a research group at the National Institute of Neurological Disorders and Stroke, National Institutes of Health in the United States (Adams and Bischof, 1994). ESTs are generated through transcription cloning from both ends of a cDNA sequence, through what is called incomplete *unedited single-pass sequencing reads* of cDNA, resulting in frequent errors (Marra et al., 1998). EST data can be used in general evaluations of gene expression but are not considered suitable for gene expression studies that require greater detail. ESTs have been shown to be valuable in facilitating gene identification and in genome mapping, and EST data comprise the bulk of most public DNA sequence databases (Gerhold and Caskey, 1996; Marra et al. 1998; Quackenbush, 2001; Wolfsberg and Landsman, 2001). The criticism of ESTs in gene libraries has been due primarily to an overabundant representation in the data of genes that are frequently expressed, resulting in redundancies, and an absence of representation of genes that are rarely expressed. Researchers generally try to correct for the presence of redundant EST data in a gene library.

1.5   MICROARRAY TECHNOLOGY AND APPLICATION

High-density DNA microarray technology allows researchers to monitor the interactions among thousands of gene transcripts in an organism on a single experimental medium, which is often a glass microscope slide or nylon membrane. Prior to the computerization and miniaturization of this technology, researchers were limited to examinations of much smaller numbers of genetic units per experiment and were able to assess interactions among genes under changing conditions on a much smaller scale. Microarray technology is particularly useful in the evaluation of gene expression patterns in complex disorders because of its ability to observe the expression of the same genes in different samples at the same time and in response to the same stimuli.

The use of microarrays in biomedical research is equivalent to some of the technological advancements found in the computer science industry, such as that of parallel distribution. Distributing the "work" of an experiment in a parallel fashion facilitates solving computationally complex problems and becomes more than the equivalent of running thousands of experimental steps at the same time. Microarrays are generally designed to provide parallel distribution of the work of an experiment. Each microarray can represent thousands of separate biochemical assays performed in a much shorter time period.

Microarrays can be used to evaluate the dynamic expression of genes in response to normal cellular activity (for example, changes in gene transcription, cell division) or in response to external stimuli (for example, a toxic substance, viral infection). The ability to simulate a large variety of cellular conditions and then translate and process the resulting large quantities of data, provides a systematic way to evaluate cellular function and genetic variations and may be particularly important in testing for genetic susceptibility to diseases and disorders as well as genetic susceptibility (or the ability to respond effectively) to specific therapies or interventions.

The biostatistician's concern lies in the statistical methods and computations that are required to appropriately normalize, analyze, and interpret the vast amounts of data obtained from gene expression studies using microarrays. It is important, however, for the biostatistician to develop a basic understanding of the procedures involved in production of the arrays and in the experiments that generate gene expression data. An understanding of how the data will be applied in a biomedical context is also an important factor. A biostatistician's initial task is to consider the appropriate statistical normalization [3] procedures that may need to be performed on data that are generated from microarray experiments. Understanding the components of the microarray experiments and the levels of sample processing are the crucial preliminary requirement that will guide the biostatistician (Nguyen et al., 2002; Kerr, 2003; Simon et al., 2002; Dobbin et al., 2003). Chapter 2 focuses on data normalization techniques and relevant controversies. The present chapter will provide the biostatistician with some basic principles underlying the microarray technology, beginning with a review of some terms and concepts common to studies that use DNA microarrays.

1.5.1 History of Microarray Development

Microarray technology developed through the application of advanced technologies from the fields of biology and physiochemistry to the analysis of ligand assays, particularly those involving immunoassays. Assays are determinations of the amount of a particular substance within a mixture of different substances. For example, assays have been in use for decades to identify blood proteins; to test for chemical exposure; to perform urinalyses; to screen for drugs; to screen for certain congenital mutations (such as α-fetoprotein); to test for blood clotting disorders; to measure antibody titers; and to test for enzymes specific to injury to the heart muscle or liver tissue. Immunoassays help determine the amount of antibodies present in a biological sample that are involved in the very specific antibody–antigen binding that occurs in immunologic response processes. Researchers in this field were among the first to introduce microarray technology. Labeling techniques implemented in immunoassays included fluorescent labeling of either the antibody or the antigen to detect its presence, as well as radioactive labeling and enzyme-linked immunosorbent assay (ELISA).

Immunoassay technology, as developed in the 1950s and 1960s, involved the attachment of antibodies to solid supports and relied on the specificity of target molecules binding to the antibody (Polsky-Cynkin et al. 1985; Ekins, 1998). These same techniques would subsequently be adapted for DNA analysis. Early assays utilized macroarray technology, whereby the samples were applied or "spotted" manually onto a test surface, creating sample spot sizes of 300 μm or more. Once arrays were designed to support "sample spots" of less than 200 μm in diameter; however,

---

[3] Normalization is the process of standardizing the data so that reasonable data comparisons can be made.

the use of specialized robotics and imaging equipment became a requirement. Further development of the technology reduced the spot size to 20 to 25 μm (and even significantly smaller in some of the current technologies), allowing researchers to observe and evaluate changes in much greater numbers of spots (or individual assays) per experiment. Early labeling methods used radioactive labeling of the known sequences. Additional detection methods have involved electric signal transduction or electron transfer reactions between the known and unknown samples. The simplest and most common detection method used in immunoassays, which is still in use in DNA microarray technology, is the direct labeling of biological molecules in the sample solution with fluorescent dyes such as fluorescein isothiocyanate.

The *Southern blot* (named after E. M. Southern, a British biologist) was the first array of genetic material, and it is still commonly used today. The Southern blot is a technique for the transfer of denatured DNA fragments to a nitrocellulose filter for detection by hybridization to radioactively labeled probes (now commonly replaced by nonradioactive materials). It was based on the principle that DNA and RNA strands could be labeled for detection and used to probe other nucleic acid molecules that were attached to a solid surface. After using gel electrophoresis to separate the DNA fragments, E. M. Southern found that single strands of DNA could form strong covalent bonds with a solid surface, such that the strands would not reassociate with each other but would be available to bond with a complementary segment of RNA (Southern, 1975). This array technique used porous surfaces as the solid support for the DNA strands. The advance to using a glass surface for genetic arrays facilitated the application of fluorescent dyes for labeling, greatly decreased the chemical reaction time since the substances did not diffuse into a porous surface, and accommodated miniaturization, as well.

In the 1980s, the group of R. P. Ekins in the Department of Molecular Endocrinology at the University College London were the first to use simple *microspotting* techniques to manufacture arrays for immunoassay studies with high sensitivity (Ekins and Chu, 1999). Although Ekins' work and patents in the construction and use of microarray-based assays were specific to the analysis of antibodies in the field of immunodiagnostics, his research group expected that the technology would have applications for all types of biological binding assays. Researchers at Boehringer Mannheim (Germany; acquired as part of Corange in 1998 by Hoffman–La Roche Ltd., Basel, Switzerland) led the way in mechanizing the construction of the solid supports and methods for microspotting, decreasing the production costs and industrializing the technology. Ekins and Chu (1991) first named the product of this technique *multianalyte microspot immunoassays*. They utilized the ratio of the fluorescent signals to the absorbance to measure the radiometric intensity and also used a dual-channel scanning-laser confocal microscope to provide an image of the fluorescent signals in the experiment.

Numerous groups of researchers have furthered the technology introduced by Ekins, Chu, and their colleagues. In the United States, notable research has been accomplished by Stephen P. A. Fodor and his colleagues at Affymetrix, Inc. (Santa Clara, California) (Fodor et al., 1991), as well as by groups at Stanford University, particularly Patrick O. Brown, in the Department of Biochemistry and Biophysics

(Stanford, California) (Schena et al. 1995), David Botstein, Ph.D. (now director of the Lewis-Sigler Institute for Integrative Genomics at Princeton University), and among groups at the National Center for Genomic Research at the National Institutes of Health. Brown and his colleagues at Stanford are credited with engineering the first DNA microarray chip, while Stephen Fodor and colleagues at Affymetrix, Inc., created the first patented DNA microarray wafer chip, the GeneChip®. Numerous commercial entities and academic groups have since contributed to advancements in DNA microarray technology. A small selection of these are outlined at the end of the chapter.

1.5.2 Tools of Microarray Technology

The following is a simplification of the complicated biochemical processes and detailed protocols involved in the preparation of nucleic acid materials and microarrays and in the conduct of gene expression studies in the biology laboratory. It is provided as an introduction to the technology, and readers are encouraged to consult reference publications and specialists in the field to improve their understanding of the technology and of the experimental processes that create the data they will subsequently analyze.

Array

The array is a solid base on which a grid of "spots" or droplets of genetic material of known sequence is arranged systematically. The array is commonly a small piece of glass or nylon (similar to a microscope slide), with thousands of spots or wells that can each hold a droplet representing a different cDNA sequence. Array sizes commonly vary from that of a microscope slide (2.5cm × 7.5cm) (Schermer, 1999) to square silicon chips of 0.5cm × 0.5cm (Warrington et al., 2000). Every spot in the grid of the array can represent an independent experimental assay for the presence and abundance of a specific sequence of bases in the sample polynucleotide strand.

The selection of material for the array depends on the cost, density, accuracy, and form of polynucleotide to be fixed to the slide (Sinclair, 1999). Glass, silica chips, and charged nylon are common forms of array materials currently in use (Kricka and Fortina, 2001). Coated glass microscope slides possess relatively low inherent fluorescence, which makes them a good choice for the arrays. A form of silicon hydride that is used in the manufacture of semiconductor devices is one of the coatings added to microarray slides. The coating repels water and helps the DNA stick to the surface of the slide while preventing the spread of each spot or deposit of DNA material (Cheung et al. 1999; Duggan et al., 1999; Schermer, 1999).

An important difference between DNA microarrays and the dot blot techniques (such as the Southern blot) is the use of a nonporous rigid test surface such as glass. In dot blotting, polynucleotide products must first diffuse into pores in the test surface before binding to the DNA strands of the probe. This requires more time to allow for diffusion prior to many of the other steps throughout the biochemical processes (Southern et al., 1999). The transparency and rigidity of the glass slide in microarrays and the precise location of each spot on the array also contribute to better image production and can result in better data acquisition (Schena and Davis, 1999).

Spot sizes of cDNA applied to microscope plates were initially at least 100 to 200 μm. Technological developments have reduced the size to 5 μm and even to 2 μm on some commercially prepared arrays, depending on the medium used, thus improving the spatial resolution of the technique (Sinclair, 1999). Microarrays containing thousands of array spots are considered high-density microarrays. They can be created in a variety of grid styles that facilitate the type of experiment to be undertaken.

The cDNA of known sequence that is fixed onto the array after microspotting is commonly referred to as the *probe*, while the polynucleotide of unknown sequence in the biological sample solution is commonly referred to as the *target*. The target will be evaluated indirectly, through its hybridization to the known polynucleotides on the array (Warrington et al., 2000; Kricka and Fortina, 2001; Quackenbush, 2001).

Spotter

The robotic machine that applies the droplets of different cDNA strands of known sequence to a well or spot on the array is called the *spotter* or *arrayer*. The product is applied to each spot in a grid to accommodate a large number of tests within each experiment. The spotter utilizes either contact or noncontact methods to apply the probe material to the array. Contact spotting is performed with an instrument similar to a fountain pen under constant pressure. Relatively new methods of noncontact spotting apply ink-jet technology or the piezoelectric capillary effect to complete the grid of probe droplets. The original spotting technology, which used a pin or needle as a spotter, is still in use in many labs. Noncontact spotting methods generally increase the speed of microarray production (Duggan et al., 1999; Theriault et al., 1999).

Piezoelectric capillary jets are thought to be superior as efficient and accurate spotters; however, the need to keep the capillary nozzle full with a larger volume of the biological sample results in the use of a greater amount of polynucleotide probe material in the manufacture of a microarray than when other types of spotters are used. Because of this characteristic, piezoelectric capillary jet spotters are best suited to microarray production of a large number of spots of the same material. Laboratories use this characteristic to their advantage by creating large numbers of microarray slides at once, and programming the spotter to place a particular polynucleotide probe in specific wells in the grids of each slide before rinsing the solution out of the capillary nozzles and then applying a probe of another kind to different wells in the same microarray slides. Manufacturers of spotters or arrayers offer a variety of solutions to match each laboratory's needs, and many laboratories customize their equipment as they improve their experimental protocols.

Immobilization of cDNA of known sequence onto the array

The cDNA that is spotted onto the array must be fixed onto the surface to maintain the integrity of each spot that is required for high image definition and also to prevent the cDNA from washing off during the various steps involved in processing the array.

Adherence to the slide coating is the first process that immobilizes the DNA on the array, followed by air drying, which is then commonly followed by ultraviolet irradiation for DNA fixation (Cheung et al., 1999; Duggan et al., 1999).

Labeling a sample for detection

To identify and measure the presence of a polynucleotide of unknown sequence (in the target sample) after it binds to the material on the microarray, it is labeled with a fluorescent dye such as fluorescein, rhodamine, or coumarin. The dye is incorporated with the molecule during reverse transcription. A different dye color is used for each sample, and generally only two dye colors are used, due to the accompanying scanning and imaging requirements to detect fluorescent light of specific wavelengths. Experimental protocols commonly call for the application of one fluorescent dye to polynucleotides from the experimental or unknown target sample, and a different fluorescent dye to the polynucleotides from a known or control target sample (Cheung et al., 1999; Schermer, 1999; Brazma et al., 2000). Figure 1.1 illustrates the preparation of target samples.

Hybridization

Molecular hybridization is the association of single strands of polynucleotides through their specific base-pairing properties to form a complementary double-stranded molecule. This is the chemical process that occurs between the labeled polynucleotide strands of target tissues (including those of unknown sequences) and their complementary strands of cDNA of known sequence among the spots on the array. Ideally, if a polynucleotide from the target sample contains a base sequence that is complementary to that of a polynucleotide at one spot of the array, it will hybridize to the molecule at that spot. The location of that spot on the array grid will then be detectable by the fluorescent light that is given off during the scanning and imaging processes. When many target polynucleotides hybridize to complementary cDNA probe strands at one spot on the array then the fluorescent signals emitted and detected at that spot will have greater intensity.

The slide or array is chemically processed to reduce a positive charge on the surface, and heat or a chemical alkali is applied to break some of the double-stranded cDNA probes into single strands for binding with the target sample. There should be a sufficient level of cDNA on each spot of the array so that two unknown sequences of polynucleotides, if complementary to the cDNA sequence at a particular spot, could hybridize to that spot at the same time without introducing interference. Excess solution containing the labeled target polynucleotides is then washed off the array prior to scanning. The hybridization signals that result are dependent on many variables, including temperature, hybridization and processing time, relative humidity, buffers and reagents added, salt concentration, and rinsing of the reagents and excess sample (Cheung et al., 1999; Duggan et al., 1999).

Ongoing improvements in labeling and hybridization technology include the development of a process called *dendrimer signal detection* (Stears et al., 2000). Rather

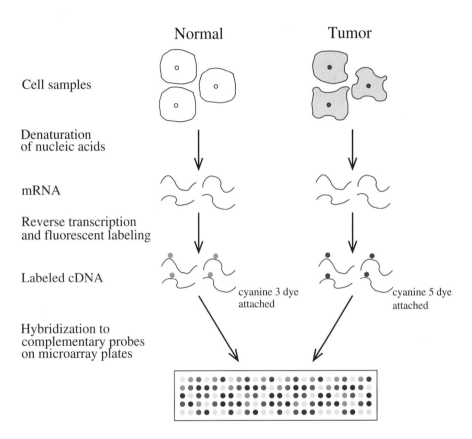

Fig. 1.1 Preparation of target samples: the process from the cell samples to the microarray. See the insert for a color representation.

than incorporating the fluorescent dyes during the reverse transcription process, resulting in modified cDNA transcripts, this process labels dendrimers, which will then specifically bind with complementary sequences along the unmodified cDNA strands. A *dendrimer* is a complex of partially double-stranded oligonucleotides that form a stable, spherical structure with a determined number of free ends. The nucleotide sequence of a cDNA probe on the surface of the microarray will specifically bind to its complementary sequence on a free arm of the dendrimer. The dendrimers, prelabeled with a fluorescent dye, have a predetermined quantified fluorescent intensity. Since each cDNA transcript binds to a single dendrimer, the amount of signal generated is directly proportional to the number of cDNA molecules detected. This labeling method reportedly maintains a low background signal over increasing amounts of RNA, as well as over increasing numbers of scans, facilitating the detection of transcripts normally present in low abundance.

Scanning/imaging the array

Many scanners use a specific frequency of light from a laser (for example, an argon laser) in the ultraviolet region to excite the fluorescent dye attached to the target samples that have hybridized to their complementary probe sequences on the array. The photons emitted by the excited dye are collected at a detector, which measures and records their levels, converting the measurements to electrical signals. A confocal microscope or charge-coupled device that records the intensity of the photons is used as the scanning detector. Since two different fluorescent labels are generally used in gene expression studies, each slide is scanned at two wavelengths, and the imager must be capable of detecting the hybridized polynucleotides and measuring their amounts at least at the two different wavelengths of light, and must possess high-resolution scanning capabilities (Evertsz et al., 2000). Cyanine 3 (Cy3) and cyanine 5 (Cy5) are fluorescent dyes that are commonly used in this technology because their emission spectra are well separated, providing a low probability of crosstalk.

Imagers commonly use a grid over the array to associate the signal from each spot with its location on the array, and thereby with its base-pairing sequence identification. With good filtration of the scattered light, the detector will record only the light from the fluorescently labeled hybridized pairs, and greater intensity of fluorescence will be detected at spots where more polynucleotides have hybridized to the array (Schermer, 1999). The lab technician should be able to adjust the sensitivity of the scanner for each batch of microarrays to correct for variations in dye intensity that may occur during hybridization and array processing. This may involve adjusting the wavelength and output of the excitation laser as well as adjusting the voltage input of the detector, which is commonly a photomultiplier tube. Scanners that automate the sensitivity adjustments are beneficial. The use of a confocal scanner will also decrease the recorded fluorescent noise by controlling the depth of focus used in scanning the true signals of hybridization while reducing the capture of background fluorescence. The ratio of the fluorescent light emissions between the two different wavelengths (corresponding to the two different dyes used to label the unknown and

Fig. 1.2 Microarray image showing differentially expressed genes. Red spots: gene transcripts of high expression in target labeled with cyanine 5 dye. Green spots: gene transcripts of high expression in target labeled with cyanine 3 dye. Yellow spots: gene transcripts with similar expression in both target samples. See the insert for a color representation.

control polynucleotide target samples) is the indirect measurement of the relative gene transcript expression levels.

Final image processing

A digitized scanned array image is obtained from the microarray scanner/imager and is displayed on a monitor. False coloration of the fluorescent intensities, translated on the computer monitor as pixel intensities, is applied to the image to produce a color image for the analyst to read. If the biochemist tagged the polynucleotide from the unknown experimental sample with a red dye and the control polynucleotide sample with a green dye, and the false colorations mimic the fluorescent tagging, then visualization of a red spot on the final array grid indicates that the unknown polynucleotide hybridized abundantly to the cDNA affixed at that location on the microarray slide. A final green spot indicates that the control polynucleotide hybridized abundantly to the cDNA affixed at that location, a yellow spot indicates that the unknown and the control polynucleotides hybridized in relatively equal amounts at that location on the microarray, and a black spot indicates that neither sample of polynucleotides hybridized at that location (Brazma et al., 2000). An alternative image coloration scheme is to apply one false color (such as red) to represent up-regulation (increased transcription) of a gene in the experimental sample on a microarray image, while a second color (such as green) is applied to represent down-regulation of a gene in the final image (Sherlock et al., 2001). The standard image format for microarray images is a 16-bit tagged image file format (TIFF) (Schermer, 1999). Figure 1.2 illustrates a cDNA microarray image.

Verification of microarray data

Laboratory researchers and manufacturers of commercial arrays will use other techniques to confirm the findings in a microarray experiment. Northern blot analysis and reverse transcriptase polymerase chain reaction (RT-PCR) are techniques commonly used to make such verifications. The Northern blot technique analyzes samples of RNA on a nylon membrane but is otherwise analogous to the Southern blot technique (Lockhart and Winzeler, 2000; King and Sinha, 2001).

Carrying out a cDNA microarray experiment in a biology laboratory is a lengthy, complex, multistep process. In brief, it involves the mechanical preparation of solutions of cDNA clones in a 96-well grid pattern on a plate; PCR amplification of the cDNA; verification of the clone sequences by gel-filtration chromatography; purification and assembly of clones; mechanical conversion of the plate format to a configuration of 3,072 assays; verification of the probe sequences; spotting of the microarray slides; hybridization; cleaning, scanning, and imaging the slides; and database analysis and verification of the experimental results using an alternative laboratory method. It is after this complex process that the experimental results can be analyzed statistically. Figure 1.3 presents a flowchart illustrating the production and processing of a cDNA microarray experiment in the Cancer Genomics Core Laboratory of Wei Zhang at the University of Texas M. D. Anderson Cancer Center.

Analysis of data from the array experiment

Analysis of the cDNA microarray data is a new challenge for the biostatistician. It is covered in depth in subsequent chapters of the book, so receives only a brief introduction in this chapter. Microarray laboratory analysis requires the application of a filter to remove gene transcripts from the analysis that do not contribute information to the experimental outcome, such as transcripts that were not measured accurately, and those that do not change across the series of experiments. This step will probably be undertaken by the biology laboratory personnel during the scanning and visualization processes, but its effect on the raw data should be communicated to the biostatistician (Meltzer, 2001). Data analysis by the biostatistician may require the application of a normalization procedure, the simplest method being a linear transformation, to the data from each experiment to correct for variables within the experimental processes (Kellam, 2001; Quackenbush, 2001). Nonlinear normalization will require the application of more sophisticated statistical methods. There are many factors of experimental variability that should be taken into account, such as the amount and purity of the polynucleotides, changes in temperature and relative humidity of the experimental environment, fluorescent labeling efficiency, hybridization results and saturation effects, and increased background fluorescence intensity levels. Normalization frequently requires the use of housekeeping genes or reference mRNA strands (added to a sample at a specific, measurable level) during the experiment (Hollon, 2001; Kellam, 2001; Wu, 2001). The biologists conducting the experiments can provide information about the particular housekeeping genes that are used. Some researchers are proponents of repeating a microarray experiment on replicated samples to assist the biostatistician in correcting for variability across experimental samples (Wu, 2001). Following normalization of the data, a computer visualization analysis is performed to identify similarities or patterns in gene expression profiling. The overall

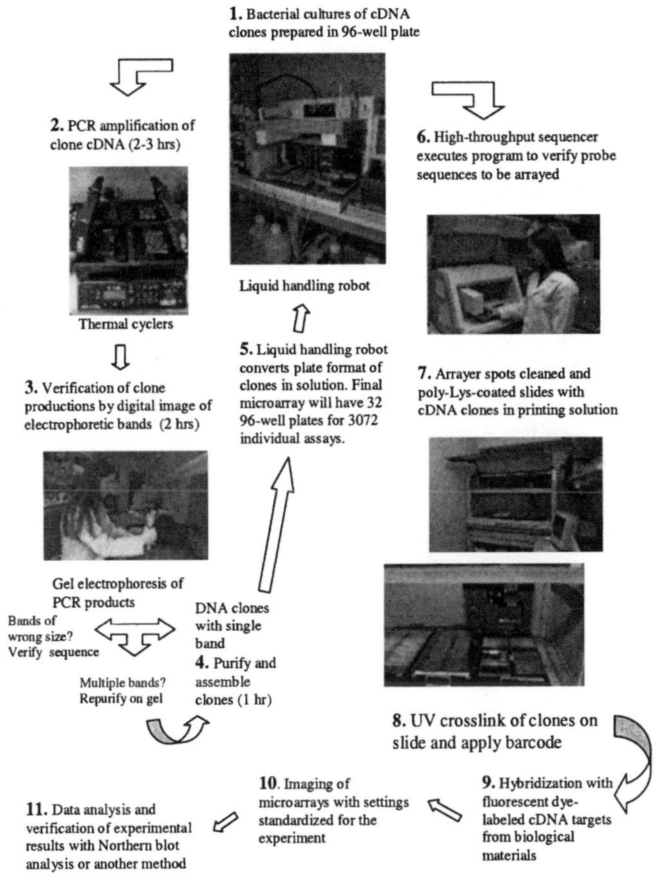

Fig. 1.3 cDNA microarray production and processing flowchart in the Cancer Genomics Core Laboratory at the M. D. Anderson Cancer Center.

1. Bacterial cultures of cDNA clones prepared in 96-well plate

2. PCR amplification of clone cDNA (2-3 hrs)

Thermal cyclers

6. High-throughput sequencer executes program to verify probe sequences to be arrayed

Liquid handling robot

3. Verification of clone productions by digital image of electrophoretic bands (2 hrs)

5. Liquid handling robot converts plate format of clones in solution. Final microarray will have 32 96-well plates for 3072 individual assays.

7. Arrayer spots cleaned and poly-Lys-coated slides with cDNA clones in printing solution

Gel electrophoresis of PCR products

Bands of wrong size? Verify sequence

DNA clones with single band

4. Purify and assemble clones (1 hr)

Multiple bands? Repurify on gel

8. UV crosslink of clones on slide and apply barcode

10. Imaging of microarrays with settings standardized for the experiment

11. Data analysis and verification of experimental results with Northern blot analysis or another method

9. Hybridization with fluorescent dye-labeled cDNA targets from biological materials

results must then be applied to the biological model of interest for its meaningful and appropriate interpretation.

Temporal and end-result studies

Microarray technology in gene expression studies provides the researcher with the choice of taking measurements at one point in time as an end-result study, or at several points in time as a serial study. The researcher can expose the cells to a change and then take successive samples over time. This method allows the researcher to observe the overall changes in gene transcript expression as well as the order of change. It also allows the researcher to identify which gene transcripts are actively guiding the responses, as well as which transcripts change expression as a result of an initial change in a guiding gene's activity.

1.5.3   Limitations of Microarray Technology

There are many limiting factors to the accuracy and application of DNA microarray technology, the most basic of which is that the technology does not measure gene expression levels or mRNA abundance directly. The technology uses an indirect measurement of gene transcript levels through capture of the intensity of fluorescent dyes bound to the polynucleotides that hybridize to the array's experimental probes (Wu, 2001). Many of the limiting factors will be corrected through ongoing and future research and the resulting advancements in the technology. Other limitations should be understood by the biostatistician so that corrections and normalization methods can be adequately performed and so that the data gained from the experiments may be interpreted appropriately.

Limitations concerning DNA and RNA samples

The following list summarizes some of the limitations of microarray experiments with which the biochemists deal directly and which the biostatistician should also keep in mind.

1. The most basic limitation regarding microarray experiments in the laboratory is the availability of clones and/or tissue samples in sufficient quantity.

2. The quality of the RNA and cDNA samples, depending on the purity and con-centrations of the polynucleotides, the storage and maintenance of the samples, the spotting process, and the experimental protocol, will limit the accuracy of the resulting data (Hollon, 2001).

3. Different molecules of mRNA undergo reverse transcription to varying degrees of efficiency, resulting in what is known as *reverse transcription bias*.

4. The fluorescent dyes typically have a greater binding affinity to one type of nucleotide, such as guanine (G); therefore, cDNA strands that contain more

guanine in their sequence will appear brighter upon detection of the microarray's fluorescence. This is known as *sequence bias*.

5. Fluorescence is a nonlinear phenomena; it is linear only over a limited range.

6. Measurements of gene expression using cDNA microarrays currently provide only relative expression levels which gene transcripts are more abundantly expressed in one sample in relation to the same gene transcripts in another tissue sample or in one experiment in comparison to another experiment (Brazma and Vilo, 2000).

7. Gene expression can provide only partial information about activities in a cell. There are many variations in expression, and a gene's product (that is, the protein) may become more or less active because it is being produced at a faster rate, is being degraded by other proteins, or is being chemically modified. Therefore, it is usually not possible to make negative claims saying that some genes are not involved in certain biological processes.

8. DNA/RNA hybridization is very sensitive to temperature and ionic strength in solution, and these characteristics depend on the base sequences in the DNA or RNA strand. No set of experimental conditions is optimal for all genes. Thus, some genes may be nondetectable because the intended hybridization simply does not happen under the experimental conditions chosen.

9. The measurements of microarrays are averages of the expression of many cell types over a period of time. Thus, the technology has a limited space-time resolution to detect transient molecular events in certain types of cells.

Limitations concerning fixed DNA on the array

Preparation of the microarray and hybridization are not perfect processes, and some variability in the results will occur. Examples include cross-linking of the fixed cDNA strands into double-stranded forms that remain even after the thermochemical processes are applied to separate the probe molecules into single strands. This will decrease the number of strands available to hybridize with the target polynucleotides. Additionally, during the drying phase of array preparation, the cDNA molecule may adhere to the glass slide at various places along its strand, also decreasing its ability to hybridize with a complementary target.

Detecting fluorescent light as a means of identifying complementary DNA strands

Light from labeled target molecules that *hybridize* to the glass slide and are not washed from the array will also be detected during scanning procedures, and this will form background light as *noise*. The development and use of array materials with higher signal-to-noise ratios will greatly improve the capability of scanning methods to detect specifically fluorescence that represents complementary hybridization over the fluorescent background. Image processing is a problematic area. Insufficient concentration or labeling of the target polynucleotides, insufficient exposure time, or too little cDNA on a spot of the array to capture adequate signal during hybridization can occur (Cheung et al., 1999; Wu, 2001). Additional misreading of the results during scanning and visualization processes may occur when foreign organic particles are present on the array, such as dust, clothing fibers, skin, or oil from direct human contact; and when the solid support material of the array has an inherent fluorescence similar to that of the labeling dyes (Schermer, 1999). The choice of method for spot segmentation when reading the scanned microarray image has also been shown to cause variability in the resulting microarray data (Ahmed et al., 2004).

Limitations to microarray data storage, retrieval, and shared communication and analysis

DNA microarray data are obtained on any number of different tissue samples under diverse experimental conditions. It is the experimental variability that makes microarray data much more complex than the data created in the human genome sequencing projects. Efforts to standardize microarray data have been under way for some time through work groups spanning international research organizations. Work groups have been actively supportive of developing standards for describing, storing, and using microarray data and of the creation of centralized public repositories for data that can be shared among researchers. Most work groups advocate the deposit of "raw" research data into a central repository before it has undergone normalization techniques, and the development of standard sets of control probes and samples to be used as reference points for common normalization methods (Brazma et al., 2000). Researchers agree that standardization of microarray data will require the adoption of standard descriptors to identify the specific tissue, cell type, and state of pathophysiology at the time a sample is taken, as well as general ontologies to describe the experimental environment and protocol from which data were derived. Ongoing efforts of the work groups of the Microarray Gene Expression Database (MGED), in collaboration with the Object Management Group (OMG), among others, have resulted in models for standardization, including (1) Minimal Information About a Microarray Experiment (MIAME 3.0); (2) Microarray Gene Expression — Object Model (MAGE-OM), and (3) Microarray Gene Expression — Markup Language (MAGE-ML). Additional work groups of the MGED include those for standards in normalization and for ontologies. The work of the MGED is accessible online through the support of the European Bioinformatics Institute at http://www.mged.org/.

1.5.4 Oligonucleotides versus cDNA Arrays

Use of the term *DNA microarrays* usually refers to cDNA arrays, whereas *DNA chips* or *oligo chips* are terms commonly used to refer to oligonucleotide arrays. In this book, however, we use the term *microarrays* in a more general sense to refer to both cDNA and oligonucleotide arrays. Oligonucleotides are shorter sequential base-pair segments, ranging from 15 to 70 nucleotides in length, taken from the hundreds of nucleotides in a DNA segment that function as a gene (Aitman, 2001, Jordan, 2002).

DNA chips use oligos as the probing material on the array. Affymetrix, Inc. was the first company to develop commercially a DNA chip (using the trade name GeneChip®). Oligos are synthesized by standard methods and then spotted onto the chip, or they can be synthesized directly on the chip (in situ or in silico) through a process of photolithography, which is the array production process used by Affymetrix. Glass or polypropylene supports are commonly used for oligo arrays (Elder et al., 1999). Treatment of the support with chemical linkers before spotting or in situ synthesis of the oligo chains promotes adherence of the oligos and improves the hybridization efficiency of the oligo probes (Green et al., 1999); (Southern et al., 1999). Improved methods for adhering probes to the array include UV-irradiation processes that facilitate increased signal intensities (Kimura et al. 2004). Lockhart and colleagues introduced the use of DNA chip technology in gene transcript expression studies with a GeneChip® array made of over 16,000 different oligo probes (Lockhart et al., 1996). Some of the original products developed by Affymetrix included a 1.28cm × 1.28cm chip onto which 450,000 individual oligo probes were synthesized, and provides an array chip with sufficient probes for the evaluation of over 12,000 target "genes" in gene transcript expression profiling studies. Commercial arrays and their accompanying analytical systems are very costly. Some commercial arrays provide protocols for stripping the arrays following scanning and imaging in order to use them in another experiment (Elder et al. 1999).

Researchers will want to consider issues such as specificity and efficiency of hybridization, and accuracy and reproducibility of resulting gene transcript expression levels when assessing the advantages and disadvantages of using oligo arrays versus cDNA arrays. Probe oligos may be more accessible for hybridization than the probe cDNA strands, due to their much shorter chains with single terminal points for attachment to the slide or chip. Another advantage of oligos is their use to detect a subregion of a gene, which is a valuable tool when there is a "family" of a gene with a high similarity of sequence. With a shorter oligo, a region that is different within the family can be detected. Additionally, having uniform lengths for the oligo probes enhances the chance of finding optimal hybridization, and it is easier to engineer. Single strands of cDNA must be spotted onto the array as complete molecules in order to promote fixation to the surface and subsequent accurate hybridization (Sinclair, 1999); however, one spot on this type of array may be sufficient to identify a specific gene transcript. Oligo probes, on the other hand, may undergo cross-hybridization with several genes, requiring the use of the same oligo sequence in many spots on the array in order to identify a specific gene. Affymetrix uses 32 to 40 probes in each probe set of its oligo arrays, and the identification of a gene is made only if positive hybridization can be detected in the majority of the probes in the set. (The automated normalization techniques used in the Affymetrix system are explained further below.)

There are also disadvantages to the use of oligo arrays. The potential for cross-hybridization mentioned previously, due to the use of only about 25 - 70 nucleotides in the oligo strands versus hundreds in cDNA strands, results in a loss of specificity in an oligo array experiment. Additionally, the oligos that are generated directly on the chip do not undergo a purification process – all products and by-products created during the photolithographic synthesis of the oligos may be fixed to the chip (Green et al., 1999; Theriault et al., 1999.) Some commercially manufactured oligo chips have been found to have irregularities in the fluorescent intensities that are detected from the arrays, including bright edges and fluorescing streaks, sometimes caused by the packaging processes; as well as regions known as dark spots where the fluorescent signal is artificially low (Schadt et al. 2000). Of course, artifacts may occur on glass surfaces as well. An additional disadvantage to overcome when using the automated analytical programs that are part of the oligo chip system is that the default parameters in the analytical software are preset estimates based on data from the commercial entity; thus, they may not be updated automatically based on the user's experimental findings and normalization criteria. (Affymetrix responded to this issue by developing "tunable" parameters allowed researchers to prioritize sensitivity versus specificity and to test for these using rigorous statistical methods (Foster and Huber, 2002, and their next generation products include some of these improvements). Commercial software programs include many automatic corrections, such as automatic correction of background fluorescence by detecting and subtracting the background intensity before analyzing the intensities of hybridization signals (Warrington et al., 2000). For a recent comparison of commercial software products, see Liu et al., 2004.

Automated normalization techniques in the oligonucleotide array systems of Affymetrix, Inc.

To improve the signal-to-noise ratios and the specificity of gene transcript identification, Affymetrix incorporates two forms of probe redundancy in its high-density oligonucleotide arrays. Firstly, the arrays are manufactured with multiple oligos of varying sequence that will hybridize to different parts of the same target polynucleotide; and secondly, the arrays are manufactured with probes that form a perfect match (PM) and a mismatch (MM) with the target polynucleotide of interest. Generally, 11 to 20 PM and MM probe pairs are used in each probe set. The perfect match oligo probe will contain a segment of a wild-type allele (creating a perfect complementary match with a segment of the target polynucleotide of interest), while the mismatch oligo probe will be a copy of the PM oligo that has been altered by one base at a central position, usually the thirteenth position (Lipshutz et al., 1999; Schadt et al., 2000; Chudin et al., 2001). The altered base is created by reversing the base pairs at that site – replacing adenine with thymine (A → T) and guanine with cytosine (G → C). Mismatches of base pairs at the center of an oligonucleotide (as opposed to mismatches at the end of the oligo strand) weaken the hybridization bonds enough that the image detector can discern a difference in signal intensity between a mismatch pair and a perfect match pair following hybridization. The mismatch probes therefore serve as controls, helping to discern a true hybridization signal from that produced by nonspecific hybridization. The PM–MM contrast is designed to subtract out cross-hybridization (that is, nonspecific binding to the probes). The assumption is that the nonspecific binding, which by definition does not have a detailed match, should not be different between PMs and MMs. The concept seems to work at the probe set level but not on the level of probe pairs, since about 33% of probe pairs have fewer PMs than MMs.

Affymetrix's GeneChip® software aligns an image grid automatically, matching signals to their region of origin on the array; performs image segmentation; corrects for background noise; normalizes the data; and performs statistical calculations for the presence of a gene transcript and for its differential expression. The presence of a gene transcript is determined on the basis of a consistent pattern of hybridization that will occur between the sample target and the set of oligo probes on the array that are perfect matches and mismatches to the target. Originally, the presence of a gene transcript in the target solution was characterized by the signal intensity of the hybridized target–probe pair, derived from the average difference in fluorescent intensity between the PM and MM probe–target hybrid pairs (Schadt et al., 2000; Warrington et al., 2000); Zarrinkar et al., 2001.) The presence of a specific gene transcript in the target solution was determined by an overall "positive" signal from the probe—target hybridized pairs. Affymetrix has more recently updated their software programs to reflect ongoing statistical research and improved methodologies. According to Affymetrix's automatic analysis, a probe pair is positive when the averaged PM–MM intensity is greater than a calculated difference threshold, and the PM–MM ratio is greater than a set ratio threshold. A probe pair is "negative" when the averaged PM - MM intensity is less than the difference threshold or the PM/MM intensity is less than the ratio threshold. GeneChip® system software sets default values for the difference and ratio thresholds that can be changed by the user (Warrington et al., 2000).

Many researchers applying oligo chip technology have preferred to develop and apply their own statistical algorithms, and have found that the functional relationship between the paired PM and MM probe intensities is not always linear (Schadt et al., 2000; Chudin et al., 2001). Suggestions for improvements in the technology have included the use of higher numbers of replicates, longer hybridization times, examining the PM/MM sums as well as the differences, using quality scores from arrays that have already been analyzed to evaluate the performance of new probes; and developing and applying more sophisticated algorithms (Schadt et al., 2000). These methods are reviewed in detail in Chapter 2.

1.5.5 SAGE: Another Method for Detecting and Measuring Gene Expression Levels

Commonly used methods for the measurement of gene expression levels are divided into two general categories: analog and digital. Analog methods are based on sample hybridization to cDNA clones or oligonucleotides on arrays (cDNA microarrays and chips), while digital methods are based on the generation of sequence tags (EST generation and serial analysis of gene expression) (Audic and Claverie, 1997).

Serial analysis of gene expression (SAGE) involves the sequencing of very short, unique sequence *tags* of nucleotides from a sample, with each tag representing a transcription product (Velculescu et al., 1995). Whereas EST methods have utilized segments that are 100 to 300 nucleotides in length, SAGE methods are based on segments of only 9 to 11 nucleotides that are located precisely within the gene. The abundance of the sequenced tags in a sample is then analyzed to represent the level of the gene transcript expression in the sample. SAGE technique runs a high risk of error when two or more genes share the same tag and when a gene has more than one tag, due to nonspecific assignment of tags or polymorphism. A tag of 9 or 10 base pairs is not a complete representation of a gene's entire transcribed sequence, but it should be enough to identify the gene unambiguously. SAGE does not involve hybridization, and evidence of relative gene expression is not dependent on the sequences affixed to the microarray, as is the case in cDNA microarray technology. Many public databases from large-scale gene expression studies use SAGE data (Audic and Claverie, 1997; Meltzer, 2001; Wolfsberg and Landsman, 2001). However, cDNA microarray technology may still be more appropriate than SAGE for the analysis of large numbers of samples (Meltzer, 2001).

1.5.6 Emerging Technologies

Emerging mRNA/cDNA amplification techniques include the in vitro transcription reaction (IVT), which is thought to be particularly important for the amplification of very limited genetic samples, such as samples of tumor tissue obtained by laser-capture microdissection. The IVT method uses a linear amplification process, as opposed to the exponential amplification that is possible through the PCR technique. The advantage of using the IVT technique is the reduction in a quantitative PCR bias. This is a bias that occurs through application of the PCR technique in which a relative abundance of cDNA clones that are not truly representative of the mRNA levels in the original tissue sample are generated in PCR (Lockhart and Winzeler, 2000; Nallur et al., 2001).

Among the DNA microarray technologies in various levels of development are the DNA microfluidics chip, also known as the lab-on-a-chip, whole cell arrays, electronic chip activation technology (Heller et al., 1999; Ramsey, 1999), and microsphere-based fiber optic microarrays (Epstein et al., 2003). Increased miniaturization, together with sample preparation and hybridization technologies that facilitate the parallel analysis of greater numbers of samples on multiple arrays is leading to the approximation of an entire study of gene transcript expression profiles for a specific pathophysiological state (Zarrinkar et al., 2001, Weeraratna et al., 2004), as well as whole genome studies for various organisms.

1.6 SAMPLING OF RELEVANT RESEARCH ENTITIES AND PUBLIC RESOURCES

Public repositories for array-based gene expression data

International sequence databases include the GenBank®, [National Center for Biotechnology Information (NCBI), National Institutes of Health, Bethesda, Maryland], the DNA DataBank of Japan (DDBJ), at the Center for Information Biology in Mishima, Japan; and the European Molecular Biology Laboratory (EMBL), supported by the European Bioinformatics Institute in Cambridgeshire, UK. Through an agreement known as the International Nucleotide Sequence Database Collaboration, the three organizations conduct daily exchanges of data through the Internet. GenBank contains nucleotide sequences from more than 140,000 organisms. As of August 2003, GenBank® reported that the collaboration held approximately 33.9 billion nucleotide bases from 27.2 million individual sequences (Benson et al. 2004).

Notable contributors to the EST database of the U.S. National Center for Biotechnology (dbEST) and to supporting its public availability include Washington University Genome Sequencing Center (St. Louis, Missouri) through its support by the Howard Hughes Medical Institute; members of the I.M.A.G.E. Consortium (see below); and Merck & Co., Inc. (Whitehouse Station, New Jersey).

The Institute for Genomic Research (TIGR) in Rockville, Maryland, is a not-for-profit research institute with academic partnerships throughout the world. This group focuses on the analysis of genomes and gene products from a wide variety of organisms, including many viruses and bacteria. TIGR researchers have completed the genome sequencing of many pathogens. Published EST sequence data are available through the TIGR Web site, `http://www.tigr.org`.

GeneX Database, supported by the U.S. National Center for Genomic Research (NCGR), and ArrayExpress, supported by the European Bioinformatics Institute (EBI) in the UK, are compliant with current recommendations of standardization (Brazma et al., 2000).

I.M.A.G.E. Consortium (Integrated Molecular Analysis of Genomes and their Expression), brought together four researchers and their colleagues, with financial support from the U.S. Department of Energy, in the creation of an extensive cDNA library. Researchers from the University of Iowa (Iowa City, Iowa), Centre National de la Recherche Scientifique (Villejuif, France), Novartis Corporation (Hagerstown, Maryland), the National Institutes of Health (Bethesda, Maryland), and Lawrence Livermore National Laboratory (Livermore, California) were the original contributors to the I.M.A.G.E. research product, which is Internet accessible at `http://image.llnl.gov/`.

Institutes of the Human Genome Project:

- Whitehead Institute for Biomedical Research in Cambridge, Massachusetts

- The Wellcome Trust Sanger Institute in Hinxton, Cambs, United Kingdom

- Baylor College of Medicine in Houston, Texas

- Washington University in St. Louis, Missouri

- Department of Energy's Joint Genome Institute (JGI) in Walnut Creek, California

The U.S. Department of Energy's Joint Genome Institute (JGI) is a consortium of researchers from the Department of Energy's Lawrence Berkeley, Lawrence Livermore, and Los Alamos National Laboratories. Partner institutions include Oak

Ridge National Laboratory (genome annotation), Brookhaven National Laboratory (molecular biology), Pacific Northwest National Laboratory (proteomics), and Stanford Genome Center (finishing).

Peter Lemkin, a computer scientist at the Laboratory of Experimental and Computational Biology, National Cancer Institute (NCI), developed a program (MicroArray Explorer) for the quantitative analysis of cDNA expression profiles across a group of microarrays (Hollon, 2001).

The Stanford Microarray Database (SMD), which is Internet accessible (`http://genome-www5.stanford.edu/`) is implementing annotations developed by the Array SML working group at Stanford University, Laboratory of Patrick O. Brown, Department of Biochemistry and Biophysics.

Some commercial entities in the United States

Affymetrix, Inc. (Santa Clara, California) uses silica chips for the support structure of an array. Their product requires the use of specialized equipment for scanning and processing data from each study. Affymetrix developed the process of in situ synthesis of oligonucleotides of specific sequences on a silica surface, known as the GeneChip®.

The oligos in the GeneChip® have the characteristics of photospecificity and varying sequences in predetermined locations on an array. The National Institutes of Health established an agreement with Affymetrix, Inc., known as the Academic Access Program, through which Affymetrix offers volume discount pricing of its array products to academic researchers. The company agreed in early 2002 to provide public access to the sequence data of their oligonucleotide probes (Foster and Huber, 2002). New Affymetrix products include the Human Genome U133A array, which all researchers to test gene expression in 96 biological samples at one time, and the GeneChip® Mapping 100K Array Set, a two-microarray set that can be used to genotype over 100,000 SNPs and is the first in a family of commercial products that will facilitate large-scale whole-genome association studies, which had been previously unaffordable or impractical.

Agilent Technologies, Inc. (Palo Alto, California) developed a whole genoma on a single microarray chip in 2003.

Amersham Biosciences (Piscataway, New Jersey) provides products and services for gene and protein research. CodeLink™ prearrayed slides are among the products distributed by Amersham Biosciences.

Applied Biosystems (Foster City, California) developed the first automated DNA sequencer in 1986 that labeled different nucleotide bases with fluorescent dyes, eliminating the need for radioactivity in gene studies.

BioDiscovery (Marina del Rey, California) develops software products for automated microarray work flow.

BioTrove, Inc. (Woburn, Massachusetts) has developed the Living Chip™ system, a process for rapid and parallel nanoliter-scale liquid processing for ultra high-throughput analysis of molecular, biochemical and cellular samples. Its array support consists of 25,000 isolated nanoliter reaction containers in individually stackable plates the size of a standard microplate, and is designed for the analysis of much smaller volumes of sample more quickly.

CLONTECH Laboratories (Palo Alto, California) produces cDNA arrays on nylon membranes and offers some arrays on glass slides. CLONTECH offers two general types of arrays: (1) arrays containing genes that are grouped according to their function in the cell (that is, apoptosis, oncogenesis, normal cellular regulation), and (2) arrays specific to the area of application (for example, immunology, hematology) (De Francesco, 1998).

CombiMatrix Corporation (Burlingame, California) manufactures ArrayChips using programmable in situ synthesis of oligonucleotides on a highly porous membrane overlying a semiconductor chip. Array densities range from 1,000 to 500,000 individual assay sites per square centimeter (Montgomery, 1999).

Compugen (Tel Aviv, Israel) has developed a computer platform (LEADS) that uses advanced proprietary algorithms to create a complete view of the transcriptomes of complex organisms, and to model complex biological phenomena.

Eurogentec(Liège, Belgium) produces DNA microarrays and provides custom oligo synthesis.

Exigon (Vedback, Denmark) produces DNA analogues (LNA) for oligonucleotides, as well as nucleotide embolization products.

Gene Logic, Inc. (Gaithersburg, Maryland) has created GeneExpress databases for specific research applications through an analytical system known as *Restriction Enzyme Analysis of Differentially-expressed Sequences* (READS). They use GeneChip® arrays licensed from Affymetrix, Inc., to generate gene expression data from a variety of tissue samples. Commercial databases that they have created include BioExpress, ToxExpress, and PharmExpress. Gene Logic demonstrated accurate applications of gene transcript expression profiling in hepatotoxicity analysis at a conference on microarray data analysis in November 2001 that was presented by the Cambridge Healthtech Institute (Newton Upper Falls, Massachusetts) (Foster and Huber, 2002).

Genisphere, a developer and manufacturer of array labeling and hybridization products, is a division of Datascope Corporation (Montvale, New Jersey). Genisphere produces dendrimer labeling products.

Genometrix, Inc. (The Woodlands, Texas) manufactures microarrays specifically for the study of gene expression in the mouse as well as for experiments in cancer and toxicity, using a glass or microplate substrate (De Francesco, 1998).

Illumina, Inc. (San Diego, California), in competition with manufacturers of other high-density arrays, has developed the Sentrix® BeadChip and BeadArray technology for whole-genome studies. Each BeadChip contains over ten million features, distributed across a number of discrete array regions. The SNP genotyping BeadChip will contain over 200,000 sequence-specific bead types (each locus requires two allele-specific probe sequences), with greater than 30 times the average redundancy of each bead, or feature.

Invitrogen Life Technologies (Carlsbad, California) is a leading supplier of molecular biology reagents and kits. Invitrogen produced the first complete kit for making cDNA libraries (The Librarian), in 1987. Invitrogen launched a pre-cast electrophoresis gel in 1996 that is the only room-temperature stable gel to offer rapid and accurate protein electrophoresis.

Link Technologies, Ltd. (Lanarkshire, Scotland) manufactures SynBase™ controlled-pore glass supports for arrays.

Matrix Science, Inc. (London, UK) offers proprietary algorithms to be used with its search engines to search mass spectrometry data against sequence databases.

Micralyne (Edmonton, Alberta, Canada), has developed micro-electro-mechanical-systems (MEMS)-based components. Micralyne's MEMS solutions include lab-on-a-chip devices and microfluidics.

Motorola Life Sciences (Pasadena, California) manufactures high-performance microarray tools for genetic analysis. This company developed CodeLink™ prearrayed slides, subsequently purchased by Amersham Biosciences.

MWG Biotech AG (Ebersberg, Germany) has developed HPSF(R) synthesis technology for microarray applications. This technology facilitates rapid production of purified, salt-free oligonucleotides.

Nanogen (San Diego, California) has developed electronic chip activation technology, a process wherein the DNA solution is poured over a microchip and a row, column, or spot on the chip is activated electrically, introducing a positive charge onto the chip's surface. The single strands of negatively charged DNA in solution migrate in microfluidic channels toward the charge, where they undergo chemical bonding to the chip's surface. The chip is then washed off, and another DNA solution is poured over it. After hybridization, the charge is reversed. Electronic activation at the chip surface can also increase hybridization efficiency between the DNA strands of known and unknown sequences.

NextGen Sciences, Ltd. (Cambridgeshire, UK) developed the first bench-top automated 2-D electrophoresis system.

RZPD (Berlin, Germany) provides high-throughput technology and automation solutions. RZPD's portfolio includes clones of genomic and cDNA libraries, expression clones with full open reading frames, siRNA resources, high-density colony, DNA, and protein arrays, custom microarrays, expression profiling, Affymetrix service, high-throughput PCR amplification, and cDNA library generation.

Sequenom Genetic Systems (San Diego, California) has developed the MassARRAY™ system of hardware, software, and reagent products for large-scale, high-throughput DNA analysis.

SIRS-Lab GmbH (Jena, Germany) produces pre-defined and custom biochips and arrays from polynucleotides and from 6,000 sequenced cDNA clones in their databases.

Stratagene (La Jolla, California) develops and manufactures amplification and microarray products.

Among the companies manufacturing microarray spotters are BioRobotics, Inc. (Woburn, Massachusetts); Cartesian Technologies, Inc. (Irvine, California); and Genomic Solutions (Ann Arbor, Michigan).

Tecan Group, Ltd. (Maennedorf, Switzerland) offers a suite of products for gene expression and analysis.

Zeptosen AG (Witterswil, Switzerland) has developed the SensiChip™ Microarray Bar, which consists of six different or identical microfluidic arrays, using planar waveguide technology.

2

Cleaning and Normalization

2.1 INTRODUCTION

It is crucial for any high-throughput technology to have sufficient quality control for each operation or step in the study, especially at the data acquisition level. The technology for microarray studies is still evolving, and many researchers are conducting their studies using different types of customized microarrays, including home-made array chips. In general, the current technology does not consistently generate robust and reliable data when used in the average laboratory. Ideally, reliable microarray data should exhibit the qualities of *accuracy*, assessed by the probable error of a measurement, and *precision*, defined by the reproducibility of a measurement. The images from the hybridized arrays constitute the essential raw data for microarrays, where intensities of the signals are measured using specialized imaging software. The intensities of the signal represent the amount of fluorescent DNA bound to microarrays and is subject to considerable uncertainty because of large- and small-scale intensity fluctuations within spots, nonadditive background, and fabrication artifacts (Brown et al. 2001), contributing to poor-quality images.

Other sources of systematic variation, internal or external to the sample, include fluctuations in the physical properties of the dyes, efficiency of dye incorporation, probe coupling and processing procedures, target and array preparation in the hybridization process, background and overshining effects, and scanner settings, among others. Critical first steps in any analysis of gene expression data include an attempt to clean the data by automatic procedures that can improve image quality, by separating signal from noise, and by handling missing values. Subsequently or simultaneously, the removal of the various sources of variation can also be accomplished by performing appropriate *normalization* procedures. The literature that describes normalization methods for different types of cDNA, oligonucleotide, and other arrays is expansive and constantly evolving. Researchers distinguish DNA microarray data analysis into primarily two types: (1) one-channel DNA data that reflect absolute intensities and are derived from technologies that employ the hybridization of individual labeled cDNA probes to a microarray; versus (2) two-channel DNA data that represent relative intensities or ratio data derived from the simultaneous competitive hybridization of two distinct cDNA probes, each labeled with a different fluorescent dye. Many technologies use background subtraction methods that precede both global and local normalizations. In particular, one-channel radioactivity-based technologies may use a low number of background measurements to generate a single background intensity that can be subtracted from all array element signal intensities. In contrast, two-channel fluorescence-based technologies may subtract the local measurement of background intensity from each array element individually.

In this chapter we focus mainly on describing standard cleaning processes and on summarizing common normalization methods for oligonucleotide arrays and cDNA arrays. The different normalization procedures may be described as global or local (with intensity and location dependence), linear or nonlinear, and by the characteristic of whether they are applicable to single or multiple array analyses. The importance of replication and limitations of the different normalization methods is also addressed.

2.2 CLEANING PROCEDURES

Gene spots are often composed of characteristic imperfections such as irregular contours, donut shapes, artifacts, and low or heterogeneous expression. The simplest initial cleaning attempt is to perform background correction. It is often assumed that the signal observed is a combination of the true signal (from the specific hybridization of interest) and the background signal (due to nonspecific hybridization and/or contamination). The standard approach is simply to subtract the background estimate directly from the spot intensity. However, the background signal may increase due to dust, fibers, fingerprints, autofluorescence of the coated glass, hybridization problems resulting from dehydration near the edge of coverslips, or residual effects from inadequate washing (Hess et al. 2001). Quality assurance is required in the initial step of extracting numerical foreground and background intensities (that is, the image processing step). Further, physical contaminations can cause missing values in the extracted image; thus, appropriate methods to handle missing values are also considered.

2.2.1 Image Processing to Extract Information

Many image analysis methods have been adapted to deal with the specific problems of microarrays. Two issues of great importance in obtaining good data are determining the background signal and reducing the impact of poor-quality spots on the data set.

Target Patch
⟶

Target Mask

Target Site

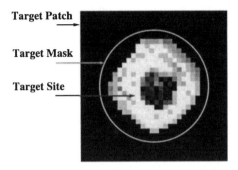

Fig. 2.1 Target patch, mask, and site.

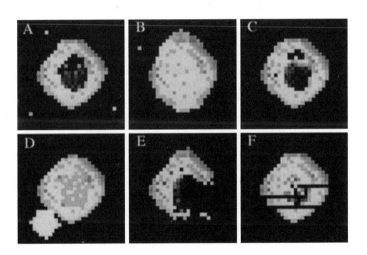

Fig. 2.2 Examples of spot imperfections. A. donut shape; B. oval or pear shape; C. holey heterogeneous interior; D. high-intensity artifact; E. sickle shape; F. scratches.

Good reviews of existing methods can be found in Bozinov and Rahnenführer (2002) and Smyth et al. (2002). For cDNA arrays, the image processing phase consists of three steps: (1) The *addressing* step identifies the target areas or the combined area of a spot and its background; (2) the *segmentation* step partitions the target area into foreground and background areas; and (3) the *reduction* step extracts the red (R) and green (G) intensities and assigns the ratio R/G to represent the relative abundance of each spot. The image needs to be segmented into target patches corresponding to predetermined cDNA targets positioned by the robot. Depending on how the robot finger places the cDNA on the slide or how the slide is treated, the target site may exhibit between-image variability and even between-target variability. Ideally, every spot on a microarray has the shape of a circle, and all spots should have consistent diameters. A diagrammatic representation of target patch, mask, and site is given in Figure 2.1. Problems arise when the observed spots present with variable diameters, variable contours (sickle shape, donut shape, oval or pear shape, scratched or interrupted shape), black holes inside spots, high background and/or low foreground, or spatial artifacts due to dirt on the slide or slide treatment (Figure 2.2). Image analysis procedures may try to rectify the spatial problems by capturing the true shape of the spots. Other image methods may use the distributional properties of the pixel intensity values, such as their histogram, to discriminate between foreground and background areas. There are also hybrid image analysis approaches that combine the spatial and distributional methods.

Basic implementations that assume circular spots and ignore irregular shapes are included in most of the common software packages, such as ScanAlyze (Eisen, 1999), GenePix (Axon Instruments Inc., 1999), and QuantArray (GSI Lumonics, 1999). Both ScanAlyze and GenePix allow estimation of the spot diameter for each circle individually.

QuantArray also implements two distributional methods for image analysis. The first method, described in Chen et al. (1997), relies heavily on first defining a good background region from which eight pixels are selected at random. It is assumed that the foreground pixels from within the target mask form a sorted list. The analysis starts with a target window consisting of the lowest eight foreground intensities in the list. Pixels in the target window are compared to the background pixels via the Mann–Whitney rank sum statistic at some fixed critical value ($\alpha = 0.05$ or 0.01). If the test statistic is not significant, the target window is slid up the sorted list by x (perhaps 1) pixel(s) to form an updated target set. This procedure is iterated until statistical significance is reached. The true spot signal thus consists of the final target set of eight pixels and all lighter ones in the sorted list. The second method implemented by QuantArray relies on the observed histogram of the pixels and defines the background as pixels with intensities between the 5th and 20th percentiles while those with intensities between the 80th and 95th percentiles form the foreground. This method is not adaptive since it does not account for the variable size of the spot.

Two adaptive segmentation methods that do not assume spot circularity are the watershed method (Beucher and Meyer, 1993) and seeded region growing (Adams and Bischof, 1994). Both methods require the specification of starting points and the spot detected area is enlarged in a stepwise manner until some criterion is reached.

Seeded region growing is implemented in the Spot software (Buckley, 2000), based on the R programming language. Details can be found in Yang et al. (2002b) and at `http://www.stat.berkeley.edu/users/terry/zarray/`.

For background estimation, the Spot software uses a nonlinear filter, the morphological opening (Soille, 1999), to smooth out all local peaks and artifacts over the entire slide image, thus extracting only the background intensities. The nonlinear filter combines a local minimum filter called *erosion* with a local maximum filter called *dilation*. The advantages of this method of background estimation are that the background estimates derived are obtained at the actual spot locations and are not influenced directly by the bright pixels belonging to the actual spots. A comparison study of various segmentation and background estimation methods was carried out by Yang et al. (2002). They concluded that morphological opening is a more reliable background estimation method than others and that the choice of background correction method has a larger influence on the log ratios of intensities than does the choice of segmentation method.

The methods described above do not work well when an artifact falls into the background area, resulting in an overestimation of the background value, or when the signals are so weak that there is no marked transition between foreground and background. ImaGene (BioDiscovery Inc., 1997) attempted to circumvent these problems with a hybrid solution of shape segmentation and distributional methods. Initially, a circular shape is selected to form the basis of the separation of pixels into foreground and background. Subsequently, the distribution of the pixels is depicted from which an arbitrary fixed interval is chosen. All pixels with intensities that lie outside this interval are considered as extreme and eliminated, thus minimizing the effects of outliers and artifacts.

A more recent approach was proposed by Bozinov and Rahnenführer (2002) based on the use of pixel clustering methods to discriminate a target area of one gene spot into foreground and background pixels. The clustering procedure is simplified since the number of clusters is known to be two (foreground and background). They implemented two pixel extraction methods: (1) Partitioning Around Medoids (PAM) (Kaufman and Rousseeuw, 1990), and (2) k-means (MacQueen, 1967; Bock, 1992). Let R_i and G_i denote the expression values for the ith pixel of a particular spot corresponding to the red and green dyes, respectively, and let $\boldsymbol{w}_i = (R_i, G_i)^T$. Also, let $\boldsymbol{w}_1^*$ and $\boldsymbol{w}_2^*$ denote the medoids (data points) that are representative of the foreground and background clusters. The PAM pixel extraction algorithm is applied, using the Manhattan distance function $d(\boldsymbol{w}_i, \boldsymbol{w}_j) = |R_i - R_j| + |G_i - G_j|$ to measure the distance between the ith and jth pixels. This algorithm solves for $\boldsymbol{w}_1^*$ and $\boldsymbol{w}_2^*$ by finding a local minimum of the objective function

$$\sum_i \min_{k=1,2} d(\boldsymbol{w}_i, \boldsymbol{w}_k^*).$$

Alternatively, the k-means pixel extraction algorithm uses an objective function based on the sum of the squared Euclidean distances to the two cluster centers $\boldsymbol{w}_1^*$ and $\boldsymbol{w}_2^*$ and looks for minimum variance partitions. Both the PAM and k-means methods require some random starting points, and the optimization step is iterated until convergence according to some predefined criterion. Exact algorithmic details can be found in Bozinov and Rahnenführer (2002). Their proposed methods can cope with exemplary spots that contain one or more imperfections commonly encountered in practice. The methods proved to be robust with respect to variability in target area size, shape, or pixel intensity; and even artifacts encapsulated within the spots could be isolated dynamically. However, one of the shortcomings of these methods manifests itself in the case of low expression spots with large bright artifacts, thus misleading the clustering algorithm to classify the artifact as foreground while merging the actual gene spot with the real background. Possible remedies to this problem are still being investigated.

Other researchers suggested more sophisticated methods of background adjustment to produce positive adjusted intensities when the resulting background estimate is larger than the foreground estimate, or for highly irregular spots. A simple measure of spot irregularity was proposed by Brown et al. (2001) where the normalized standard deviation of the ratio measurement y_i, termed the *spot ratio variability* (SRV), can be calculated as $\mathrm{SRV}_i = \sigma_{y_i}/y_i$. This SRV can subsequently be employed in assigning significance estimates to expression ratios through the calculation of robust confidence limits. Theilhaber et al. (2001) proposed a Bayesian algorithm of putting a prior distribution on the foreground intensity before estimating the fold change based on expression ratios on a regular scale. Kooperberg et al. (2002) proposed a Bayesian method for background correction and the computation of log-expression ratios from glass spotted arrays. They assumed that there are two additive effects: (1) the effect of RNA attaching to the unprocessed array within the background region, and (2) the effect of RNA hybridizing to the target cDNA or to the glass medium. Thus, the intensity of background and foreground pixels may be assumed to be, respectively, random variables with mean μ_b and $\mu_f = \mu_t + \mu_b$, where both μ_t and μ_b are nonnegative. They formulated relationships between the background and foreground intensities observed, x_b and x_f, with their respective true intensities, μ_b and μ_f, and appropriate prior beliefs on the true intensities. Subsequently, a posterior distribution can be calculated from which a better estimate of μ_t can be derived. The main advantage of this Bayesian method is that it reduces the variation of the estimates of the expression ratio when the expression levels are low, while maintaining unchanged estimates for expression ratios corresponding to higher expression levels.

2.2.2 Missing Value Estimation

Another problem often encountered in practice is that gene expression microarray experiments can generate data sets with multiple missing expression values. Missing values occur for diverse reasons, including insufficient resolution, image corruption, or due simply to dust or scratches on the slide. Missing data may also occur systematically as a result of the robotic methods employed in generating the microexpression arrays.

Some simple strategies often employed in practice to handle missing or suspicious data are to flag their positions manually and exclude them from subsequent analysis, or to replace missing log-transformed data by zeros or by an average expression over the rows (samples or experiments) (Alizadeh et al. 2000). Such approaches are not optimal since the correlation structure of the data is not taken into account. An expensive alternative is to repeat the experiment.

More sophisticated imputation methods of missing data that take advantage of the correlation structure of the data have been proposed by Troyanskaya et al. (2001) based on k-nearest neighbors (k-NN) or a singular value decomposition (SVD approach). These methods are summarized briefly below.

Imputation based on k-nearest neighbors

The k-NN imputation algorithm uses a gene similarity measure (Euclidean distance, Pearson correlation, variance minimization) to impute missing values. Suppose that for the jth sample or experiment, the expression value y_{ij} is missing for gene i. A weighted average is calculated from the k genes with nonmissing values for experiment j that have closest expression profiles to gene i in the remaining samples or experiments; the weight may be proportional to the similarity measure.

Imputation based on Singular Value Decomposition

As a preliminary step, all missing values in matrix A are imputed using the row average method. The SVD algorithm produces a set of mutually orthogonal expression patterns that can be combined linearly to approximate the gene expressions in the $N \times M$ microarray data matrix A of N genes and M conditions/samples/experiments. The singular value decomposition of A is

$$A = U_1 \Lambda U_2^T,$$

where U_1 and U_2 are orthogonal matrices. The columns of U_2 form the eigenvectors or eigengenes of $A^T A$, corresponding to the eigenvalues on the diagonal of matrix Λ. The largest k significant eigengenes are selected empirically to form the basis for the imputation process. Gene i with missing value for a sample j is regressed against the k eigengenes (while ignoring all expression values corresponding to experiment j). An estimate of the missing y_{ij} is obtained from a linear combination of the k eigengenes weighted by the regression coefficients. The SVD imputation process is iterated until the total change in the matrix A converges to an sufficiently small arbitrary value.

Troyanskaya et al. (2001) compared the performance of their proposed methods to the row-average algorithm in terms of computational complexity and accuracy measured via normalized root mean square (RMS) errors as a function of the fraction of missing values. They concluded that although row averaging is the fastest method, it does not perform well in terms of accuracy. They recommend the k-NN imputation method as the most robust against the increasing fraction of missing data. However, they also extended a cautionary note in drawing critical biological conclusions from partially imputed data. Researchers should flag locations with missing data and assess carefully any significant biological results suggested by the corresponding imputed data.

2.2.3 Sources of Nonlinearity

A key assumption in the analysis of microarray data is that the quantified signal intensities are related linearly to the expression levels of the corresponding genes. A recent study by Ramdas et al. (2001) examined this relationship experimentally for two types of microarrays commonly encountered: radioactively labeled cDNAs on nylon membranes and fluorescently labeled cDNAs on glass slides. They uncovered two discrepancies: signal quenching associated with excessive dye concentrations, and a nonlinear (square-root) transformation of the raw data introduced by the scanner. These nonlinearities were revealed by serial dilution experiments, which is recommended as a quality control step. The problem of nonlinearity has been recognized by many other researchers, and there are normalization methods proposed to correct for nonlinearity, as described below.

2.3 NORMALIZATION AND PLOTTING PROCEDURES FOR OLIGONUCLEOTIDE ARRAYS

A major concern with oligonucleotide arrays is the problem of *saturation*, where both PM and MM probe intensities reach the maximum intensity allowed by the scanner. Much of the important information may lie in the high values of PM and MM. Saturation may result in missing out on differential expressions between chips due to the artificial maximum tableau applied to intensities at the high end. A simple way to circumvent this problem is to tune down the scanner. Others perform normalization procedures in a simultaneous attempt to reduce the effects of multiple sources of variation. Exploratory plots are helpful in detecting obscure sources of variation. For example, one may consider direct chip-to-chip comparison of PM values via box plots of $\log_2(\text{PM})$, $\log_2(\text{MM})$, $\log_2(\text{PM/MM})$, or PM-MM. Alternatively, one can explore intensity-related biases for each pairwise chip comparison via *M–A plots* of $M = \log_2(\text{PM}_k/\text{PM}_l)$ versus abundance $A = \log_2 \sqrt{\text{PM}_k \times \text{PM}_l}$ for two different chips k and l. If the M–A plots exhibit any obvious curvature deviating from the horizontal line at zero, normalization is recommended.

2.3.1 Global Approaches for Oligonucleotide Array Data

Most classical normalization procedures for oligonucleotide arrays are global approaches, based on normalization of the overall mean or median array intensity to a common standard, such as those implemented in the Affymetrix GeneChip software (Affymetrix Inc., Santa Clara, California). Detailed descriptions of Affymetrix normalization methods can be found in the Version 5.0 Affymetrix Microarray Suite User Guide. Normalization methods implemented are similar to scaling and enable comparison analysis of an expression and baseline array. The main goal is to minimize discrepancies between an experiment and baseline array due to variation in sample preparation, hybridization conditions, staining, or probe array lot. Microarray Suite offers three types of normalization:

- *User-Defined Normalization.* User-defined normalization multiplies the signal of each probe set on an array by a user-specified normalization value, where a value of 1 is equivalent to no normalization.

- *All Probe Sets Normalization.* This method adjusts or *normalizes* the trimmed mean signal (TMS) of the experiment to the trimmed mean signal of the baseline. A normalization value is computed such that

$$\text{TMS}_{\text{baseline}} = (\text{normalization value}) \times \text{TMS}_{\text{experiment}}.$$

- *Selected Probe Sets Normalization.* For this normalization option, Microarray Suite utilizes user-selected probe sets to compute the trimmed mean signal of the experiment and baseline and to derive a normalization value analogous to that shown above. User-selected probe pairs may be excluded or *masked* from an expression analysis by creating a probe mask file that specifies the probe pairs to be excluded from an analysis. Microarray Suite allows the user manually to include or exclude specific probe set names or to utilize more advanced features that automatically generate three types of probe masks: cross-hybridization, hybridization, or spike probe mask. The *cross-hybridization* probe mask specifies probe pairs that have a perfect match (PM) or mismatch (MM) probe cell with an intensity that exceeds a user-specified limit. The *hybridization* probe mask method analyzes one or more user-specified cell intensity files to generate a hybridization probe mask composed of probe pairs that meet either of the following criteria: $\text{PM} - \text{MM} \leq \delta_D$ or $\text{PM/MM} \leq \delta_R$; where δ_D and δ_R are threshold values that are user-modifiable with defaults of 30 and 1.5, respectively. The *spike* probe mask method compares two user-specified cell intensity files, one derived from a *spiked* target that contained known amounts of a control transcript, and one derived from an unspiked target. A hybridization probe mask is generated comprising probe pairs that meet either of the following criteria: $(\text{PM-MM})_{\text{spike}} - (\text{PM-MM})_{\text{unspike}} < \alpha_D$ or $(\text{PM-MM})_{\text{spike}}/(\text{PM-MM})_{\text{unspike}} < 1 + \alpha_R$, where α_D and α_R are threshold values that are user-modifiable with defaults of 30 and 1, respectively.

2.3.2 Spiked Standard Approaches

Normalization procedures implemented in the Affymetrix GeneChip software are referred to as *global* or *scaled average difference*. There are several limitations to global normalization. First, global normalization does not absolutely quantify mRNA abundances. Second, global normalization implicitly assumes that the mean expression level of all monitored mRNAs is constant. The validity of this assumption depends on the number and biological characteristics of genes monitored by an array and does not hold for smaller arrays, where only a limited set of mRNAs is monitored. Third, global normalization does not deal well with low-abundance transcripts that are present at levels below the detection sensitivity, typically about 1:100,000 mRNAs for Affymetrix GeneChip assays. Such mRNAs induce noisy and sometimes negative intensity values, which cannot be log-transformed automatically.

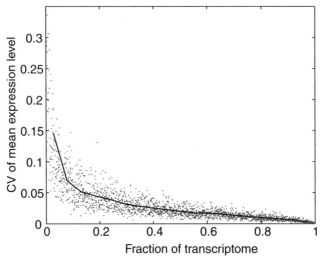

Fig. 2.3 Plot to assess constant-mean normalization. The overall data set represents 19,031 mRNAs for samples covering widely divergent developmental stages of the nematode *Caenorhabditis elegans*. The mRNAs were monitored by three array designs with 13 hybridizations for each design. The subsets were chosen randomly and ranged in size from 20 to 19,031 genes or 0.05% to 100% of this transcriptome, respectively. From Hill et al., (2001).

Hill et al. (2001) evaluated two alternative approaches to the standard global normalization scheme, which they coined *frequency* and *scaled frequency* normalization based on the use of spiked standards. These procedures require the presence of a common pool of biotin-labeled transcripts of known concentrations spiked into each hybridization. Appropriate plotting procedures can assist in deciding if the global normalization method is adequate; that is, whether the assumption that the mean expression level on an array should be the same for all samples and all arrays. To evaluate the constant-mean assumption, Hill et al. (2001) constructed hypothetical arrays of their overall data by choosing random subsets of genes with varying sizes represented by fractions of transcriptome. One can examine the associated coefficient of variation (CV) of the mean expression level via plots such as Figure 2.3. This plot indicates that the constant-mean assumption for global normalization may be acceptable for arrays measuring more than 10% of the genes, but the CV is substantially increased for smaller subsets and can be more dramatic if there is bias in the selection of genes on the array.

Spike-in frequency normalization: This procedure consists of two main steps:

Step 1. Calibration of the arrays: This step allows investigators to transform average differences (ADs) to cRNA frequency estimates. First, a number of control scripts are spiked into the hybridization solution at known concentrations. From a single hybridization one can produce a plot of the AD values for the spike-in controls versus the transcript frequency in units of transcripts per million. A generalized model may be fitted to the plotted points, which can subsequently be used to calibrate the AD values of the other genes on the array.

Step 2. Estimation of the minimum detectable frequency on the array. Logistic regression can be applied to define chip sensitivity as the frequency at which the estimated probability of a gene being *present* is at least $p\%$, where p is often greater than 50%.

Hill et al. (2001) noted that frequency estimates might be biased by experimental limitations on the accuracy with which control transcripts can be spiked into cRNA. cRNA impurities can lead to *spike-skew* normalization effects where all readouts from one array are systematically higher or lower than those from another array. To mitigate the effects of spike-skew, Hill and colleagues developed the hybrid scaled frequency normalization method described below.

Spike-in scaled frequency normalization: The principle here is to remove technical variation in the ratio of spiked transcripts to cRNAs. The process consists of two main steps:

Step 1. Initially, assume constant mean and compute globally scaled average differences for all arrays in a set. Pool together all spiked cRNAs on all the arrays and fit a single linear model to their corresponding scaled average differences to produce a calibration function.

Step 2. Individual array sensitivities are computed similarly to step 2 of frequency normalization, where low-end frequencies can be damped out based on the sensitivity values for each array.

Hill et al. (2001) also compared the performance of their proposed methods to the globally normalized approach in terms of reproducibility and accuracy (in measuring true biological variation). To assess reproducibility, the basis of comparison was experimental data from sets of three or four replicated hybridizations of the same array design, augmented by simulations. Reproducibility performance for each method was measured by the median absolute coefficient of variation (MEDACV) of probe sets across the replicated hybridizations. Perfect agreement or reproducibility of all transcript readouts in the set of hybridizations is reflected by a zero value for MEDACV. Accuracy assessment was based on calculating the fraction of computed fold changes between the modulated condition and a chosen baseline, under different levels of spike-skew. The spike-in scaled frequency normalization method consistently exhibits the lowest MEDACV values, is the most effective in eliminating spike-skew effects, and possesses high accuracy levels of above 99% regardless of spike-skew.

2.3.3 Geometric Mean and Linear Regression Normalization for Multiple Arrays

Geometric mean normalization is similar to the common standardization technique often used by statisticians. First, for each array, all data values on the original scale that are considered *outliers*, that is, are more than 3 standard deviations from the mean, are flagged. Next, the mean and standard deviation of the logged data are calculated, ignoring the outliers. Finally, the logged expression data are transformed linearly such that the normalized (logged) values lie on a standard scale with mean zero and unit variance. This method acquired its name from the fact that normalization with respect to the mean of the logged data is equivalent to normalization with respect to the geometric mean in the unlogged data.

For any two arrays, the simplest approach is to fit a curve or a *linear regression* with the intensities of array A on the x-axis versus the intensities of array B on the y-axis, or vice versa. The intensities of array A (or B) are then normalized simply by subtracting the difference between the fitted curve and the line through the origin of coordinates with slope 1. This approach is simple enough to apply to a multiple of k arrays where a baseline array is chosen. It is easily extendable to additional arrays without affecting the analysis of the original set of arrays by using the baseline array of the original k arrays to normalize the new arrays. However, this approach has the obvious drawback of being baseline-array dependent, which may affect subsequent analyses of the raw intensities, such as computing expression indices. The use of standard linear regression therefore leads to an asymmetrical method in which the result of normalization is not equivalent for different choices of the baseline array.

2.3.4   Nonlinear Normalization for Multiple Arrays Using Smooth Curves

Åstrand (2001) proposed an alternative approach where all arrays are treated uniformly. Briefly, the raw feature intensities (PM and MM intensities) of all the M arrays, denoted by a $N \times M$ matrix $\boldsymbol{A}$, are logged and transformed using an orthonormal matrix $\boldsymbol{H}$ to produce a transformed matrix $\boldsymbol{A}_{\mathrm{proj}}$. The first row of $\boldsymbol{H}$ has each element equaling $\sqrt{1/M}$, and the remaining rows form a set of orthonormal contrast; $\boldsymbol{H}$ is unique only when $M = 2$. One can choose a specific column of the transformed matrix $\boldsymbol{A}_{\mathrm{proj}}$ as a baseline array and fit a smooth curve, such as a *loess curve*, for each of the other columns. A curve-fitting technique is proposed which uses a re-descending M-estimator and a biweight function with an appropriate modification to ensure that the same set of robust weights is employed for each of the $M - 1$ contrast vectors. The weights and fitted curves are invariant to the choice of orthonormal contrasts. In the special case of $M = 2$ the transformation reduces to a transformed basis of the difference versus the mean of the two arrays. To add a new set of arrays to an original set of M original arrays that have been normalized and further analyzed, the new set is normalized separately first and scale-transformed to match the scale of the original set. The scales for the original and the new set of arrays may be determined by their respective geometric means.

Other approaches using nonlinear smooth curves have been proposed in Schadt et al. (2000, 2001) and Li and Wong (2001a,b). For many experimental situations, plots of the PM/MM ratios between a baseline and an experimental array often show substantial slope change from the low-intensity region to the high-intensity region (10 to 50% in slope value differences). An immediate extension of the linear method is to fit a *nonlinear regression* of the baseline array values on the experimental array values. One such procedure using smoothing splines with generalized cross-validation (GCVSS) was described in Schadt et al. (2000). However, such a procedure is inadequate if the expression profiles of the two arrays are very different. Schadt et al. (2001) extended the nonlinear normalization method to an invariant difference selection algorithm (IDS) with the GCVSS procedure defined previously. A set of probes is defined to be invariant if the orderings of these probes (based on the PM/MM ratios) are identical for both the baseline and experimental arrays. IDS is an iterative process that updates M, the total number of ordered PM/MM ratios at the current step. Let the ranks for the ith PM/MM ratio corresponding to the baseline and experimental arrays be denoted by B_i and E_i, respectively. For the ith PM/MM ratio, one calculates R_i, the threshold for ratio intensity i, by interpolating between a low-ratio-intensity threshold (L) and a high-ratio-intensity threshold (H) with the following formula:

$$R_i = \frac{1}{2M}[L(B_i + E_i) + H(2M - B_i - E_i)].$$

To determine if the ith ratio belongs to the invariant set, a rank ratio test statistic, D_i, is calculated where $D_i = 2|B_i - E_i|/(B_i + E_i)$. The ith ratio is considered approximately invariant if $D_i < R_i$. The nonlinear normalization curve is constructed by applying the GCVSS technique to the final approximately invariant set. This particular approach, among others, is implemented in the software DNA-chip Analyzer (dChip) (Li and Wong, 2001a,b) and is available at the Web site http://www.dchip.org.

Quantile normalization was recently discussed by Bolstad (2001) and Bolstad et al. (2003). This approach aims to synchronize the quantiles over the entire set of chips available and assumes that there is an underlying common distribution of intensities across chips. For multiple arrays, one can use pairwise quantile–quantile plots or overlaying density plots of cell intensities across chips to visualize the intensity distributions. The intensities of N genes from M arrays may be arranged in a $N \times M$ matrix $\boldsymbol{A}$, where each column corresponds to intensities in one array. Each column of $\boldsymbol{A}$ is sorted; then each row of $\boldsymbol{A}$ is projected onto the vector $(1/\sqrt{M}).\boldsymbol{1}_N$, to yield $\boldsymbol{A}_{\mathrm{proj}}$. This projection is equivalent to replacing each individual element in a particular row of matrix $\boldsymbol{A}$ with the quantile values averaged over the columns or arrays. The normalization process is achieved by rearranging each column of $\boldsymbol{A}_{\mathrm{proj}}$ according to the ordering of the original $\boldsymbol{A}$. This method can be improved further to allow greater differentiation between chips in the tails of the distributions. Further investigation is warranted on scaling and centering extreme tail values appropriately without affecting the corresponding quantiles in the other chips.

Bolstad et al. (2003) also discussed a *cyclic loess* approach based on initially plotting the difference in log expression values versus the average of the expression values, termed the *M versus A plot* (Yang et al. 2001; Dudoit et al. 2002a). Let y_{i1} and y_{i2} denote probe intensities for the ith probe for arrays 1 and 2, respectively. A *loess* (local regression, described by Cleveland and Devlin, 1988) normalization curve is fitted to the M_i versus A_i plot where $M_i = \log(y_{i1}/y_{i2})$ and $A_i = \frac{1}{2}\log(y_{i1}\,y_{i2})$ to produce fitted values $\hat{M}_i$. The normalization correction is derived from the residual $M_i - \hat{M}_i$ and is applied with equal weights to the probe intensities from the two arrays. For more than two arrays an *orthonormal contrast-based method* can be used for normalization. The cyclic loess approach can also be applied to paired, two-color-channel cDNA data (Yang et al., 2001, 2002), and the contrast-based method has also been used for multiple cDNA array normalization (Åstrand, 2001). We discuss these in further detail in a subsequent section. Bolstad et al. (2003) conducted a careful comparison study of these three normalization methods (quantile, cyclic loess, orthonormal contrast). They established that all three methods are able to reduce the variation of a probe set measure across a set of arrays to a greater degree than two other methods that make use of a baseline array (Affymetrix scaling method; nonlinear method by Schadt et al., 2001). Specifically, the quantile method performed favorably in terms of speed as well as bias and variability measures, and thus is the recommended normalization method for high-density oligonucleotide arrays. Software implementing these three normalization methods (in R) is available via the package Affy, part of the Bioconductor project, and can be found at http://www.bioconductor.org. Accompanying literature include those by Cope et al. (2004) and Gautier et al. (2004). A very recent development called *fastlo* was discussed by Ballman et al. (2004). It is a model-based normalization technique that yields normalized values similar to cyclic loess and quantile normalization, but is at least an order of magnitude faster than cyclic loess. A particularly innovative model of molecular interactions on short oligonucleotide microarrays proposed by Zhang, Miles, and Aldape (2003) is the *positional-dependent-nearest-neighbor (PDNN) model*. Such a model reveals how probe signals depend on probe sequences and indicates that the amount of nonspecific binding can be estimated from a simple rule. In particular, a simple free-energy model was developed for the formation of RNA–DNA duplexes on short oligonucleotide microarrays based on the nearest-neighbor model. It takes into account two different modes of binding on the probes — gene-specific binding versus nonspecific binding — and assigns a different weight factor at each nucleotide position on a probe to reflect that different parts of the probe may contribute differently to the stability of binding. A key advantage of the PDNN model is that it offers a means to check data quality and appropriateness of probe design, allowing problematic probe signals to be detected directly from the model fitting. Since the determination of NSB and GSB requires only PM probe signals in the PDNN model, the capacity of the array can be doubled through replacement of MM probes with additional PM probes. Thus, the model provides a practical guide for microarray design in terms of probe selection.

2.4 NORMALIZATION METHODS FOR cDNA MICROARRAY DATA

For cDNA microarrays, normalization methods focus on balancing the fluorescence intensities of green (Cy3) and red (Cy5) dyes, allowing comparison of the expression levels across slides. When two identical mRNA samples labeled with different dyes are hybridized on the same slide, one often observes higher intensities for the green dye. This phenomenon, referred to as *dye bias*, may result from a variety of internal and external factors to the sample, such as the physical properties of the dyes, efficiency of dye incorporation, fluctuations in processing procedures, or scanner set-

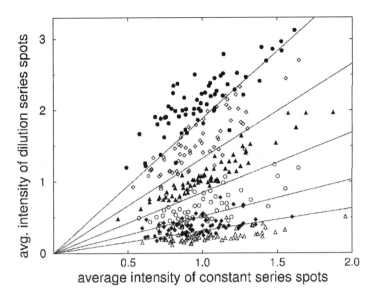

Fig. 2.4   Plot for normalization. From Schuchhardt et al. (2000).

tings. Plots of the log ratio intensities versus the overall spot intensity may reveal dye biases that are spot-intensity dependent.

Schuchhardt et al. (2000) compared normalization strategies in the context of cDNA microarrays, spotted on glass slides and hybridized with a radioactively labeled probe. The experiment was based on several mouse tissues with a fixed amount of spike-in *Arabidopsis thaliana* cDNA. A grid of 384 blocks (36 spots per block) was spotted with a 384-pin gridding head. All clones were spotted twice within a grid, and altogether, nine slides with an identical spotting pattern were produced. The sources of variation in this study could have been attributable to double spotting, constant *Arabidopsis* controls versus *Arabidopsis* dilution control series (six steps, each corresponding to a twofold dilution), and pin-tip and slidewise fluctuations, among others. A graphical approach to assess background noise is to plot the intensity of background spots versus the average signal intensity of k nearest neighbors, overlaid with a regression line. To gain insight into the magnitude of fluctuations and correlations between double-spotted signals, scatter plots of the 384 doubly spotted pairs were produced for every block within each slide. These scatter plots should conform to a straight-line pattern along the diagonal; where deviations from this expected behavior essentially reflect random fluctuations in target volume. Figure 2.4 illustrates how the presence of a control signal can help to distinguish the various classes of the dilution series. The intensity of the averaged diluted control signal is plotted versus the intensity of the averaged constant control signal. Figure 2.4 shows that the six classes of the dilution series are reasonably well separated. With the presence of a reference control signal, investigators can decide whether an observed strong signal intensity is due to high task concentration in the target clone or rather, is a consequence of systematically excessive volume spotted by the corresponding pin. The most direct normalization method with a constant control signal is to calculate the ratio of the dilution signal and the constant control signal. Four normalization strategies (none, slidewise, pinwise, average pinwise) were compared in terms of classification and prediction performance. The prediction task was to predict the correct dilution class of a given spot using the nearest mean classifier trained on a different slide. The quality of classification was measured by the percentage of correct assignments. This experiment led to the conclusion that the average pinwise normalization strategy, by first averaging intensities over several slides and then calculating the ratio of the averaged quantities, was superior to the other methods considered. In particular, this strategy minimizes the effect of a strong fluctuation in the denominator.

Yang et al. (2002) present a detailed review of existing normalization methods and propose improved methods that can account for intensity and spatial dependence in the dye biases for different experimental setups. They distinguish between three situations: (1) within-slide normalization (Figure 2.5), (2) paired-slide normalization for dye-swap experiments (Figure 2.6), and (3) multiple-slide normalization. For each situation, one needs to consider the set of genes to use for normalization. For example, if only a small proportion of genes are expected to be expressed differentially or there is symmetry in the expression levels of the up- or down-regulated genes, then all genes on the array may be used for normalization. Alternatively, a smaller subset of *housekeeping genes* that have constant expression across a variety of conditions may be used for normalization purposes. One can also use spiked controls or a titration series of control sequences for normalization purposes. For spiked control methods, DNA sequences from an organism different from the one being studied are spotted on the array and included equally in the two different mRNA samples. This would induce equal dye intensities expressed by these spotted control sequences. This desirable property can also be achieved by the titration series approach, where spots consisting of different concentrations of the same gene or EST, such as genomic DNA that are known to express constant expressions, are printed on the array.

2.4.1 Single-Array Normalization

Global normalization methods are the most widely used mainly because they are mathematically and computationally easy to implement. For single-slide cDNA microarray experiments, *location normalization* can be performed by subtracting a location parameter c from the log-intensity ratios. Letting R and G represent red and green intensities, respectively, and assuming that $R/G = k$,

$$\text{Normalized Expression} = \log_2(R/G) - c,$$

where $c = \log_2 k$ may be estimated as the gene-specific median/mean of the log-intensity ratios. Others estimated k along with cutoffs for the intensity values by an iterative method (Chen et al., 1997) or by constraining the arithmetic mean of the intensity ratios of all the genes on a specific microarray to be equal to 1 (GenePix).

When there is evidence of spot intensity dependence, one can perform an *intensity dependent normalization* by calculating

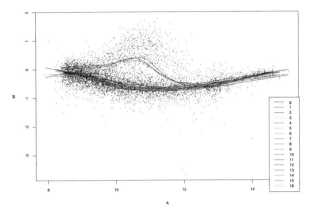

Fig. 2.5 M versus A plot for within-slide normalization. The plot depicts loess curves corresponding to 16 print tips and for the entire data set (labeled g). Data was collected from mice (eight treatment mice with the apo AI gene knocked out, and a control group of eight normal mice). Data details are given in Callow et al. (2000). See the insert for a color representation. From Yang et al. (2001a).

$$\text{Normalized Expression} = \log_2(R/G) - c(I),$$

where I represents overall spot intensity. There are a variety of ways to estimate $c(I)$. Yang et al. (2001) suggested performing a robust locally linear fit using the Splus function *lowess()* of $\log_2(R/G)$ versus I based on 20% of the the data used for smoothing at each point. A more general approach was proposed by Kepler et al. (2002), based on a different local regression method, to estimate the normalized expression levels as well as the expression-level dependent error variance. Their method is based on two main assumptions: (i) that a large majority of genes will not exhibit significant differences in relative expression levels between treatment groups; and (ii) that departures of response from linearity are small and slowly varying. Sapir and Churchill (2000) considered a more constrained approach by fitting a robust regression line of $\log_2(R/G)$ versus I.

Analogously, Yang et al. (2002b) also proposed using the *loess* fit to handle systematic differences that may exist between the print tips due to different physical tip properties (length, opening) or just tip degradation over time. *Print tip normalization* depends on both the print tip and the overall spot intensity. Let k represent the kth grid in an array that shares the same print tip; then normalization is performed by calculating

$$\text{Normalized Expression} = \log_2(R/G) - c_k(I),$$

where, as before, $c_k(I)$ is the normalization function obtained by the loess fit of $\log_2(R/G)$ versus I for the kth print tip. This process recenters the log-intensity ratios from the various print tip groups. However, if print tip differences also show

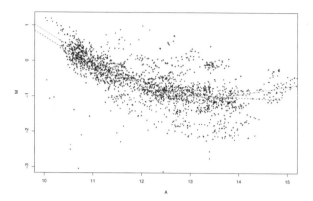

Fig. 2.6 M versus A plot for paired-slide normalization. The plot depicts loess curves obtained from a follow-up experiment of the apo AI mice study. Details of the experiment are described in Yang et al. (2001a). The dots represent log ratios for slide C3K5 with a corresponding solid loess curve. The crosses represent log ratios for slide C5K3 with a corresponding dotted loess curve. See insert for color representation. From Yang et al. (2001a).

evidence of different spreads, *scale normalization* is also performed. Yang et al. (2002) assume that for the same print tip group k, $\log(R/G) \sim N(0, a_k^2\sigma^2)$, where σ^2 is the true variance of the log-intensity ratios and a_k^2 is the scale factor for the kth print tip group. Estimates for a_k can be calculated using the maximum likelihood method or a robust alternative based on the median absolute deviation approach.

2.4.2 Multiple Slides Normalization

The simplest multiple slides experiment is the dye-swap experiment when the slides are paired. Reverse assignment of dyes is applied to two hybridizations of two mRNA samples. Any within-slide normalization may be applied to each of the paired slides to adjust for location. One would expect the normalized log-intensity ratios on the paired slides to be of equal magnitude with opposite signs. Thus, assuming that the normalization constant c is approximately the same on the two slides, one can write

$$c \approx \tfrac{1}{2}[\log_2(R_1/G_1) + \log_2(R_2/G_2)],$$

where c can be estimated, using all the genes, by a loess curve of $\tfrac{1}{2}[\log_2(R_1/G_1) + \log_2(R_2/G_2)]$ versus $\tfrac{1}{2}(\log_2\sqrt{R_1G_1} + \log_2\sqrt{R_2G_2})$.

When one wishes to perform experimentwise comparisons, multiple unpaired slides are involved, which introduce scale heterogeneity. This problem can analogously be rectified by using scale adjustment methods similar to those described above for within-slide scale normalization.

2.4.3 ANOVA and Related Methods for Normalization

A common practice with microarray data is to estimate a differential expression based on ratios of the raw signals. This simple approach is inadequate because simple ratios do not necessarily account for the differential behavior of dyes or variations between samples and arrays. In addition, ratios expressing fold change in fluorescence do not correspond directly to fold changes in the raw expressions.

Many researchers use ANOVA models for microarray data that can account for experiment-wide systematic effects that could bias inferences made on the data from the individual genes. The ANOVA models are designed to effectively normalize the data without the need to introduce preliminary data manipulation by combining the normalization process with downstream data analysis.

In particular, Kerr et al. (2000) identified four basic experimental factors: (1) Array main effects (A) measure overall variation in fluorescent signal that arise when, for example, arrays are probed under inconsistent conditions that increase or reduce hybridization efficiencies of labeled cDNA; (2) dye main effects (D) measure differences in the two-dye fluorescent labels; (3) gene main effects (G) occur when certain genes emit a consistently higher or lower fluorescent signal compared to other genes (these effects may be due to the natural property of these specific genes or because of differential hybridization efficiency and differential labeling efficiency for different sequences); and (4) variety main effects (V) occur when the categories or varieties of the factor of interest induce a consistently higher or lower trend in expression levels for the genes spotted on the arrays. Gene-specific interaction effects due to array × gene, variety × gene, or dye × gene can also be included in the model. For experiments in which each gene is spotted only once on each array, Kerr et al. (2000) proposed an ANOVA model based on the logarithms of the original fluorescence measurements that can incorporate such main effects and gene-specific interaction effects. Let

$$\log(y_{ijkl}) = \mu + G_i + A_j + D_k + V_l + (GA)_{ij} + (GD)_{ik} + (GV)_{il} + \epsilon_{ijkl}, \quad (2.1)$$

where μ is the overall average signal; i, j, k, and l index genes, arrays, dyes, and varieties; and ϵ_{ijkl} represents independently, identically distributed error terms with mean zero. If one is only interested in the interactions between genes and varieties (GV) that can give evidence of differential expression for a given gene, one does not need to include the terms GA or GD to obtain a more parsimonious model. Parameter estimates for the ANOVA model can be computed by using a least-squares fit with constraints that all main and interaction effects sum to zero over their respective indices. From the resulting ANOVA table, the sums of squares reflect the relative contribution of each set of effects. The data can be adjusted or *normalized* to remove the overall effects of uninteresting factors. For example, if one is interested in testing for array × gene interactions, one can fit model (2.1) without the $(GA)_{ij}$ terms and create a new data set from the residuals, considered as the normalized values. Subsequently, a resampling-based approach, such as permutation or the bootstrap, based on the residuals and the full model (2.1) can be carried out to test for the interactions in question and also to construct relevant confidence intervals. The properties of ANOVA estimates are tied to the experimental design. For microarray experiments, it is impossible to design full factorial designs. However, Kerr et al. (2000) suggested the use of designs that are balanced across the samples of interest since they provide the greatest efficiency: for example, the Latin square design. An extended model discussed in Kerr et al. (2002) allows for the incorporation of a spot effect when there are duplicated spots within an array. In more recent work, Wu et al. (2003) developed a software package called MicroArray ANalysis Of VAriance (MAANOVA), implemented in both Matlab and R environments.

In contrast to Kerr et al. (2000), who considered only additive linear effects, Rattray et al. (2001) allowed a multiplicative effect. For each experiment j, let a_j and b_j denote the parameters determining the systematic linear effects, and let α_i denote the true expression value for gene i corrupted by noise ϵ_{ij}. Rattray et al. (2001) proposed the linear normalization model

$$y_{ij} = \alpha_i a_j + b_j + \epsilon_{ij}.$$

The multiplicative effect in the first term can account for variance fluctuations in distribution of the logged expression levels for different arrays or experiments. One cannot incorporate a general parameter with gene and experiment indices in this model since this will result in overparameterization; thus, not all parameters are identifiable. For n experiments, the probability density for $\mathbf{y}_i = (y_{i1}, y_{i2}, ..., y_{in})^T$ may be expressed as an appropriate multivariate normal density from a distribution with mean $\alpha_i \mathbf{a} + \mathbf{b}$ where $\mathbf{a} = (a_1, a_2, ..., a_n)^T$ and $\mathbf{b} = (b_1, b_2, ..., b_n)^T$. Assuming equal variance for every gene, two maximum likelihood approaches (least squares or latent variables) can be used to estimate $\mathbf{a}$ and $\mathbf{b}$ as solutions of

$$\mathbf{C}\mathbf{a} = \lambda_1 \mathbf{a}, \qquad \mathbf{b} = \bar{\mathbf{y}},$$

where $\bar{\mathbf{y}}$ and $\mathbf{C}$ denote, respectively, the mean vector (dimension n) and covariance matrix (dimension $n \times n$) estimated from the observed data, and λ_1 is the largest eigenvalue of $\mathbf{C}$. Thus, $\mathbf{a}$ is the first principal component of $\mathbf{C}$ and may be referred to as an *eigenarray*. Finally, the normalized data are obtained by linearly transforming the data so that they are distributed about the 45° line through the origin of the two-dimensional principal subspace where the axes are formed by the first and second principal components. Effectively,

$$\text{normalized } y_{ij} = \frac{y_{ij} - b_j}{a_j}.$$

2.4.4   Mixed-Model Method for Normalization

Wolfinger et al. (2001) extended the ANOVA model to include a mixture of fixed and random effects. Specifically, they considered fitting the normalization mixed model to explain the fixed treatment (T) main effect and random effects representing array (A) effects and array × treatment interaction (AT) effects. For the ith gene, let

$$y_{ijk} = \mu + A_j + T_k + (AT)_{jk} + \epsilon_{ijk},$$

where A_j, $(AT)_{jk}$, and ϵ_{ijk} are assumed to be independently normally distributed with mean zero and variance components σ_A^2, σ_{AT}^2, and σ_ϵ^2, respectively. To fit this normalization model, the method of restricted maximum likelihood (REML) can be used where REML parameter estimates are found by the usual numerical methods, such as Newton–Raphson. Once again the residuals from this model form the normalized data and can be used as input data to a subsequent gene model that allows gene-specific inferences to be made using separate estimates of variability. Wolfinger et al. (2001) advocated the use of mixed models for their additional flexibility. Researchers are free to select fixed and random effects in the models to portray scientifically reasonable patterns of variability and covariability in the data, or to include additional random effects, such as those arising from biological replicates. Further, the simple mixed model described above can be extended to accommodate more complex covariance structures.

Related work includes an application of the mixed-model ANOVA to a complex microarray described by Jin et al. (2001). Currently, MAANOVA by Wu et al. (2003) can carry out computations for the fixed effects model, but future releases will include functions for mixed-model analysis.

2.4.5 SNOMAD

Colantuoni et al. (2002) developed SNOMAD, a collection of general approaches and specific algorithms that aim to refine the meaning and validity of differential gene expression ratios, thus facilitating the identification of differentially expressed genes and a comparison of expression data across diverse microarray technologies, experimental paradigms, and biological systems. Since much artifactual variation present in gene expression data is not constant across the range of element signal intensities, global normalization is not sufficient to address this variation. SNOMAD focuses on local normalization processes that address bias and variance that are distributed nonuniformly across absolute signal intensity. The majority of the transformations within SNOMAD are directed at the refinement of paired microarray data that derive from either (1) two sets of element signal intensities generated in two individual hybridizations using a one-channel microarray technology (mostly radioactivity based), or (2) two sets of intensities generated in a single two-channel fluorescent experiment (from simultaneous two-color hybridizations).

In a single two-channel hybridization, one can use SNOMAD to draw standard scatter plots, such as (1) a scatter plot of normalized raw intensities for control versus experimental conditions; (2) a scatter plot of the logarithm of the normalized raw intensities for control versus experimental conditions; (3) a scatter plot of the logarithm of geometric mean intensity or mean log(intensity) (x-axis) versus the ratio of intensities (y-axis). Plots of type 1 will display similarity in expression levels across the two experiments but cannot depict proportional change well. For plots of type 2, one can look for patterns that deviate from the line of slope 1; however, these plots still accentuate similarity rather than difference in expression levels, an attribute that is easily captured by type 3 plots.

Colantuoni et al. (2002) discussed *local mean normalization*, called *balancing* of gene expression ratios, based on calculating a mean intensity locally (via the loess function) across the range of mean expression levels. Practically, the residuals from the loess fit are obtained by subtracting the fitted values from the original log ratio values. A plot of the residuals (y-axis) versus the mean log intensity (x-axis) can depict the points that are at least twofold significant judged by their distance from the line (parallel to the x-axis) that crosses y at zero. Another method discussed by the same authors is the *local variance correction*, which performs a variance-stabilizing function by dividing each log(ratio) value by the corresponding locally calculated standard deviation (also via a loess function). The resulting values are called *local Z-scores*. A subsequent plot of the derived local Z-scores (y-axis) versus the mean log intensity values (x-axis) can highlight the significant differential genes. SNOMAD is implemented in the R statistical language and allows users flexibility to input their own choice of window span for the loess function and the appropriate fold size for differential determination. SNOMAD can be found at the following Web site
`http://pevsnerlab.kennedykrieger.org/snomadinput.html`

2.5 TRANSFORMATIONS AND REPLICATION

2.5.1 Importance of Replication

The importance of replication in microarray gene expression studies has been addressed by Lee et al. (2000). They conducted a controlled experiment involving replication of cDNA hybridizations on a single microarray to investigate inherent variability in gene expression data and the extent to which replication in an experiment can affect consistent and reliable findings. Their study was based on a biological assumption. For a gene to be detected on a slide, the following assumptions must be met: (1) The mRNA must be contained in the target sample tissue; (2) some of the mRNA in the sample must be converted to probe; and (3) some of the probe must be detected by the cDNAs deposited on the slide as an observed gene expression. Statistical models were developed for these three events, and subsequently, a mixed normal distribution was used to model the distribution of the gene expressions observed. From the mixture distribution, a posterior probability can be calculated from the gene expression observed that quantifies the likelihood that the gene is truly expressed in the tissue and thus can be used to classify whether the gene transcript is present. The results of this study showed that (1) any single microarray output is subject to substantial variability; (2) false positive expressions may be prevalent in microarray studies; and (3) the precision of gene classification (as expressed or unexpressed) varies with the number of replicates but seems to level out with three replicates. However, the optimal number of replicates in a general microarray study will depend on many factors, including array equipment type, laboratory technique, and the condition and preparation of samples. Another study that emphasizes the importance of replicate microarray experiments has been reported by Pritchard et al. (2001) based on mouse gene expression data collected from different tissues, such as the kidney, liver, and testis. They demonstrated that even for genetically identical mice of the same age housed under the same conditions, there were still genes that expressed significant variation at the mouse level. In particular, their data suggest that both specific genes and functional classes of genes will be consistently variable, even in multiple tissue types. Genetically diverse populations such as humans are likely to show even greater variability in gene expression. A more recent study by Pan et al. (2002b) discussed how to calculate the number of replicates (arrays or spots) in the context of applying a normal mixture model approach to detect changes in gene expression. Their estimation depends on several factors, including a given magnitude of expression change, a desired statistical power to detect it, a specified type I error rate, and the statistical method being used to detect it. In general, although most researchers agree that replication is desirable, there is still no clear answer as to whether replication of spots, or arrays, or subjects is the most important. However, as microarray technology advances, perhaps this will become a moot point.

2.5.2 Transformations

We have seen that there are a multitude of normalization strategies. A similar situation exists for detecting differentially expressed genes. There are two main approaches to distinguish real signals from noise in a chip-to-chip comparison: thresholding or replicate analysis. The latter approach is desirable but expensive. The former approach involves imposing an arbitrary threshold of signal difference, or fold-change ratio, between experimental and control samples, above which differences are considered to be real. Often, differential expression is assessed by taking ratios of expression levels of different samples at a spot on the array and flagging those where the magnitude of the fold difference exceeds some threshold, incorporating the common fact that the variability of these ratios is not constant. For example, an increase or decrease of at least twofold may be considered significant. The use of fold-change ratios can be inefficient and erroneous. The uncertainty associated with dividing two intensity values further increases overall errors (Miles, 2001; (Newton et al., 2001; Yang et al. 2002b). The methods are often variants of Student's t-test; earlier simple methods were discussed by Schena et al. (1995, 1996), and DeRisi et al. (1996). Chen et al. (1997) considered a less arbitrary threshold by using replicated housekeeping genes. More recently, methods that implicitly assume a nonconstant coefficient of variation were discussed by Hughes et al. (2000), Baggerly et al. (2001), Newton et al. (2001), and Rocke and Durbin (2001), among others. Parametric tests often rely on the assumption that the data are normally distributed with equal variances across experimental conditions. Such assumptions are not usually met with real data. In contrast, a simple alternative such as the Mann–Whitney nonparametric test does not rely on such strong assumptions, but a large number of replicate experiments is required.

Most methods for normalization or for the detection of differentially expressed genes often employ some form of data transformation. Log transformations provide good variance stabilization at high levels of gene expression, but inflate the variance of near-background observations, particularly in data that have been background corrected. This result was proven by Rocke and Durbin (2001) via asymptotic variance results from a straightforward delta-method approach based on a two-component error model of the expression levels

$$y = \alpha + \mu e^{\eta} + \epsilon,$$

where y is the measured expression level for a single array, α is the mean background noise, μ is the true expression level, and η and ϵ are normally distributed error terms with means 0 and variance σ_{η}^2 and σ_{ϵ}^2, respectively. This model implies that at low expression levels the measured expression $y \approx \alpha + \epsilon$ and thus is approximately normally distributed with mean α and constant variance σ_{ϵ}^2. A promising transformation was introduced independently by Munson (2001), Durbin et al. (2002), and Huber et al. (2002), of the form

$$g(y) = \log[(y - \alpha) + \sqrt{(y - \alpha)^2 + c)}] \tag{2.2}$$

where

$$c = \hat{\sigma}_{\epsilon}^2 / \hat{\sigma}_{\eta}^2.$$

Such a transformation can stabilize the asymptotic variance of microarray data across the full data range and symmetrize the data simultaneously. One main advantage of this transformation is that low-level observations need not be removed before performing subsequent analyses. A variant of the procedure of Rocke and Durbin (2001) was discussed in Geller et al. (2003), the main differences being use of the median and interquartile range (IQR) instead of the mean and standard deviation for robustness against outliers. Baggerly et al. (2001) explored different models for the variability to be expected of a log ratio under different types of replication: (1) within-sample (channel) replication; and (2) between-sample (arrays and/or channels) replication. For within-sample replication, a beta distribution was used to model p, the probability of a single labeled target strand binding to one of the two replicated spots. Thus, the distribution of X, the number of target strands that will bind to a particular spot, follows a beta-binomial distribution with parameters $\alpha = \beta = 0.5$, assuming symmetry about $p = 0.5$. The variance function of the log ratio of replicate densities could be calculated asymptotically and a plot of the variance of the log ratio against the log intensity can depict an exponential decay curve displaced vertically from zero. An immediate extension to between-sample replication (assuming two samples come from two independent experiments but measure the same thing) is to model n_i, the number of labeled strands of gene i as a Poisson random variable with parameter λ_i; variation can be incorporated by letting λ_i be a random variable from a gamma distribution. Other scenarios discussed by Baggerly et al. (2001) include replicating ratios between samples on a single array and replicating ratios between samples on two arrays. In general, their models can be used to predict how the variance of a log ratio changes with the intensity of the signal at the spot, independent of the identity of the gene. The variance estimable from the replicate ratios can be used to define the precision with which a ratio measurement can be made, thus allowing the differences to be scaled or *studentized*, with the aid of a robust loess fit to the absolute values of the differences between replicate ratios. A case study was thoroughly described by Baggerly et al. (2002). The experiment used spotted two-channel (red and green) cDNA arrays involving 1,152 genes spotted in duplicate, together with 96 positive controls, 96 negative controls, and 576 blanks. Spot intensities were measured as the sums of all pixel intensities within a fixed-radius circle with some local background correction. Figure 2.7 demonstrates how appropriate plots can typically be used for quality control, to identify outliers, as well as to fit the standard deviation of replicate ratios, and subsequently, identify the differentially expressed genes. To assess the degree that a gene is differentially expressed, one can locally studentize values by subtracting the fitted loess value from the observed channel difference and then divide it by the local standard deviation estimated by one of the models discussed above.

The problem of detecting genes that are differentially expressed is to be covered in Chapter 5.

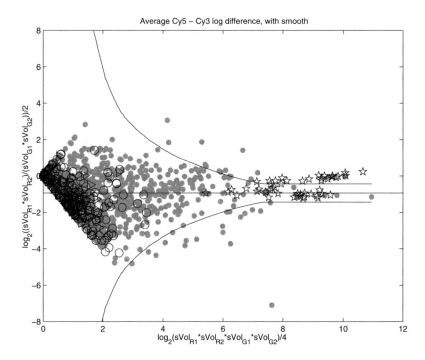

Fig. 2.7 Average differences in log intensities (sVol) for the Cy5 and Cy3 channels, with local standard deviation bounds. Genes are represented by light grey spots, negative controls by triangles, positive controls by stars, and blanks by circles. Spots outside the bands correspond to differentially expressed genes. There is a clear extreme case at the bottom, corresponding to p53 which had been transfected into one of the two samples used in this experiment. Note that the positive controls appear to have some differences. From Baggerly et al. (2002).

2.6 ANALYSIS OF THE ALON DATA SET

Alon et al. (1999) used Affymetrix oligonucleotide arrays to monitor absolute mea-surements on expressions of over 6,500 human genes in 40 tumor and 22 normal colon tissue samples. These samples were taken from 40 different patients, so that 22 patients supplied both a tumor and a normal tissue sample. Alon et al. (1999) focused on the 2,000 genes with highest minimal intensity across the samples.

We performed quantile normalization on the logged expression values obtained from the Alon data set, using the 2,000 genes with the highest minimal intensities. Figure 2.8 displays MA plots and fitted loess curves to illustrate the before and after effects on the gene intensities from a select subset of 12 tumor samples corresponding to patients 1, 5, 6, 8, 9, 12, 13, 25, 29, 30, 37, and 38, using the data on patient 1 as the baseline array. It can be seen that after normalization, the point clouds are all centered around $M = 0$ and all nonlinear relationships between arrays have been removed. Figure 2.9 displays the normalized densities for all genes for the 40 tumor and 22 normal samples.

2.7 COMPARISON OF NORMALIZATION STRATEGIES AND DISCUSSION

After normalization, the data measured should be adjusted in such a way that subse-quent comparison or analysis should reveal only biological differences relevant to the scientific question being addressed. The separation of normalization and modeling into two distinct tasks is convenient but may be a controversial issue. Normalization that is carried out blindly may result in removing large components of biological sig-nal, treated as noise in the prior step, which are then lost in subsequent analysis. Some researchers may even consider this as "unprincipled" (Kerr et al., 2000) and would prefer to combine these two steps of the analysis within a unified probabilistic frame-work. The question of how different normalization strategies and parameterizations may affect subsequent analysis, such as clustering, does not have a straightforward answer. Different researchers have attempted to carry out comparison studies. For example, Rattray et al. (2001) investigated the effects of normalization and prepro-cessing in a reduced dimension projection based on principal components analysis.

Kroll and Wölfl (2002) recently introduced a new ranking diagram to compare global normalization methods based on mean or median values. They proposed use of the rank intensity plot (RIP) to plot intensity values versus their corresponding ranks. When experimental data are available from different experiments with the same gene set, the RIP should display a separate curve for each experiment, and each curve would feature the global distribution of the values. A measure for the extent of normalization is provided by the *relative rank deviation* (RRD) curve superimposed on RIP curves, defined as the standard deviation of the intensities of a given rank divided by the mean of intensities of that given rank.

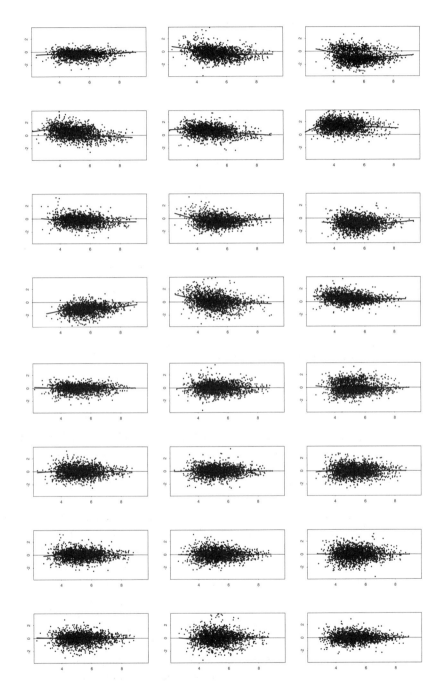

Fig. 2.8 Effect of normalization on the Alon data set. Upper four panels: pairwise MA plots using unadjusted data from 12 randomly selected tissue samples. Lower four panels: pairwise MA plots for these same 12 tissues after normalization.

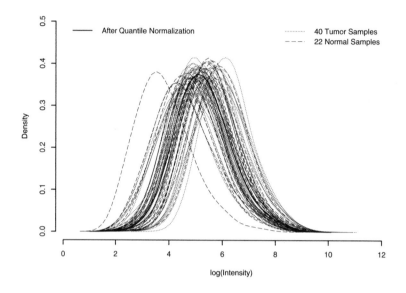

Fig. 2.9 A plot of the densities of the logged expression values for the Alon data set, 40 tumor and 22 normal tissue samples, after quantile normalization.

Hoffmann et al. (2002) used a specific prototype data set to study the impact of global scaling and different invariant-set normalization methods (either taking the raw fluorescence values at the feature level or the average difference at the probe set level as input) and all possible combinations with three different statistical algorithms for detection of differentially expressed genes (parametric ANOVA using the F-test, nonparametric ANOVA using the Kruskal–Wallis test, and the SAM procedure (Tusher, Tibshirani, and Chu, 2001)). Thus, three different statistical criteria were considered for filtering the data to detect differential expression: confidence, fold change, or absolute difference. The general findings of this study indicate that the normalization method has a very high influence, even more so than the choice of statistical criterion, on the final set of differentially expressed genes that were detected. In particular, the number of genes detected as differentially expressed varies by a factor of almost 3. In comparing any two different combinations of methods, the percentage of genes detected simultaneously by both methods vary between 41 and 100%. In particular, under the same normalization scheme, the genes detected by SAM generally are a much smaller subset of those detected by nonparametric ANOVA which, in turn, form a subset of slighter smaller size than those genes detected by parametric ANOVA.

Human gene expression data present great challenges for microarray-based studies due to their genetically diverse nature. In addition, environmental conditions cannot be controlled carefully in humans. Meaningful interpretation of global gene expression in humans will require an extensive characterization of normal variability. Some researchers have suggested the creation of a comprehensive database of normally variable genes for human tissues and organs as well as highly variable genes (Pritchard et al., 2001).

3

Some Cluster Analysis Methods

3.1 INTRODUCTION

In bioinformatics, much attention is centered on the cluster analysis of the tissue samples and also the genes. Cluster analyses have demonstrated their utility in the elucidation of unknown gene function, the validation of gene discoveries, and the interpretation of biological processes; see Alizadeh et al. (2000), Eisen et al. (1998), and Iyer et al. (1999) for examples. The main goal of microarray analysis of many diseases, in particular of unclassified cancer, is to identify novel cancer subtypes for subsequent validation and prediction, and ultimately to develop individualized prognosis and therapy. Limiting factors include the difficulties of tissue acquisition and the expense of microarray experiments. Thus, many experiments attempt to perform a cluster analysis of a small number of tumor samples on the basis of a large number of genes, often resulting in gene-to-sample ratios of approximately 100-fold. Many researchers have explored the use of clustering techniques to arrange genes in some natural order, that is, to organize genes into groups or clusters with similar behavior across relevant tissue samples (or cell lines). Although a cluster does not automatically correspond to a pathway, it is a reasonable approximation that genes in the same cluster have something to do with each other or are directly involved in the same pathway.

The clustering techniques that have been used in the past to cluster tissue samples or the genes include hierarchical agglomerative clustering, k-means clustering, and the self-organizing map, among many others. A natural generalization of these methods is to extend to two-way clustering procedures that simultaneously cluster both genes and samples. These procedures seek global organization of genes and samples. A nice review of these methods with application to DNA microarray experiments can be found in Tibshirani et al. (1999).

In this chapter, we briefly outline in a general context some methods for the cluster analysis of multivariate data consisting of n independent observations $y_1, \ldots, y_n$ taken on a p-dimensional feature vector Y associated with the random phenomenon of interest. The focus is on model-based methods of clustering. These methods will be applied in the next two chapters to the classification of microarray data. Before we proceed to describe model-based clustering, we shall very briefly outline some other methods of cluster analysis that have been commonly used to cluster microarray data. For a more detailed account of cluster analysis, the reader is referred to the many books that either consider or are devoted exclusively to this topic; for example, Hartigan (1975), Hastie et al. (2001b, Chapter 14), Ripley (1996), and Seber (1984, Chapter 7).

Some of the clustering methods to be considered here, like the hierarchical agglomerative methods, can be applied directly to cluster gene expression data. However, there is no reason why the clusters should be hierarchical for microarray data. Also, there are other disadvantages with these hierarchical methods, as to be discussed in Section 3.4. As advocated by Marriott (1974, Page 67), "it is better to consider the clustering problem *ab initio*, without imposing any conditions."

In recent times, model-based clustering has become very popular in the statistical literature. As to be discussed in some detail in Section 3.7.2, an advantage of model-based clustering is that it provides a sound mathematical framework for clustering. In particular, it provides a principled statistical approach to the practical questions that arise in applying clustering methods, namely, the question of what metric (distance function) to adopt and the question of how many clusters there are in the data (Fraley and Raftery, 1998).

3.2 REDUCTION IN THE DIMENSION OF THE FEATURE SPACE

We shall focus on normal mixture models for the clustering of continuous data, which means that a covariance matrix Σ_i has to be estimated for each group G_i. We shall see in Section 3.9 that a mixture model with unrestricted group-covariance matrices in its normal component distributions is a highly parameterized one with $\frac{1}{2}p(p+1)$ parameters for each component-covariance matrix Σ_i $(i = 1, \ldots, g)$. Thus for the clustering of tissue samples on the basis of the available genes, a normal mixture model with unrestricted component-covariance matrices cannot be applied directly to the data.

One way to handle this dimensionality problem is to ignore the correlations between the genes and to cluster the tissue samples by fitting mixtures of normal component distributions with diagonal covariance matrices. Elliptical clusters can be obtained under this restricted model, but their axes must be aligned with the axes of the feature space. Thus it is proposed to use mixtures of factor analyzers. This approach enables a normal mixture model to be fitted to a sample of n data points of dimension p, where p is large relative to n. The number of free parameters is controlled through the dimension of the latent factor space. By working in this reduced space, it allows a model for each component-covariance matrix with complexity lying between that of the isotropic and full covariance structure models.

Even with the use of factor analysis to reduce the number of parameters in the normal mixture model, it is still not feasible to fit mixtures of factor analyzers directly to microarray data without first reducing the dimension of the feature vector. That is, the dimension of the expression-signature vector in the clustering of the tissues and the dimension of the expression profile vector in the clustering of the genes. With the EMMIX-GENE clustering procedure of McLachlan et al. (2002), which has been developed specifically for the clustering of tissue samples, the number of genes are reduced first by a combined screening and clustering approach. The EMMIX-GENE procedure is to be described in the next chapter.

The mixture of factor models discussed above provides a global nonlinear approach to dimension reduction as it postulates a finite mixture of linear submodels (factor models) for the distribution of the full observation vector given the (unobservable) factors. Thus, it is a local dimensionality reduction method.

A more straightforward and commonly used approach to dimension reduction is principal component analysis (PCA), which is to be discussed in Section 3.19 of this chapter. But it is only a global linear method, and so is not as effective as a mixture of factor models. As a consequence, the leading principal components need not necessarily reflect the direction in the feature space best for revealing the group structure of the tissues. Thus a potential problem with a PCA is the determination of an appropriate number of principal components (PCs) useful for clustering. A common practice is to choose the first few leading components. But it is not always clear where to stop and whether some of these components are caused by some artifact or noises in the data unrelated to the clustering task (Liu et al., 2003).

Finally, in the last section of this chapter we consider linear projections of the feature data that are able to incorporate the class structure of the data when available, as in supervised classification (discriminant analysis).

3.3 CLUSTER ANALYSIS

In discriminant analysis (supervised classification), the existence of the different classes is known and there are observations of known class origin (training data) for the purposes of forming a prediction rule. In contrast, cluster analysis (unsupervised classification), is concerned with multivariate techniques that can be used to create groups amongst the observations, where there is no *a priori* information regarding the underlying group structure, or at least where there are no available data from each of the groups if their existence is known. The problem of discriminant analysis is to be considered in Chapter 6.

Available methods of cluster analysis can be categorized broadly as being hierarchical or nonhierarchical. The former category is one in which every cluster obtained at any stage is a merger or split of clusters at other stages. Thus it is possible to visual not only the two extremes of clustering, that is, n clusters with one observation per cluster and a single cluster with all n observations, but also a monotonically increasing strength of clustering as one goes from one level to another. A hierarchical strategy always optimizes a route between these two extremes. An hierarchical method may proceed by progressive fusions, beginning with n single observation groups (agglomerative hierarchy) or it may proceed by progressive divisions, beginning with a single group of n observations (divisive hierarchy). The route may be defined by progressive fusions, beginning with n entities (agglomerative), or by progressive divisions, beginning with a single group and decomposing it finally into individual entities (divisive hierarchy). In practice, divisive methods that seek to optimize some criterion over all possible subdivisions present impossible computational difficulties unless n is very small. For example, there are $2^{n-1} - 1$ ways of making the first subdivision, and to compute a statistic for each of them and choose the optimum quickly is a prohibitive task.

In nonhierarchical procedures, new clusters are obtained by both lumping and splitting of old clusters. Thus the intermediate stages of clustering are different, although the two extremes of clustering are the same as with hierarchical techniques.

3.4   SOME HIERARCHICAL AGGLOMERATIVE TECHNIQUES

Most of the hierarchical agglomerative clustering procedures used in practice, can be implemented with the data represented by a matrix D of proximities, where $D_{jk} = (D)_{jk}$ is the proximity between the jth and kth observations. Usually, D will be symmetric for the adopted method of defining the proximities; see, for example, Section 2.2 of Jain and Dubes (1988) for the properties that proximities must satisfy. The proximity D_{jk} is either a similarity or dissimilarity measure. The closer the features of the observations j and k resemble each other, the larger or smaller D_{jk} is, depending on whether it is a measure of similarity or dissimilarity. For example, if D_{jk} is given by the Euclidean distance between the feature vectors of observations j and k,

$$
\begin{aligned}
D_{jk} &= \|\boldsymbol{x}_j - \boldsymbol{x}_k\| \\
&= \{(\boldsymbol{x}_j - \boldsymbol{x}_k)^T (\boldsymbol{x}_j - \boldsymbol{x}_k)\}^{1/2},
\end{aligned}
\tag{3.1}
$$

then D_{jk} is a dissimilarity measure.

In average linkage clustering, the distance between two clusters is the average of the pairwise distances between two observations, one from the first cluster and the other from the second cluster. With single-link (nearest neighbor) clustering, the distance between two clusters is defined as the distance between their two closest members. For complete-link (farthest neighbor) clustering, the distance between two clusters is defined as the distance between their two farthest members.

It is the different ways of defining distances between clusters that gives rise to the variety of hierarchical clustering techniques. One of the more commonly employed ways of defining the dissimilarity between two expression signatures in microarray analysis is $1 - \rho$ or $1 - |\rho|$, where ρ is the sample correlation. It is invariant under location and scale transformations of the data. So it is invariant under standardization of the columns of the microarray data matrix A. As to be discussed further in Section 3.6, hierarchical agglomerative methods specify in advance the metric for defining distances between points and clusters without sufficient prior knowledge about the problem.

Suppose that cluster r and cluster s are being combined to form the single cluster v. Then Lance and Williams (1967) showed that the distance $d(v, t)$ between this new cluster v and any currently existing cluster t can be expressed as

$$d(v, t) = \alpha_r d(r, t) + \alpha_s d(s, t) + \beta d(r, s) + \gamma |d(r, t) - d(s, t)|, \qquad (3.2)$$

where the coefficients α_r, α_s, β, and γ characterize the particular technique. This formula (3.2) applies to most of the commonly referenced hierarchical clustering techniques. For example, for single linkage, $\alpha_r = \alpha_s = \frac{1}{2}$, $\beta = 0$, and $\gamma = -\frac{1}{2}$, while for complete linkage, the values of the coefficients are the same apart from γ which is now $\frac{1}{2}$.

Under a hierarchical clustering, the data points are thus fashioned into a phylogenetic tree (a dendrogram) whose branch lengths represent the degree of similarity between the sets. Strict phylogenetic trees are best suited to situations of true hierarchical descent (such as in the evolution of species) and are not designed to reflect the multiple distinct ways in which expression patterns can be similar; this problem is exacerbated as the size and complexity of the data set grows (Tamayo et al., 1999).

The results of a hierarchical clustering can be conveniently represented by a dendrogram. It is a special type of tree structure that depicts the nested structure of the clusters. A seminal paper in the analysis of microarray data is Eisen et al. (1998), in which the authors propose hierarchical clustering of genes as a means to identify patterns in the high-dimensional data generated by microarrays.

Bittner et al. (2000) used an average-linkage(agglomerative) hierarchical clustering procedure to cluster $M = 31$ cutaneous melanoma samples on the basis of $N = 3,613$ genes. The distance was based on the correlation ρ (the dissimilarity measure was taken to be 1-ρ). A replica of their dendrogram from Goldstein, Ghosh, and Colon (2002) is given in Figure 3.1. It can be seen in Figure 3.1 that each node of the dendrogram represents a cluster.

Another example of a dendrogram for this data set in Bittner et al. (2000) is given in Figure 3.2, but this time for single linkage with correlation-based distance. Based on cutting the dendrogram in Figure 3.1 at a value of 0.54, Bittner et al. (2000) found a cluster of 19 samples. It can be seen that if we were to cut the dendrogram in Figure 3.2 at the same cutoff point of 0.54, then we would obtain a cluster of 27 tissue samples. Ideally, it would be nice to have an objective function for choosing the cutoff point other than by visual inspection of the dendrogram. With a normal mixture model-based approach to clustering to be discussed shortly, we have an objective function, namely the likelihood function, for choosing a clustering for a specified number of clusters g and for deciding on the value of g. This model-based approach is applied to this data set in Section 4.13.3.

A commonly used hierarchical technique is the agglomerative method of Ward (1963). With this method the objective function to be minimized is the increase in

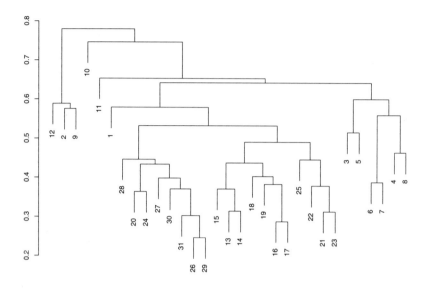

Fig. 3.1   An example of a dendrogram. From Goldstein, Ghosh, and Conlon (2002).

the pooled within-cluster sum of squares as the clustering proceeds hierarchically from g to $g - 1$ clusters, commencing with $g = n$. It can be implemented using the formula (3.2) for the revised distance between two clusters as a consequence of the fusion of two clusters. The appropriate values of the coefficients are given by $\alpha_r = (n_r + n_t)/n_{rst}$, $\alpha_s = (n_s + n_t)/n_{rst}$, $\beta = -n_t/n_{rst}$, and $\gamma = 0$, where $n_{rst} = n_r + n_s + n_t$ and n_r, n_s, and n_t denote the number of members of clusters r, s, and t, respectively. Several of the comparative studies discussed in Jain and Dubes (1988) (Section 3.5.2) have concluded that Ward's method outperforms the other hierarchical clustering methods.

Strictly hierarchical techniques of clustering have proven to be popular, mainly because of their simple implementation in an agglomerative manner, particularly for the clustering of tissue samples containing thousands of genes. As argued in Marriott (1974, Chapter 8), strictly hierarchical techniques may have serious disadvantages, unless there is some special reason for imposing the nested structure of the dendrogram. This structure is not a natural one to impose if the primary purpose of the clustering is to find a natural grouping of the data. It is true that if there is a genuine grouping of the data set, with little or no overlap between the groups, then it will be revealed by any method of cluster analysis. But in less obvious cases, a typical question is whether a partition of the data into $g + 1$ clusters gives a better representation of the data than a partitioning into g clusters for some g. To answer this it is necessary

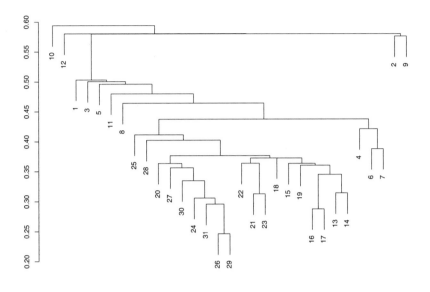

Fig. 3.2   Another example of a dendrogram. From Goldstein, Ghosh, and Conlon (2002).

to compare the best division into g clusters with the best division into $g + 1$ clusters. But hierarchical methods will not usually give both. The deterministic nature of hierarchical clustering can cause points to be grouped based on local decisions, with no opportunity to reevaluate the clustering. In particular, there is the problem of chaining, the risk of which increases with the dimensionality of the feature space. We give in Figure 3.3 a two-dimensional scatter plot, which shows fairly clearly that there are two clusters. Most divisive methods of cluster analysis would produce these two clusters. However, most agglomerative hierarchical methods would join up the chain between them at an early stage in the process, and it would never be broken thereafter. Also, hierarchical clustering has been noted by statisticians to suffer from lack of robustness, nonuniqueness, and the inversion problems that complicate interpretation of the hierarchy (Morgan and Ray, 1995).

Note that care must be taken in interpreting the results of experiments which have been undertaken to compare hierarchical clustering methods with model-based methods. For example, if Euclidean distance is used to assess the quality of the clusters as in Datta and Datta (2003), then methods that use Euclidean distance such as several of the hierarchical ones, will be favored over normal-based methods that use Mahalanobis distance.

Fig. 3.3   Two-dimensional scatter plot illustrating possible chaining of clusters.

3.5 k-MEANS CLUSTERING

A commonly used clustering method is k-means, where k refers to the number of clusters to be imposed on the data. It can be implemented by randomly selecting k observations to be the initial k seeds (cluster centers). The observations are then visited in turn in some prespecified order with an observation being assigned to the ith cluster if it is closest (in Euclidean distance) to the current mean of the ith cluster (the ith seed point in the first instance). After an observation has been assigned to a cluster, its mean is updated and the next observation is visited. The process is terminated when there is no change in the cluster memberships of the observations. As to be discussed further in Section 3.9.3, the k-means method of clustering attempts to impose spherical clusters on the data. However, clusters tend to be elliptical in many problems in practice. Also, the use of squared Euclidean distance is sensitive to atypical observations that can produce very large distances.

The procedure can be made more robust but more computationally expensive as, for example, with k-medoids, as proposed by Kaufman and Rousseeuw (1990). They refer to this approach as PAM (partitioning around medoids). With this approach, a cluster center is restricted to being one of the observations.

Self-Organizing Maps (SOMs) are implemented in a similar manner to k-means clustering, but now the cluster centers (prototypes) are represented as the nodes (initially at random) on a grid usually in two-dimensional space. During training, an observation is selected randomly. The prototype closest to it is identified. The identified prototype is and its neighbors are adjusted to look similar to the observation. This process is repeated until it converges. Although SOMs are favored by some biologists, it may be hard to come up with an initial spatial structure in real problems. The reader is referred to Hastie et al. (2001b, Chapter 14) for further details on the SOM and PAM methods.

In addition to the clustering methods discussed above, many *ad hoc* methods of clustering have been proposed for microarray data.

3.6   CLUSTER ANALYSIS WITH NO *A PRIORI* METRIC

It has been seen with the clustering methods presented in the previous sections that they assume a similarity measure or metric is known *a priori*. Often the Euclidean metric is used as with k-means clustering. However, it is more appropriate to use a distance function (metric) that depends on the shape of the clusters. For example, if a cluster is multivariate normal with mean μ and covariance matrix Σ, the appropriate distance between a point y and the center μ of the cluster is the squared Mahalanobis distance

$$d(y, \mu) = (y - \mu)^T \Sigma^{-1} (y - \mu) \tag{3.3}$$

between y and μ. The difficulty is that the shape of the clusters is not known until the clusters have been identified, and the clusters cannot be effectively identified unless the shapes are known. Indeed, as noted by Hansen and Tukey (1992), "The shakiest part of any clustering procedure is the choice of the metric."

To avoid reliance on any *a priori* metric, Coleman et al. (1999) advocate the use of affine invariant clustering algorithms. This means that the clustering produced on the transformed data $Cy + a$ is the same as on the untransformed data y. Here C is a nonsingular matrix. It means that the clustering is invariant under location (translations of the data), scale (stretchings of the data), and rotation (orientations of the data). Thus affine-invariant metrics are particularly appropriate for use in clustering, since the results do not depend on irrelevant factors such as the units of measurement or the orientation of the clusters in space. Hartigan (1975, Page 63) has commented that "Invariance under this general class of linear transformations seems less compelling than invariance under the change of measuring units of each of the variables."

Essentially, affine invariance of clustering is equivalent to assuming that the metric is quadratic but otherwise unspecified; that is, the distance between any two points y_1 and y_2 is given by

$$d(y_1, y_2) = (y_1 - y_2)^T B^{-1} (y_1 - y_2) \tag{3.4}$$

with B a positive-definite symmetric matrix. Quadratic metrics can arise naturally in a number of ways, such as with mixture models with component distributions such as the multivariate normal or other elliptically symmetric distributions (the t-distribution). Note that Euclidean distance corresponds to the use of (3.4) with B equal to the $p \times p$ identity matrix.

3.7   CLUSTERING VIA FINITE MIXTURE MODELS

3.7.1   Definition

In recent times much attention has been given in the statistical literature to the use of finite mixture models as a device for clustering; see, for example, McLachlan and Basford (1988) and McLachlan and Peel (2000). With this approach, the observed data $\boldsymbol{y}_1, \ldots \boldsymbol{y}_n$, are assumed to have come from a mixture of a finite number, say g, of groups $G_1, \ldots, G_g$ in some unknown proportions $\pi_1, \ldots, \pi_g$. The mixing proportions π_i lie between zero and one, and sum to one. The feature vector $\boldsymbol{Y}$ is taken to have the density $f_i(\boldsymbol{y})$ in group G_i $(i = 1, \ldots, g)$. Thus unconditionally with respect to its group of origin, the feature vector $\boldsymbol{Y}$ has the mixture density

$$f(\boldsymbol{y}) = \sum_{i=1}^{g} \pi_i f_i(\boldsymbol{y}). \tag{3.5}$$

In this mixture framework, the posterior probability that an observation with feature vector $\boldsymbol{y}_j$ belongs to the ith component of the mixture is given by

$$\tau_i(\boldsymbol{y}_j) = \pi_i f_i(\boldsymbol{y}_j) / f(\boldsymbol{y}_j) \tag{3.6}$$

for $i = 1, \ldots, g$.

On specifying a parametric form $f_i(\boldsymbol{y}_j; \boldsymbol{\theta}_i)$ for each component density, we can fit this parametric mixture model

$$f(\boldsymbol{y}_j; \boldsymbol{\Psi}) = \sum_{i=1}^{g} \pi_i \, f_i(\boldsymbol{y}_j; \boldsymbol{\theta}_i) \tag{3.7}$$

by maximum likelihood via the expectation-maximization (EM) algorithm of Dempster, Laird, and Rubin (1977); see also McLachlan and Krishnan (1997). Here $\boldsymbol{\Psi} = (\boldsymbol{\omega}^T, \pi_1, \ldots, \pi_{g-1})^T$ is the vector of unknown parameters, where $\boldsymbol{\omega}$ consists of the elements of the $\boldsymbol{\theta}_i$ known *a priori* to be distinct. In order to estimate $\boldsymbol{\Psi}$ from the observed data, it must be identifiable. This will be so if the representation (3.5) is unique up to a permutation of the component labels.

The actual fitting of finite mixture models by maximum likelihood via the EM algorithm is to be be described shortly in Section 3.11. Let $\hat{\boldsymbol{\Psi}}$ denote the estimate of $\boldsymbol{\Psi}$ so obtained. Then

$$\tau_i(\boldsymbol{y}_j; \hat{\boldsymbol{\Psi}}) = \hat{\pi}_i \, f_i(\boldsymbol{y}_j; \hat{\boldsymbol{\theta}}_i) / \sum_{h=1}^{g} \hat{\pi}_h \, f_h(\boldsymbol{y}_j; \hat{\boldsymbol{\theta}}_h) \tag{3.8}$$

is the estimated posterior probability that the jth observation with feature vector $\boldsymbol{y}_j$ belongs to the ith component of the mixture $(i = 1, \ldots, g; j = 1, \ldots, n)$. The mixture approach gives a probabilistic clustering in terms of these estimated posterior probabilities of component membership. An outright partitioning of the observations into g nonoverlapping clusters $C_1, \ldots, C_g$ is effected by assigning each observation to the component to which it has the highest estimated posterior probability of belonging. Thus the ith cluster C_i contains those observations assigned to group G_i. That is, C_i contains those observations j with $\hat{z}_{ij} = (\hat{\boldsymbol{z}}_j)_i = 1$, where

$$
\begin{aligned}
\hat{z}_{ij} &= 1, && \text{if } \tau_i(\boldsymbol{x}_j; \hat{\boldsymbol{\Psi}}) \geq \tau_h(\boldsymbol{x}_j; \hat{\boldsymbol{\Psi}}) && (h = 1, \ldots, g; h \neq i), \\
&= 0, && \text{otherwise.} &&
\end{aligned} \tag{3.9}
$$

As the notation implies, $\hat{z}_{ij}$ can be viewed as an estimate of z_{ij} which, under the assumption that the observations come from a mixture of g groups $G_1, \ldots, G_g$, is defined to be one or zero according as the jth observation does or does not come from G_i ($i = 1, \ldots, g; j = 1, \ldots, n$).

Although the estimated posterior probabilities $\tau_i(\boldsymbol{y}_j; \hat{\boldsymbol{\Psi}})$ may have limited reliability in small samples, they may well give a satisfactory outright assignment of the data. If the maximum of the fitted posterior probabilities $\tau_i(\boldsymbol{y}_j; \hat{\boldsymbol{\Psi}})$ over $i = 1, \ldots, g$ is near to one for most of the observations $\boldsymbol{y}_j$, then it suggests that the mixture likelihood-based approach can put the n observations into g distinct clusters with a high degree of certainty. Conversely, if the maximum is generally well below one, it indicates that the components of the fitted mixture model are too close together for the n observations to be clustered with any certainty. Hence these estimated posterior probabilities of component membership can be used to provide a measure of the strength of the clustering (Basford and McLachlan, 1985).

The mixture likelihood-based approach to clustering, which is a divisive technique, is usually applied in a nonhierarchical manner. However, initially it may be applied hierarchically, in that any obvious group structure would be removed first before proceeding to investigate any further grouping in the data set. For example, if initially fitting a two-component mixture model splits the data into two widely separated clusters, then the search for further clusters would proceed by fitting mixture models to each of these two clusters considered separately.

In the above, there is a one-to-one correspondence between the mixture components and the groups. For multivariate data of a continuous nature, attention has concentrated on the use of multivariate normal components because of their computational convenience. In those cases where the underlying population consists of g groups in each of which the feature vector is able to be modeled by a single normal distribution, the number of components g in the fitted normal mixture model corresponds to the number of groups. However, when the distribution of a group is unable to be modeled adequately by a single normal distribution but rather needs a normal mixture distribution, the components in the fitted g-component normal mixture model and the consequent clusters will correspond to g subgroups rather than to the smaller number of actual groups represented in the data.

3.7.2 Advantages of Model-Based Clustering

It can be seen that this mixture likelihood-based approach to clustering is model based in that the form of each component density of an observation has to be specified in advance. Hawkins, Muller, and ten Krooden (1982) commented that most writers on cluster analysis "lay more stress on algorithms and criteria in the belief that intuitively reasonable criteria should produce good results over a wide range of possible (and generally unstated) models." For example, the trace $\boldsymbol{W}$ criterion, where $\boldsymbol{W}$ is the pooled within-cluster sums of squares and products matrix, is predicated on normal groups with (equal) spherical covariance matrices; but as they pointed out, many users apply this criterion even in the face of evidence of nonspherical clusters or, equivalently, would use Euclidean distance as a metric. They strongly supported the increasing emphasis on a model-based approach to clustering. Indeed, as remarked by Aitkin, Anderson, and Hinde (1981) in the reply to the discussion of their paper, "when clustering samples from a population, no cluster method is, *a priori* believable without a statistical model." Concerning the use of mixture models to represent nonhomogeneous populations, they noted in their paper that "Clustering methods based on such mixture models allow estimation and hypothesis testing within the framework of standard statistical theory." Previously, Marriott (1974) had noted that the mixture likelihood-based approach "is about the only clustering technique that is entirely satisfactory from the mathematical point of view. It assumes a well-defined mathematical model, investigates it by well-established statistical techniques, and provides a test of significance for the results." In the present context of the analysis of gene expression data, Yeung et al. (2001a) commented that "in the absence of a well-grounded statistical model, it seems difficult to define what is meant by a 'good' clustering algorithm or the 'right' number of clusters." A mixture model-based approach to the clustering of microarray data has been adopted also by Ghosh and Chinnaiyan (2002), Medvedovic and Sivaganesan (2002), and Liu et al. (2003), among others.

In finite mixture models, each component corresponds to a cluster. Thus the problem of choosing an appropriate clustering method can be recast as statistical model choice. Outliers are handled by adding one or more components representing a different distribution for outlying data. It also allows the important question of how many clusters there are in the data to be approached through an assessment of how many components are needed in the mixture model. These questions of model choice can be considered in terms of the likelihood function.

In recent times, microarray experiments are being carried out with replication. With a model-based approach to clustering, the model is able to be adjusted to allow for repeated measurements, as to be discussed in Section 5.15.

3.8   FITTING MIXTURE MODELS VIA THE EM ALGORITHM

We consider now the fitting of the mixture model (3.5) by the method of maximum likelihood. The log likelihood for $\boldsymbol{\Psi}$ formed from the observed feature data

$$y = (y_1^T, \ldots, y_n^T)^T \tag{3.10}$$

is given by

$$\log L(\boldsymbol{\Psi}) = \sum_{j=1}^{n} \log f(y_j; \boldsymbol{\Psi}), \tag{3.11}$$

assuming the observed data represents an observed random sample. As discussed in McLachlan and Peel (2000, Chapter 3), an estimate $\hat{\boldsymbol{\Psi}}$ of $\boldsymbol{\Psi}$ is provided by an appropriate root of the likelihood equation

$$\partial \log L(\boldsymbol{\Psi})/\partial \boldsymbol{\Psi} = \mathbf{0}. \tag{3.12}$$

It is straightforward, at least in principle, to find solutions of (3.12) using the EM algorithm of Dempster et al. (1977). For the purpose of the application of the EM algorithm, the observed data are regarded as being incomplete. The complete data are taken to be the observed feature vectors $\boldsymbol{y}_1, \ldots, \boldsymbol{y}_n$, along with their component-indicator vectors $\boldsymbol{z}_1, \ldots, \boldsymbol{z}_n$, which are unobservable in the framework of the mixture model being fitted. Consistent with the notation introduced in the last section, the ith element z_{ij} of $\boldsymbol{z}_j$ is defined to be one or zero, according as the jth with feature vector $\boldsymbol{y}_j$ does or does not come from the ith component of the mixture, that is, from group G_i $(i = 1, \ldots, g; j = 1, \ldots, n)$.

For this specification, the complete-data log likelihood is

$$\log L_c(\boldsymbol{\Psi}) = \sum_{i=1}^{g} \sum_{j=1}^{n} z_{ij} \log f_i(y_j; \boldsymbol{\theta}_i) + \sum_{i=1}^{g} \sum_{j=1}^{n} z_{ij} \log \pi_i. \qquad (3.13)$$

3.8.1 E-Step

The EM algorithm is easy to program for this problem and proceeds iteratively in two steps, E (for expectation) and M (for maximization). The addition of the unobservable data to the problem (here the $\boldsymbol{z}_j$) is handled by the E-step, which takes the conditional expectation of the complete-data log likelihood, $\log L_c(\boldsymbol{\Psi})$, given the observed data $\boldsymbol{y}$, using the current fit for $\boldsymbol{\Psi}$. Let $\boldsymbol{\Psi}^{(0)}$ be the value specified initially for $\boldsymbol{\Psi}$. Then on the first iteration of the EM algorithm, the E-step requires the computation of the conditional expectation of $\log L_c(\boldsymbol{\Psi})$ given $\boldsymbol{y}$, using $\boldsymbol{\Psi}^{(0)}$ for $\boldsymbol{\Psi}$, which can be written as

$$Q(\boldsymbol{\Psi}; \boldsymbol{\Psi}^{(0)}) = E_{\boldsymbol{\Psi}^{(0)}} \{ \log L_c(\boldsymbol{\Psi}) \mid \boldsymbol{y} \}. \qquad (3.14)$$

The expectation operator E has the subscript $\boldsymbol{\Psi}^{(0)}$ to explicitly convey that this expectation is being effected using $\boldsymbol{\Psi}^{(0)}$ for $\boldsymbol{\Psi}$.

It follows that on the $(k + 1)$th iteration, the E-step requires the calculation of $Q(\boldsymbol{\Psi}; \boldsymbol{\Psi}^{(k)})$, where $\boldsymbol{\Psi}^{(k)}$ is the value of $\boldsymbol{\Psi}$ after the kth EM iteration. As the complete-data log likelihood, $\log L_c(\boldsymbol{\Psi})$, is linear in the unobservable data z_{ij}, the E-step (on the $(k + 1)$th iteration) simply requires the calculation of the current conditional expectation of Z_{ij} given the observed feature observation $\boldsymbol{y}_j$, where Z_{ij} is the random variable corresponding to z_{ij}. Now

$$
\begin{aligned}
E_{\boldsymbol{\Psi}^{(k)}}(Z_{ij} \mid \boldsymbol{y}) &= \mathrm{pr}_{\boldsymbol{\Psi}^{(k)}}\{Z_{ij} = 1 \mid \boldsymbol{y}\} \\
&= \tau_i(\boldsymbol{y}_j; \boldsymbol{\Psi}^{(k)}),
\end{aligned}
\qquad (3.15)
$$

where, corresponding to (3.6),

$$
\begin{aligned}
\tau_i(\boldsymbol{y}_j; \boldsymbol{\Psi}^{(k)}) &= \pi_i^{(k)} f_i(\boldsymbol{y}_j; \boldsymbol{\theta}_i^{(k)}) / f(\boldsymbol{y}_j; \boldsymbol{\Psi}^{(k)}) \qquad (3.16) \\
&= \pi_i^{(k)} f_i(\boldsymbol{y}_j; \boldsymbol{\theta}_i^{(k)}) / \sum_{h=1}^{g} \pi_h^{(k)} f_h(\boldsymbol{y}_j; \boldsymbol{\theta}_h^{(k)})
\end{aligned}
$$

for $i = 1, \ldots, g; j = 1, \ldots, n$. The quantity $\tau_i(\boldsymbol{y}_j; \boldsymbol{\Psi}^{(k)})$ is the posterior probability that the jth member of the sample with observed value $\boldsymbol{y}_j$ belongs to the ith component of the mixture. Using (3.15), we have on taking the conditional expectation of (3.13) given the $\boldsymbol{y}_j$ that

$$Q(\boldsymbol{\varPsi}; \boldsymbol{\varPsi}^{(k)}) = \sum_{i=1}^{g} \sum_{j=1}^{n} \tau_i(\boldsymbol{y}_j; \boldsymbol{\varPsi}^{(k)})\{\log \pi_i + \log f_i(\boldsymbol{y}_j; \boldsymbol{\theta}_i)\}. \qquad (3.17)$$

3.8.2 M-Step

The M-step on the $(k+1)$th iteration requires the global maximization of $Q(\boldsymbol{\varPsi}; \boldsymbol{\varPsi}^{(k)})$ with respect to $\boldsymbol{\varPsi}$ over the parameter space Ω to give the updated estimate $\boldsymbol{\varPsi}^{(k+1)}$. For the finite mixture model, the updated estimates $\pi_i^{(k+1)}$ of the mixing proportions π_i are calculated independently of the updated estimate $\omega^{(k+1)}$ of the parameter vector ω containing the unknown parameters in the component densities.

If the z_{ij} were observable, then the complete-data ML (maximum likelihood) estimate of π_i would be given simply by

$$\hat{\pi}_i = \sum_{j=1}^{n} z_{ij}/n \qquad (i = 1, \ldots, g). \qquad (3.18)$$

As the E-step simply involves replacing each z_{ij} with its current conditional expectation $\tau_i(\boldsymbol{y}_j; \boldsymbol{\varPsi}^{(k)})$ in the complete-data log likelihood, the updated estimate of π_i is given by replacing each z_{ij} in (3.18) by $\tau_i(\boldsymbol{y}_j; \boldsymbol{\varPsi}^{(k)})$ to give

$$\pi_i^{(k+1)} = \sum_{j=1}^{n} \tau_i(\boldsymbol{y}_j; \boldsymbol{\varPsi}^{(k)})/n \qquad (i = 1, \ldots, g). \qquad (3.19)$$

Thus in forming the estimate of π_i on the $(k+1)$th iteration, there is a contribution from each observation $\boldsymbol{y}_j$ equal to its (currently assessed) posterior probability of membership of the ith component of the mixture model.

Concerning the updating of ω on the M-step of the $(k+1)$th iteration, it can be seen from (3.17) that $\omega^{(k+1)}$ is obtained as an appropriate root of

$$\sum_{i=1}^{g} \sum_{j=1}^{n} \tau_i(\boldsymbol{y}_j; \boldsymbol{\varPsi}^{(k)})\partial \log f_i(\boldsymbol{y}_j; \boldsymbol{\theta}_i)/\partial \omega = 0. \qquad (3.20)$$

One nice feature of the EM algorithm is that the solution of (3.20) often exists in closed form, as is to be demonstrated for the normal mixture model in Section 3.9.

The E- and M-steps are alternated repeatedly until the difference

$$L(\boldsymbol{\varPsi}^{(k+1)}) - L(\boldsymbol{\varPsi}^{(k)})$$

changes by an arbitrarily small amount in the case of convergence of the sequence of likelihood values $\{L(\boldsymbol{\varPsi}^{(k)})\}$. Dempster et al. (1977) showed that the (incomplete-data) likelihood function $L(\boldsymbol{\varPsi})$ is not decreased after an EM iteration; that is,

$$L(\boldsymbol{\varPsi}^{(k+1)}) \geq L(\boldsymbol{\varPsi}^{(k)}) \qquad (3.21)$$

for $k = 0, 1, 2, \ldots$. Hence, convergence must be obtained with a sequence of likelihood values $\{L(\boldsymbol{\Psi}^{(k)})\}$ that are bounded above. In almost all cases, the limiting value L^* is a local maximum. In any event, if an EM sequence $\{\boldsymbol{\Psi}^{(k)}\}$ is trapped at some stationary point $\boldsymbol{\Psi}^*$ that is not a local or global maximizer of $L(\boldsymbol{\Psi})$ (for example, a saddle point), a small random perturbation of $\boldsymbol{\Psi}$ away from the saddle point $\boldsymbol{\Psi}^*$ will cause the EM algorithm to diverge from the saddle point. Further details may be found in the monograph of McLachlan and Krishnan (1997, Chapter 3) on the EM algorithm.

Let $\hat{\boldsymbol{\Psi}}$ be the chosen solution of the likelihood equation. For an observed sample, $\hat{\boldsymbol{\Psi}}$ is usually taken to be the root of (3.12) corresponding to the largest of the local maxima located. That is, in those cases where $L(\boldsymbol{\Psi})$ has a global maximum in the interior of the parameter space, $\hat{\boldsymbol{\Psi}}$ is the maximum likelihood estimate of $\boldsymbol{\Psi}$, assuming that the global maximum has been located.

3.8.3 Choice of Starting Values for the EM Algorithm

The monograph of McLachlan and Peel (2000) provides an in-depth account of the fitting of finite mixture models. Briefly, with mixture models the likelihood typically will have multiple maxima; that is, the likelihood equation will have multiple roots. There are obvious difficulties with this selection in the typical cluster analysis setting, where there is little or no *a priori* knowledge of any formal group structure on the underlying population. As the likelihood equation (3.12) tends to have multiple roots corresponding to local maxima, the EM algorithm needs to be started from a variety of initial values for the parameter vector $\boldsymbol{\Psi}$ or for a variety of initial partitions of the data into g groups. The latter can be obtained by randomly dividing the data into g groups corresponding to the g components of the mixture model. With random starts, the effect of the central limit theorem tends to have the component parameters initially being similar at least in large samples. Nonrandom partitions of the data can be obtained via some clustering procedure such as k-means. Also, Coleman et al. (1999) have proposed some procedures for obtaining nonrandom starting partitions.

3.9 CLUSTERING VIA NORMAL MIXTURES

Frequently, in practice, the clusters in the data are essentially elliptical, so that it is reasonable to consider fitting mixtures of elliptically symmetric component densities. Within this class of component densities, the multivariate normal density is a convenient choice given its computational tractability.

3.9.1 Heteroscedastic Components

Under the assumption of multivariate normal components, the ith component-conditional density $f_i(\boldsymbol{y}; \boldsymbol{\theta}_i)$ is given by

$$f_i(\boldsymbol{y}; \boldsymbol{\theta}_i) = \phi(\boldsymbol{y}; \boldsymbol{\mu}_i, \boldsymbol{\Sigma}_i), \tag{3.22}$$

where $\boldsymbol{\theta}_i$ consists of the elements of $\boldsymbol{\mu}_i$ and the $\frac{1}{2}p(p+1)$ distinct elements of $\boldsymbol{\Sigma}_i$ $(i = 1, \ldots, g)$. Here

$$\phi(\boldsymbol{y}; \boldsymbol{\mu}_i, \boldsymbol{\Sigma}_i) = (2\pi)^{-\frac{p}{2}} |\boldsymbol{\Sigma}_i|^{-1/2} \exp\{-\tfrac{1}{2}(\boldsymbol{y} - \boldsymbol{\mu}_i)^T \boldsymbol{\Sigma}_i^{-1}(\boldsymbol{y} - \boldsymbol{\mu}_i)\} \qquad (3.23)$$

In this case the solution of (3.20) exists in closed form. It follows that on the M-step of the $(k+1)$th iteration, the updates of the component means $\boldsymbol{\mu}_i$ and component-covariance matrices $\boldsymbol{\Sigma}_i$ are given explicitly by

$$\boldsymbol{\mu}_i^{(k+1)} = \sum_{j=1}^{n} \tau_{ij}^{(k)} \boldsymbol{y}_j / \sum_{j=1}^{n} \tau_{ij}^{(k)} \qquad (3.24)$$

and

$$\boldsymbol{\Sigma}_i^{(k+1)} = \sum_{j=1}^{n} \tau_{ij}^{(k)} (\boldsymbol{y}_j - \boldsymbol{\mu}_i^{(k+1)})(\boldsymbol{y}_j - \boldsymbol{\mu}_i^{(k+1)})^T / \sum_{j=1}^{n} \tau_{ij}^{(k)} \qquad (3.25)$$

for $i = 1, \ldots, g$, where

$$\tau_{ij}^{(k)} = \tau_i(\boldsymbol{y}_j; \boldsymbol{\Psi}^{(k)}) \qquad (i = 1, \ldots, g; \; j = 1, \ldots, n).$$

The updated estimate of the ith mixing proportion π_i is as given by (3.19).

3.9.2 Homoscedastic Components

Often in practice, the component-covariance matrices $\boldsymbol{\Sigma}_i$ are restricted to being the same,

$$\boldsymbol{\Sigma}_i = \boldsymbol{\Sigma} \qquad (i = 1, \ldots, g), \qquad (3.26)$$

where $\boldsymbol{\Sigma}$ is unspecified. In this case of homoscedastic normal components, the updated estimate of the common component-covariance matrix $\boldsymbol{\Sigma}$ is given by

$$\boldsymbol{\Sigma}^{(k+1)} = \sum_{i=1}^{g} \pi_i^{(k)} \boldsymbol{\Sigma}_i^{(k+1)} / n, \qquad (3.27)$$

where $\boldsymbol{\Sigma}_i^{(k+1)}$ is given by (3.25), and the updates of π_i and $\boldsymbol{\mu}_i$ are as above in the heteroscedastic case.

3.9.3 Spherical Components

A further simplification is to take the component-covariance matrices to have a common spherical form, where the covariance matrix of each component is taken to be a multiple of the $p \times p$ identity matrix $\boldsymbol{I}_p$, namely

$$\boldsymbol{\Sigma}_i = \sigma^2 \boldsymbol{I}_p \qquad (i = 1, \ldots, g). \qquad (3.28)$$

The constraint (3.28) means that the clusters produced are spherical. If we also take the mixing proportions to be equal, then it is equivalent to a "soft" version of k-means clustering. It is a soft version as with k-means, the observations are assigned outright at each of the iterations. That is, the current values of the posterior probabilities $\tau_{ij}^{(k)}$ are replaced by $\hat{z}_{ij}^{(k)}$, where

$$
\begin{aligned}
\hat{z}_{ij}^{(k)} &= 1, & \text{if } \tau_{ij}^{(k)} \geq \tau_{hj}^{(k)} \quad (h = 1, \ldots, g; h \neq i), \\
&= 0, & \text{otherwise.}
\end{aligned}
\tag{3.29}
$$

3.9.4 Choice of Root

The choice of root of the likelihood equation in the case of homoscedastic normal components is straightforward in the sense that the ML estimate exists as the global maximizer of the likelihood function. The situation is less straightforward in the case of heteroscedastic normal components as the likelihood function is unbounded. It is known that as the sample size goes to infinity, there exists a sequence of roots of the likelihood equation that is consistent and asymptotically efficient. With probability tending to one, these roots correspond to local maxima in the interior of the parameter space; see McLachlan and Peel (2000, Chapter 3). Usually, the intent is to choose as the ML estimate of the parameter vector $\boldsymbol{\Psi}$ the local maximizer corresponding to the largest of the local maxima located. But in practice, consideration has to be given to the problem of relatively large local maxima that occur as a consequence of a fitted component having a very small (but nonzero) variance for univariate data or generalized variance (the determinant of the covariance matrix) for multivariate data. Such a component corresponds to a cluster containing a few data points either relatively close together or almost lying in a lower-dimensional subspace in the case of multivariate data. There is thus a need to monitor the relative size of the fitted mixing proportions and of the component variances for univariate observations, or of the generalized component variances for multivariate data, in an attempt to identify these spurious local maximizers.

3.9.5 Available Software

The reader is referred to the appendix in McLachlan and Peel (2000b) for the availability of software for the fitting of normal mixture models, including the EMMIX program of McLachlan et al. (1999). The current version of EMMIX is available from the World Wide Web address

```
http://www.maths.uq.edu.au/~gjm/emmix/emmix.html
```

Concerning the availability of mixture modeling facilities in general-purpose statistical packages, there is the MCLUST software package of Fraley and Raftery (1998), which is interfaced to the S-PLUS commercial software.

3.10 MIXTURES OF t DISTRIBUTIONS

For many applied problems, the tails of the normal distribution are often shorter than appropriate. Also, the estimates of the component means and covariance matrices can be affected by observations that are atypical of the components in the normal mixture model being fitted. McLachlan and Peel (2000, Chapter 7) and Peel and McLachlan (2000) have considered the fitting of mixtures of (multivariate) t-distributions. The t distribution provides a longer tailed alternative to the normal distribution. Hence it provides a more robust approach to the fitting of normal mixture models, as observations that are atypical of a normal component are given reduced weight in the calculation of its parameters.

The t density with location parameter μ_i, positive-definite matrix Σ_i, and ν_i degrees of freedom is given by

$$\phi_t(y_j; \mu_i, \Sigma_i, \nu_i) = \frac{\Gamma(\frac{\nu_i+p}{2})|\Sigma_i|^{-1/2}}{(\pi\nu_i)^{\frac{1}{2}p}\Gamma(\frac{\nu_i}{2})\{1 + \delta(y_j, \mu_i; \Sigma_i)/\nu_i\}^{\frac{1}{2}(\nu_i+p)}}, \qquad (3.30)$$

where

$$\delta(y_j, \mu_i; \Sigma_i) = (y_j - \mu_i)^T \Sigma_i^{-1}(y_j - \mu_i) \qquad (3.31)$$

denotes the Mahalanobis squared distance between y_j and μ_i (with Σ_i as the covariance matrix). If $\nu_i > 1$, μ_i is the mean of Y_j, and if $\nu_i > 2$, $\nu_i(\nu_i - 2)^{-1}\Sigma_i$ is its covariance matrix. As ν_i tends to infinity, Y_j becomes marginally multivariate normal with mean μ_i and covariance matrix Σ_i. Hence this parameter ν_i may be viewed as a robustness tuning parameter. It can be fixed in advance or it can be inferred from the data for each component, thereby providing an *adaptive* robust procedure (McLachlan and Peel, 2000).

The t distribution does not have substantially better breakdown behavior than the normal. The advantage of the t mixture model is that, although the number of outliers needed for breakdown is almost the same as with the normal mixture model, the outliers have to be much larger. This point is made more precise by Hennig (2002) who has provided an excellent account of breakdown points for ML estimation of location-scale mixtures with a fixed number of components g.

3.11 MIXTURES OF FACTOR ANALYZERS

As noted in Section 3.1, a normal model with unrestricted group-covariance matrices is a highly parameterized one with $\frac{1}{2}p(p + 1)$ parameters for each component-covariance matrix Σ_i $(i = 1, \ldots, g)$. Banfield and Raftery (1993) introduced a parameterization of the component-covariance matrix Σ_i based on a variant of the standard spectral decomposition of Σ_i.

Another approach for reducing the number of unknown parameters in the form for a component-covariance matrix is factor analysis, which models the covariance structure of high-dimensional data using a small number of latent variables. For clustering purposes, this model can be extended by the mixture of factor analyzers model, which effectively allows different local factor models in different regions of the feature space (McLachlan and Peel, 2000a,b). This model was originally proposed by Ghahramani and Hinton (1997) and Hinton, Dayan, and Revow (1997) for the purposes of visualizing high-dimensional data in a lower-dimensional space to explore for group structure.

With the mixture of factor analyzers model, the ith component-covariance matrix Σ_i has the form

$$\Sigma_i = B_i B_i^T + D_i \quad (i = 1, \ldots, g), \tag{3.32}$$

where B_i is a $p \times q$ matrix of factor loadings and D_i is a diagonal matrix. It assumes that the component correlations between the observations can be explained by the conditional linear dependence of the latter on q latent or unobservable variables specific to the given component. Unlike the PCA model, the factor analysis model (3.32) enjoys a powerful invariance property: changes in the scales of the feature variables in y_j appear only as scale changes in the appropriate rows of the matrix B_i of factor loadings.

If the number of factors q is chosen sufficiently smaller than p, the representation (3.32) imposes some constraints on the component-covariance matrix Σ_i and thus reduces the number of free parameters to be estimated. Note that in the case of $q > 1$, there is an infinity of choices for B_i, since (3.32) is still satisfied if B_i is replaced by $B_i C_i$, where C_i is any orthogonal matrix of order q. One (arbitrary) way of uniquely specifying B_i is to choose the orthogonal matrix C_i so that $B_i^T D_i^{-1} B_i$ is diagonal (with its diagonal elements arranged in decreasing order). Assuming that the eigenvalues of $B_i B_i^T$ are positive and distinct, the condition that $B_i^T D_i^{-1} B_i$ is diagonal as above imposes $\frac{1}{2}q(q-1)$ constraints on the parameters. Hence then the number of free parameters for each component-covariance matrix is

$$pq + p - \tfrac{1}{2}q(q-1).$$

In our experience with microarray data sets, we have found that the choice of the number of factors q is not crucial in the clustering of the tissue samples. A formal test for q can be undertaken using the likelihood ratio λ, as regularity conditions hold for this test conducted at a given value for the number of components g. For the null hypothesis that $H_0 : q = q_0$ versus the alternative $H_1 : q = q_0 + 1$, the statistic $-2 \log \lambda$ is asymptotically chi-squared with $d = g(p - q_0)$ degrees of freedom. However, in situations where n is not large relative to the number of unknown parameters, we prefer the use of the Bayesian information criterion (BIC) of Schwarz (1978). Applied in this context, it means that twice the increase in the log likelihood $(-2 \log \lambda)$ has to be greater than $d \log n$ for the null hypothesis to be rejected.

With the factor analysis model, we avoid having to compute the inverses of iterates of the estimated $p \times p$ covariance matrix Σ_i that may be singular for large p relative to n. This is because the inversion of the current value of the $p \times p$ matrix $(B_i B_i^T + D_i)$ on each iteration can be undertaken using the result that

$$(B_i B_i^T + D_i)^{-1} = D_i^{-1} - D_i^{-1} B_i (I_q + B_i^T D_i^{-1} B_i)^{-1} B_i^T D_i^{-1}, \tag{3.33}$$

where the right-hand side of (3.33) involves only the inverses of $q \times q$ matrices, since D_i is a diagonal matrix. The determinant of $(B_i B_i^T + D_i)$ can then be calculated as

$$| B_i B_i^T + D_i | = | D_i | / | I_q - B_i^T (B_i B_i^T + D_i)^{-1} B_i | \, .$$

The estimates of the elements of the diagonal matrix D_i (the uniquenesses) will be close to zero if effectively not more than q observations are unequivocally assigned to the ith component of the mixture in terms of the fitted posterior probabilities of component membership (McLachlan et al., 2003). This will lead to spikes or near singularities in the likelihood. One way to avoid this is to impose the condition of a common value D for the D_i,

$$D_i = D \quad (i = 1, \dots, g). \tag{3.34}$$

The mixture of probabilistic component analyzers (PCAs) model, as proposed by Tipping and Bishop (1999), has the form (3.32) with each D_i now having the isotropic structure

$$D_i = \sigma_i^2 I_p \quad (i = 1, \dots, g), \tag{3.35}$$

where I_p denotes the $p \times p$ identity matrix.

The mixtures of factor analyzers model can be fitted by using the alternating expectation–conditional maximization (AECM) algorithm (Meng and van Dyk, 1997). McLachlan et al. (2002) showed how use can be made of the link of factor analysis with the probabilistic PCA model (3.35) to specify an initial value $\Psi^{(0)}$ for Ψ.

3.12   CHOICE OF CLUSTERING SOLUTION

If it were known that the data at hand came from a mixture of g normal components, then the aim is to seek a single solution of the likelihood equation to estimate the vector of parameters in the mixture model. However, in the absence of such knowledge in a clustering context, it is not suggested that the clustering of a data set be based solely on a single solution of the likelihood equation, but rather on the various solutions considered collectively. The set of plausible solutions may reveal that there are some points that always cluster together (core members). Also, often the clusters are meant to correspond to some discrete states (degrees of sickness of a patient). As there is a continuous gradation, in some patients, from one state to the other, it is reasonable to expect some overlap between the clusters. In such cases, the best one can hope to achieve is to identify the "core" patients of the states.

In applications of normal mixture models to the clustering of tissue samples, the sample size is typically too small relative to the dimension of the feature space to enable a choice to be made about the forms of the normal components on the basis of the likelihood function. In such situations, care has to be exercised in allowing the component-covariance matrices Σ_i to be completely unrestricted as then it is very difficult to distinguish between genuine and spurious local maximizers, assuming that the EM algorithm has managed to avoid some of the singularities.

For very sample sample sizes, the Σ_i may have to be constrained to be equal. For sufficiently large sample sizes, we attempt to obtain solutions with less stringent constraints on the Σ_i. One way to proceed is to fit mixtures of normal components with the component-covariance matrices Σ_i taken to be unequal but diagonal. That is, the clusters are allowed to have different shapes, but their axes must be aligned with the axes of the feature space. Call this solution S_{UD}. Solutions with the orientations of the clusters not necessarily aligned with the axes of the feature space can be obtained by (1) fitting mixtures of factor analyzers with unrestricted component-covariance matrices (that is, unequal matrices B_i of factor loadings and unequal diagonal matrices D_i), starting the EM algorithm from S_{UD}; (2) fitting mixtures of factor analyzers with unequal factor loadings B_i but equal diagonal matrices D_i, using random and k-means-based starts. Note that if mixtures of factor analyzers are fitted from random or k-means-based starts for very small sample sizes, the EM algorithm will typically not converge.

3.13 CLASSIFICATION ML APPROACH

Another likelihood-based approach to clustering besides the mixture likelihood approach is what is sometimes called the classification likelihood approach. With this approach, Ψ and the unknown component-indicator vectors $z_1, \ldots, z_n$ of the observed feature data $y_1, \ldots, y_n$ are chosen to maximize $L_c(\Psi)$, the likelihood for Ψ formed on the basis of the so-called complete-data as introduced within the EM framework for the ML fitting of the mixture likelihood. In principle, the maximization process for the classification likelihood approach can be carried out for arbitrary n, since it is just a matter of computing the maximum value of $L_c(\Psi)$ over all possible partitions of the n observations to the g components. In some situations, for example with multivariate normal component densities with unequal covariance matrices, the restriction that at least $p + 1$ observations belong to each component is needed to avoid the degenerate case of infinite likelihood. Unless n is small, however, searching over all possible partitions is prohibitive. As noted by McLachlan (1982), a solution corresponding to a local maximum can be computed iteratively by alternating a modified version of the E-step with the same M-step, as described in Section 3.8.2 for the application of the EM algorithm in fitting the mixture model (3.5). In the E-step on the $(k + 1)$th iteration, z_{ij} is replaced not by the current estimate of the posterior probability that the jth entity belongs to the ith component, but by one or zero according to whether

$$\pi_i^{(k)} f_i(y_j; \theta_i^{(k)}) \geq \pi_h^{(k)} f_h(y_j; \theta_h^{(k)}) \quad (h = 1, \ldots, g; \, h \neq i)$$

holds or not $(i = 1, \ldots, g; \, j = 1, \ldots, n)$.

The classification ML approach can be shown to be equivalent to some commonly used clustering criteria under the assumption of normal groups with various constraints on their covariance matrices, as noted originally by Scott and Symons (1971). For example, if the mixing proportions are taken to be equal or, equivalently, a separate sampling scheme is assumed for the data, then the classification ML approach with the constraint of equal covariance matrices leads to the $|\, W\, |$ criterion, as originally suggested by Friedman and Rubin (1967). If the covariance matrices are further assumed to be diagonal, then it yields the trace W criterion or, equivalently, the k-means procedure. More recently, Celeux and Govaert (1995) have considered the equivalence of the classification ML approach to other clustering criteria under varying assumptions on the component densities. From an estimation point of view, the classification ML approach yields inconsistent estimates of the parameters (McLachlan and Peel, Chapter 2, 2000).

3.14 MIXTURE MODELS FOR CLINICAL AND MICROARRAY DATA

In this section, we consider the case where, in addition to the microarray expression data, there are also available data of a clinical nature on the cases on which the tissue samples have been recorded. The tissue samples can be clustered on the basis of the clinical and microarray data considered separately. But the simultaneous use of the clinical and microarray should lead to more powerful clustering procedures in situations where the clinical data contains information beyond that provided by the microarray experiments.

Two types of mixture models (unconditional and conditional) are proposed for the simultaneous use of clinical and microarray data for the clustering of tissue samples. With the unconditional approach, the mixture distribution models the joint distribution of the clinical and microarray data, while with the conditional approach, the mixture distribution models the conditional distribution of the microarray data given the clinical data. These approaches are to be illustrated in the clustering of breast cancer tissues, as studied recently in van 't Veer et al. (2002).

It is supposed that the microarray data consist of n tissue samples, $\boldsymbol{y}_1, \ldots, \boldsymbol{y}_n$, from n microarray experiments on p genes. That is, $\boldsymbol{y}_j$ is a p-dimensional vector. It is assumed further that there is available a vector $\boldsymbol{x}_j$ of clinical measurements taken on the jth case with tissue sample $\boldsymbol{y}_j$ $(j = 1, \ldots, n)$. For the clustering of the n tissue samples (really the n cases) into g clusters, we shall fit a g-component mixture model, where the ith component represents the ith external class G_i corresponding to the ith cluster $(i = 1, \ldots, g)$. We let z_j be the (unobservable) class indicator associated with the jth tissue sample $\boldsymbol{y}_j$, where $z_j = i$ implies that the jth case is from the ith class $(i = 1, \ldots, g)$.

3.14.1 Unconditional Approach

The combined clinical and microarray data $(\boldsymbol{y}_j^T, \boldsymbol{x}_j^T)^T$ $(j = 1, \ldots, n)$ are taken to be n (independent) realizations from the mixture density,

$$
\begin{aligned}
f(\boldsymbol{y}, \boldsymbol{x}) &= \sum_{i=1}^{g} \pi_i \, f_i(\boldsymbol{y}, \boldsymbol{x}) \\
&= \cdot \sum_{i=1}^{g} \pi_i \, f_i(\boldsymbol{x}) f_i(\boldsymbol{y} \mid \boldsymbol{x}),
\end{aligned}
\tag{3.36}
$$

where $\pi_i = \mathrm{pr}\{Z = i\}$, $f_i(\boldsymbol{x})$ denotes the ith class-conditional density of the vector $\boldsymbol{x}$ of clinical features, and $f_i(\boldsymbol{y} \mid \boldsymbol{x})$ denotes the ith class-conditional density of the vector of the gene expression levels given the clinical-data vector $\boldsymbol{x}$ $(i = 1, \ldots, g)$. The symbol f is being used generically here to denote a density where, for discrete random variables, the density is really a probability function.

On specifying the forms of the densities of $f_i(\boldsymbol{x})$ and $f_i(\boldsymbol{y} \mid \boldsymbol{x})$, we can fit the mixture model (3.36) by maximum likelihood, using the EM algorithm of Dempster et al. (1977); see McLachlan and Krishnan (1997) and McLachlan and Peel (2001). In practice, the clinical features are usually nearly all discrete variables or are coded to be so. In discriminant and cluster analyses, it has been found that it is reasonable to proceed by treating discrete variables as if they are independently distributed within a class or cluster. This is known as the NAIVE assumption (Hand and Yu, 2001). Under this assumption, the ith class-conditional density of the vector of clinical features reduces to

$$
f_i(\boldsymbol{x}) = \prod_{v=1}^{p} f_{iv}(x_v),
\tag{3.37}
$$

where $f_{iv}(x_v)$ denotes the ith class-conditional density of the vth clinical feature in $\boldsymbol{x}$.

Concerning the ith class-conditional density of the vector $\boldsymbol{y}$ of gene expressions given the clinical-data vector $\boldsymbol{x}$, we can take $\boldsymbol{y}$ not to depend on the clinical-data vector $\boldsymbol{x}$ and model its marginal density $f_i(\boldsymbol{y})$ by the multivariate normal density. For clinical features that are all discrete, we can allow for some dependence between the microarray-data vector $\boldsymbol{y}$ and the clinical-data vector $\boldsymbol{x}$ by adopting the location model as, for example, in Hunt and Jorgensen (1999). With the location model, $f_i(\boldsymbol{y} \mid \boldsymbol{x})$ is taken to be multivariate normal with a mean that is allowed to be different for some or all of the various levels of $\boldsymbol{x}$.

Given that the dimension p of the vector $\boldsymbol{y}$ of gene expressions is so much greater than the number n of available tissue samples, we would not be able to use all the genes in $\boldsymbol{x}$. In the example to be presented in Section 4.15, we replace $\boldsymbol{x}$ by the vector of the means of the first 15 groups into which the genes have been clustered via the EMMIX-GENE software of McLachlan et al. (2002).

3.14.2 Conditional Approach

As an alternative to the use of the full mixture model (3.36), we may proceed conditionally on the realized values of the clinical-data vectors $x_1, \ldots, x_n$. This leads to the use of the conditional mixture model,

$$f(y \mid x) = \sum_{i=1}^{g} \pi_i(x) f_i(y \mid x), \qquad (3.38)$$

where $\pi_i(x)$ denotes the conditional probability that the class indicator takes on the value i given the vector x of clinical features. A common model for $\pi_i(x)$ is the logistic model under which

$$\pi_i(x) = \frac{\exp(\beta_{i0} + \beta_i^T x)}{1 + \sum_{h=1}^{g-1} \exp(\beta_{h0} + \beta_h^T x)} \qquad (3.39)$$

where $\beta_i = (\beta_{i1}, \ldots, \beta_{ip})^T$ for $i = 1, \ldots, g-1$, and

$$\pi_g(x) = 1 - \sum_{h=1}^{g-1} \pi_h(x).$$

3.15 CHOICE OF THE NUMBER OF COMPONENTS IN A MIXTURE MODEL

With a mixture model-based approach to clustering, the question of how many clusters there are can be considered in terms of the number of components of the mixture model being used.

3.15.1 Order of a Mixture Model

A mixture density with g components might be empirically indistinguishable from one with either fewer than g components or more than g components. It is therefore sensible in practice to approach the question of the number of components in a mixture model in terms of an assessment of the smallest number of components in the mixture compatible with the data. To this end, the true order g_o of the g-component mixture model

$$f(y; \Psi) = \sum_{i=1}^{g} \pi_i f_i(y; \theta_i) \qquad (3.40)$$

is defined to be the smallest value of g such that all the components $f_i(y; \theta_i)$ are different and all the associated mixing proportions π_i are nonzero.

3.15.2 Approaches for Assessing Mixture Order

In most of the work on inference on the number of components g in a mixture model, Bayesian or otherwise, the approach has been to separate the problem of testing for

g from the fitting of the mixture model, and hence estimation, for fixed g. However, with the Bayesian approach, increasing attention is being given to the more direct line of modeling the unknown g case by mixing over the fixed g case (Phillips and Smith, 1996; Richardson and Green, 1997).

The estimation of the order of a mixture model has been considered mainly by consideration of the likelihood, using two main ways. One way is based on a penalized form of the likelihood. As the likelihood increases with the addition of a component to a mixture model, the likelihood (usually, the log likelihood) is penalized by the subtraction of a term that "penalizes" the model for the number of parameters in it. This leads to a penalized log likelihood, yielding what are called information criteria for the choice of g; see McLachlan and Peel, (2000, Chapter 6).

The other main way for deciding on the order of a mixture model is to carry out a hypothesis test, using the likelihood ratio as the test statistic. Penalized likelihood criteria, like Akaike's AIC and the Bayesian information criterion (BIC), are less demanding than the likelihood ratio test (LRT), which requires bootstrapping in order to obtain an assessment of the P-value. However, they produce no number that quantifies the confidence in the result, such as a P-value.

3.15.3 Bayesian Information Criterion

The main Bayesian-based information criteria use an approximation to the integrated likelihood, as in the original proposal by Schwarz (1978) leading to his Bayesian information criterion (BIC). Available general theoretical justifications of this approximation rely on the same regularity conditions that break down for inference on the number of components in a frequentist framework.

In the literature, the information criteria so formed are generally expressed in terms of twice the negative difference between the log likelihood and the penalty term. This difference for the Bayesian information criterion (BIC) is given by

$$-2\log L(\hat{\boldsymbol{\Psi}}) + d\log n \tag{3.41}$$

where d is the number of parameters in the model. The intent is to minimize the criterion (3.41) in model selection, including the present situation for the number of components g in a mixture model.

3.15.4 Integrated Classification Likelihood Criterion

Another criterion in the cluster analysis context is the ICL (Integrated Classification Likelihood) criterion, which is an *a la* BIC approximation to the complete-data log likelihood. It was proposed by Biernacki, Celeux, and Govaert (1998). In its simplified form, it is given by

$$-2\log L(\hat{\boldsymbol{\Psi}}) + 2EN(\hat{\boldsymbol{\tau}}) + d\log n, \tag{3.42}$$

where

$$EN(\boldsymbol{\tau}) = -\sum_{i=1}^{g}\sum_{j=1}^{n} \tau_{ij} \log \tau_{ij}$$

is the entropy of the fuzzy classification matrix $\boldsymbol{C} = ((\tau_{ij}))$ and where

$$\boldsymbol{\tau} = (\boldsymbol{\tau}_1^T, \ldots, \boldsymbol{\tau}_n^T)^T, \tag{3.43}$$

and

$$\boldsymbol{\tau}_j = (\tau_1(\boldsymbol{y}_j; \boldsymbol{\Psi}), \ldots, \tau_g(\boldsymbol{y}_j; \boldsymbol{\Psi}))^T \tag{3.44}$$

is the vector of posterior probabilities of component membership of $\boldsymbol{y}_j$ ($j = 1, \ldots, n$).

3.16 RESAMPLING APPROACH

A guide to the final choice of g can be obtained from monitoring the increase in the log likelihood as g is increased from a single component. Unfortunately, it is difficult to carry out formal tests at any stage of this sequential process for the need of an additional component, since as is well known, regularity conditions fail to hold for the likelihood ratio statistic λ to have its usual asymptotic null distribution of chi-squared with degrees of freedom equal to the difference between the number of parameters under the null and alternative hypotheses.

A formal test of the null hypothesis $H_0 : g = g_0$ versus the alternative $H_1 : g = g_1$ ($g_1 > g_0$) can be undertaken using a resampling method, as described in McLachlan (1987). Previously, Aitkin et al. (1981) had adopted a resampling approach in the context of a latent class analysis. Bootstrap samples are generated from the mixture model fitted under the null hypothesis of g_0 components. That is, the bootstrap samples are generated from the g_0-component mixture model with the vector $\boldsymbol{\Psi}$ of unknown parameters replaced by its ML estimate $\hat{\boldsymbol{\Psi}}_{g_0}$ computed by consideration of the log likelihood formed from the original data under H_0. The value of $-2\log\lambda$ is computed for each bootstrap sample after fitting mixture models for $g = g_0$ and g_1 to it in turn. The process is repeated independently B times, and the replicated values of $-2\log\lambda$ formed from the successive bootstrap samples provide an assessment of the bootstrap, and hence of the true, null distribution of $-2\log\lambda$. It enables an approximation to be made to the achieved level of significance P corresponding to the value of $-2\log\lambda$ evaluated from the original sample. The rth-order statistic of the B bootstrap replications can be used to estimate the quantile of order $r/(B+1)$. A preferable alternative would be to use the rth-order statistic as an estimate of the quantile of order $(3r - 1)/(3B + 1)$; see Hoaglin (1985).

If a very accurate estimate of the P-value were required, then B may have to be very large (Efron and Tibshirani, 1993). Usually, however, there is no interest in estimating a P-value with high precision. Even with a limited replication number B, the amount of computation involved is still considerable, in particular for values of g_0 and g_1 not close to one. However, as noted by Smyth (2000), the process can be easily and efficiently implemented on parallel computing hardware, for example, by using B parallel processors (Smyth, 2000).

In the narrower sense where the decision to be made concerns solely the rejection or retention of the null hypothesis at a specified significance level α, Aitkin et al. (1981) noted how, analogous to the Monte Carlo test procedure of Barnard (1963) and Hope (1968), the bootstrap replications can be used to provide a test of approximate size α. The test that rejects H_0 if $-2\log\lambda$ for the original data is greater than the jth smallest of its B bootstrap replications has size

$$\alpha = 1 - j/(B+1) \tag{3.45}$$

approximately. For if any difference between the bootstrap and true null distributions of $-2\log\lambda$ is ignored, then the original and subsequent bootstrap values of $-2\log\lambda$ can be treated as the realizations of a random sample of size $B+1$, and the probability that a specified member is greater than j of the others is $1 - j/(B+1)$. For some hypotheses the null distribution of λ will not depend on any unknown parameters, and so then there will be no difference between the bootstrap and the true null distribution of $-2\log\lambda$. An example is the case of normal populations with all parameters unknown where $g_0 = 1$ under H_0. The normality assumption is not crucial in this example.

In general, the use of the estimate $\hat{\boldsymbol{\Psi}}_{g_0}$, in place of the unknown value of $\boldsymbol{\Psi}$ under the null hypothesis, will affect the accuracy of the P-values assessed on the basis of the bootstrap replications of $-2\log\lambda$. McLachlan and Peel (1997) performed some simulations to demonstrate this effect. They observed that there was a tendency for the resampling approach using bootstrap replications to underestimate the upper percentiles of the null distribution of $-2\log\lambda$, and hence overestimate the P-value of tests based on this statistic.

3.17 OTHER RESAMPLING APPROACHES FOR NUMBER OF CLUSTERS

3.17.1 The Gap Statistic

On other resampling approaches, Tibshirani et al. (2001) proposed a gap statistic as a general method for determining the number of clusters. This method compares an observed internal index, such as the within-cluster sum of squares, to its expectation under a reference null distribution. More recently, Dudoit and Fridlyand (2002) proposed a prediction-based resampling method to estimate the number of clusters in a data set. As we shall be reporting the results of some simulations in which this method is compared with a normal mixture model-based resampling approach, we now briefly describe this method.

3.17.2 The Clest Method for the Number of Clusters

The Clest method of Dudoit and Fridlyand (2002) is a prediction-based resampling method for assessing the number of clusters in the data. It is concerned with the reproducibility or predictability of the clusters. For a fixed number of clusters g, it proceeds by repeatedly dividing the original sample into two sets, a training or learning set $S_{L,b}$ and a test set $S_{T,b}$ on a given replication b. A clustering of $S_{L,b}$ is obtained and a classifier is found on the basis of this clustering as if the cluster labels were the true class labels. This classifier is then applied to the test set $S_{T,b}$ and the predicted group labels are compared using some external index a_b. This procedure is repeated B times to give $a_1, \ldots, a_B$ and their median m_g. The null distribution of m_g is approximated by the bootstrap under the uniformity hypothesis whereby the data are sampled from a uniform distribution in p-dimensional space. If $m_{g,1}^*, \ldots, m_{g,B_o}^*$ denote the B_o bootstrap values corresponding to m_g so obtained, we let $\bar{m}_g^*$ denote their sample mean and ω_g^* is taken to be the proportion of these B_o bootstrap samples that are at least as large as m_g (the assessed P-value). Finally, let $d_g^* = m_g - \bar{m}_g^*$.

To complete the definition of the Clest procedure, we need the set J, which is defined as

$$ J = \{2 \leq g \leq g_{\max}; \omega_g^* \leq \omega_{\max}, d_g^* \geq d_{\min}\}, $$

where $g_{\max}$ is the maximum value of g to be considered and $\omega_{\max}$ and $d_{\min}$ are preset thresholds. The *ad hoc* choice in Dudoit and Fridlyand (2002) for $\omega_{\max}$ and $d_{\min}$ was 0.05 each. If this set J is empty, the number of clusters is estimated as one ($\hat{g} = 1$). Otherwise, let the number of clusters be estimated by

$$ \hat{g} = \arg \max_g d_g^*. $$

Dudoit and Fridlyand (2002) applied their test procedure using the partitioning around medoids (PAM) method of Kaufman and Rousseeuw (1990), a linear normal-based classifier with diagonal group-covariance matrices, and the external index of Fowlkes and Mallows (1983). They compared the performance of their Clest procedure with six other methods using simulated data and gene expression data from four published cancer microarray studies. The six methods were the silhouette criterion of Kaufman and Rousseeuw (1990), the gap/gap PC statistics of Tibshirani et al. (2001), and the criteria proposed by Caliński and Harabasz (1974), Krzanowski and Lai (1985), and Hartigan (1985).

3.18 SIMULATION RESULTS FOR TWO RESAMPLING APPROACHES

We now report the results of some simulation experiments performed by McLachlan and Khan (2004) to compare the likelihood ratio test (LRT) under the normal mixture model with the nonparametric Clest procedure for the choice of number of clusters. They used the same eight population models as adopted by Dudoit and Fridlyand

(2002) to compare their Clest procedure with six other criteria, using the same number of replications (50) per model. From their simulations, Dudoit and Fridlyand (2002) concluded that Clest was the most robust and accurate. Hence McLachlan and Khan (2004) compared only the performance of the Clest procedure with the LRT in their simulations. Their comparison is reported here in Table 3.1.

The eight models can be described briefly as follows, where $\boldsymbol{Y}_{ij}$ $(j = 1, \ldots, n_i)$ denote the n_i observations generated independently in group G_i $(i = 1, \ldots, g)$.

Model 1 $(g = 1, p = 10, n = 200)$, where $\boldsymbol{Y}_{ij}$ is from the uniform distribution over the unit hypercube in 10 dimensions.

Model 2 $(g = 3, p = 2, n_1 = 25, n_2 = 25, n_3 = 50)$, where

$$\boldsymbol{Y}_{ij} \sim N(\boldsymbol{\mu}_i, \boldsymbol{I}_2)$$

and where

$$\boldsymbol{\mu}_1 = (0,\ 0)^T, \quad \boldsymbol{\mu}_2 = (0, 5)^T, \quad \text{and} \quad \boldsymbol{\mu}_3 = (5, -3)^T.$$

Model 3 $(g = 4, p = 10, n_i = 25$ or 50 with probability 0.5 each), where

$$\boldsymbol{Y}_{ij} \sim N(\boldsymbol{\mu}_i, \boldsymbol{I}_{10})$$

and where

$$\boldsymbol{\mu}_i = (\boldsymbol{w}_i^T, \boldsymbol{0}_7^T)^T$$

and $\boldsymbol{w}_i$ is a realization of the random variable $\boldsymbol{W}_i$ distributed as

$$\boldsymbol{W}_i \sim N(\boldsymbol{0}_3, 25\boldsymbol{I}_3).$$

Here $\boldsymbol{0}_7$ denotes a seven-dimensional vector of zeros. Any simulation where the Euclidean distance between the two closest observations belonging to different clusters is less than 1 is discarded.

Model 4 $(g = 4, p = 10, n_i = 25$ or 50 with probability 0.5 each), where

$$\boldsymbol{Y}_{ij} \sim N(\boldsymbol{\mu}_i, \boldsymbol{I}_{10})$$

and where $\boldsymbol{\mu}_i = \boldsymbol{w}_i$ and $\boldsymbol{w}_i$ is a realization of the random variable $\boldsymbol{W}_i$ distributed as

$$\boldsymbol{W}_i \sim N(\boldsymbol{0}_{10}, 3.6\boldsymbol{I}_{10}).$$

Any simulation where the Euclidean distance between the two closest observations belonging to different clusters is less than 1 is discarded.

Model 5 $(g = 2, p = 3, n_i = 100)$, where

$$\boldsymbol{Y}_{ij} \sim N(\boldsymbol{\mu}_{ij}, \boldsymbol{I}_3)$$

and where

$$\mu_{1j} = -0.5 + 0.1(j-1)/99$$

and $\mu_{2j} = \mu_{1j} + 10$.

Model 6 $(g = 2, p = 10, n_i = 100)$, where the simulated observations $Y_{ij} = (Y_{1ij}^T, Y_{2ij}^T)^T$ are formed independently by generating the Y_{1ij} as in Model 5 and by generating the Y_{2ij} as

$$Y_{2ij} \sim N(0_7, D_7),$$

and D_7 is a 7×7 diagonal matrix whose vth diagonal element is equal to $(v+3)^2$ $(v = 1, \ldots, 7)$.

Model 7 $(g = 2, p = 10, n_i = 50)$, where

$$Y_{ij} \sim N(\mu_i, I_{10})$$

and where $\mu_1 = 0_{10}$ and $\mu_2 = (2.5, 0_9^T)^T$.

Model 8 $(g = 3, p = 13, n_i = 50)$, where

$$Y_{ij} \sim N(\mu_i, \Sigma)$$

and where

$$\mu_1 = 0, \quad \mu_2 = (2, -2, 2, 0_{10}^T)^T, \text{ and } \quad \mu_3 = (-2, 2, -2, 0_{10}^T)^T,$$

$$\Sigma = \begin{pmatrix} \Sigma_{11} & O_{7.3} \\ O_{3.7} & I_{10} \end{pmatrix},$$

and

$$
\begin{aligned}
(\Sigma_{11})_{uv} &= 1.0 \quad (u = v), \\
&= 0.5 \quad (u \neq v),
\end{aligned}
$$

for $u, v, = 1, 2, 3$. Here $O_{3.7}$ denotes a 3×7 matrix of zeros.

For each simulated sample, McLachlan and Khan (2004) fitted a g-component normal mixture model, starting with $g = 1$. They kept increasing g until they reached a value of g, g_o, such that the LRT of $H_0 : g = g_o$ versus $H_1 : g = g_o + 1$ was not significant with the P-value assessed by resampling as described in Section 3.16. The component-covariance matrices were taken to be unrestricted for all but models 3 and 4 for which they were specified to be equal. For each of the eight simulation models, the value of g_o obtained in this manner on the 50 simulation trials per model are displayed in Table 3.1.

It can be seen in Table 3.1 that the relative performance of the LRT with the P-value assessed via resampling is quite encouraging for choosing the number of clusters. The good simulation results for this approach are to be expected since it is favored by having multivariate normal data.

Table 3.1 Estimating the Number of Clusters in Simulated Data

Model	Method	Number of Clusters			
		1*	2	3	4
1	Clest	48	2	0	0
1	LRT	50	0	0	0
		1	2	3*	4
2	Clest	0	1	49	0
2	LRT	0	0	49	0
		1	2	3	4*
3	Clest	0	1	20	29
3	LRT	0	0	0	47
		1	2	3	4*
4	Clest	0	0	1	49
4	LRT	0	0	0	50
		1	2*	3	4
5	Clest	0	44	0	6
5	LRT	0	50	0	0
		1	2*	3	4
6	Clest	0	43	7	0
6	LRT	0	50	0	0
		1	2*	3	4
7	Clest	26	15	6	3
7	LRT	0	46	0	0
		1	2	3*	4
8	Clest	0	16	34	0
8	LRT	0	1	48	1

Source: McLachlan and Khan (2004). The true number of groups is denoted by the asterisk.

3.19 PRINCIPAL COMPONENT ANALYSIS

3.19.1 Introduction

In exploring high-dimensional data sets for group structure, it is typical to rely on "second-order" multivariate techniques; in particular, principal component analysis (PCA). Here we briefly discuss a PCA on the sample covariance matrix

$$V = \sum_{j=1}^{n}(y_j - \overline{y})(y_j - \overline{y})^T / n, \qquad (3.46)$$

where

$$\overline{y} = \sum_{j=1}^{n} y_j / n.$$

We let $a_1, \ldots, a_p$ be the (unit) eigenvectors, corresponding to the eigenvalues $\lambda_1 \geq \lambda_2 \geq \ldots \geq \lambda_p$ of V. In the case where the variates are measured on disparate scales, we may wish to replace V by the sample correlation matrix.

If there are only a few groups and they are well separated, and the between-group variation dominates the within-group variation, then projections of the feature data y_j onto the first few principal axes should portray the group structure. However, a PCA of V may not always be useful. This point was stressed by Chang (1983), who showed in the case of two groups that the principal component of the feature vector that provides the best separation between the two groups, in terms of the Mahalanobis distance, is not necessarily the first component $a_1^T y_j$; see McLachlan (1992, Section 6.6).

To illustrate this, we generated a sample of size $n = 500$ bivariate observations from two groups G_1 and G_2 with means

$$\mu_1 = (5, 0)^T \quad \text{and} \quad \mu_2 = 0$$

and covariance matrices

$$\Sigma_1 = \mathrm{diag}(1, 10)$$

and

$$\Sigma_2 = \begin{pmatrix} 1 & 3 \\ 3 & 10 \end{pmatrix}.$$

These data are plotted in Figure 3.4, where it can be seen that these two groups are well separated. The first principal component is the diagonal line with equation roughly $y_2 = -y_1$ that separates the two groups. It accounts for most of the variation in these data with the first component accounting for approximately 90%. However, if it can be seen that if the data were to be projected onto it, then the group differences between the observations would not be revealed, as essentially all the cluster information is contained in the second principal component. Of course it could be argued that if both principal components were used, then the group differences would be revealed. But with high dimensional data sets, one cannot be be sure that an adequate number of principal components have been used for the purposes of finding the group structure in the data.

Principal component analysis has been commonly applied to microarray data; see, for example, Liu et al. (2003). In a comparison of three multivariate methods that included principal component analysis, Wouters et al. (2004) did claim that "principal component analysis has the disadvantage that the resulting principal factor scores are not very informative." However, Pittelkow and Wilson (2004) have since pointed out that suitable PC representations are possible for the data set considered in Wouters et al. (2004) if conventional practice is followed with the measurements on the genes being first mean corrected.

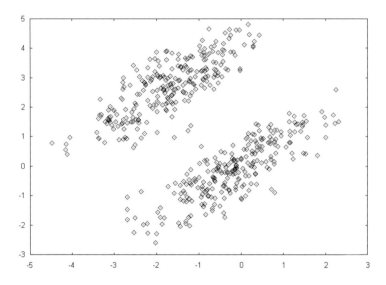

Fig. 3.4 Sample of bivariate observations.

3.19.2 Singular Value Decomposition

In the context of analyzing the tissue samples where p may be extremely large, we can avoid the computational burden of undertaking the singular value decomposition (SVD) of the $p \times p$ matrix V by working in the dual space. To see this, we note from (3.46) that the sample covariance matrix V can be written as

$$V = Y^T Y / n, \qquad (3.47)$$

where $Y = (y_1 - \overline{y}, \ldots, y_n - \overline{y})^T$ is the $n \times p$ data matrix. The singular value decomposition (SVD) of Y is given by

$$Y = U_1 \Lambda U_2^T \qquad (3.48)$$

where Λ is a diagonal matrix of singular values ordered from largest to smallest and U_1 $(n \times n)$ and U_2 $(p \times p)$ are orthogonal matrices. The columns of U_2 are the eigenvectors of V.

Now working in the dual space, we have

$$\begin{aligned} \tilde{V} &= YY^T / p \\ &= (n/p) U_1 \Lambda^2 U_1^T. \end{aligned} \qquad (3.49)$$

On noting that

$$Y^T U_1 = U_2 \Lambda, \qquad (3.50)$$

it follows from (3.49) and (3.50) that we can find Λ and U_2 via the SVD of $\tilde{V}$.

Recently, Liu et al. (2003) have developed a robust analysis for the singular value decomposition of microarray data.

3.19.3   Some Other Multivariate Exploratory Methods

There are of course other multivariate methods besides principal component analysis for exploratory analyses and dimension reduction. For example, there is multidimensional analysis (principal coordinate analysis) whereby the interpoint distances (or proximities) between the feature vectors are calculated. Then a representation of the feature data in a reduced dimensional space is formed to make the interpoint distances within the latter space as close as possible to the original interpoint distances; see Seber (1984, Section 5.5) for further details.

   We conclude this section by noting that another alternative to principal component analysis is the biplot introduced by Gabriel (1971). Pittelkow and Wilson (2003) have developed a variant of the biplot for microarray data called the GE-biplot (gene expression biplot). In the GE-biplot, standardized distances between the tissues and the variance/covariance structure of the genes are represented.

3.20   CANONICAL VARIATE ANALYSIS

3.20.1   Linear Projections with Group Structure

We consider now a reduction in the number of dimensions via linear projections in the case where the group structure of the data is known. That is, we know the classification of the data with respect to some g groups $(G_1, \ldots, G_g)$. However, the linear projections to be discussed here can be used in the unclassified case, for example, to portray clusters in a low dimensional space, by taking the clusters to represent the group structure.

   In dealing with multivariate feature observations, it often facilitates visualization and understanding to represent them in a lower-dimensional space. In particular, two- and three-dimensional scatter plots are often helpful in exploring relationships between the groups, assessing the group-conditional distributions, and identifying atypical feature observations. However, if the dimension p of the data is greater than about 7 or 10, then considerable patience and concentration are needed for a careful scrutiny of all $\binom{p}{2}$ and $\binom{p}{3}$ scatter plots of pairs and triples of the feature variables. One approach to reducing the effort involved in such an exercise is first to transform linearly the p original feature variables into a smaller number q of variables. This process is referred to in the pattern recognition literature as linear feature selection.

   For the linear projection C_q, where C_q is a $q \times p$ matrix of rank q $(q \leq p)$, there is the problem of how to choose C_q so as to best preserve the distinction between the groups, where q may or may not be specified. Often, q will be specified to be at most 2 or 3 for convenience of the subsequent analysis, in particular the graphical representations of the transformed feature data. In some situations there is interest in finding the single linear combination that best distinguishes the g groups, and so q is specified to be one.

3.20.2 Canonical Variates

A starting point in the consideration of linear projections of y is a canonical variate analysis, which expresses the differences between the means $\mu_1, \ldots, \mu_g$ in $d = \min(p, b_o)$ dimensions, where b_o is the rank of the matrix B_o defined below. A canonical variate analysis does not depend on the assumption of normality, as only knowledge of the first two moments of the group-conditional distributions is required. Let

$$B_o = \frac{1}{g-1} \sum_{i=1}^{g} (\mu_i - \bar{\mu})(\mu_i - \bar{\mu})^T,$$

where

$$\bar{\mu} = \sum_{i=1}^{g} \mu_i / g.$$

The matrix B_o is of rank $b_o \leq g - 1$, where $b_o = g - 1$ if $\mu_1, \ldots, \mu_g$ are linearly independent. In practice, we work with the sample between-group sums of squares and products matrix on its degrees of freedom,

$$B = \frac{1}{g-1} \sum_{i=1}^{g} n_i(\bar{y}_i - \bar{y})(\bar{y}_i - \bar{y})^T. \tag{3.51}$$

For mixture sampling in equal proportions from the groups, B/n converges in probability to B_o/g, as $n \to \infty$.

The canonical variates of y are defined by

$$v = \Gamma_d y, \tag{3.52}$$

where

$$\Gamma_d = (\gamma_1, \ldots, \gamma_d)^T \tag{3.53}$$

and where γ_1 maximizes the ratio

$$\gamma^T B\gamma / \gamma^T S\gamma. \tag{3.54}$$

In (3.54),

$$S = W/(n-g), \tag{3.55}$$

where W denotes the pooled within-class sums of squares and products matrix,

$$W = \sum_{i=1}^{g} \sum_{j=1}^{n} z_{ij}(y_j - \bar{y}))(y_j - \bar{y})^T \tag{3.56}$$

where z_{ij} is one or zero, according as y_j belongs to the ith group G_i or not.

For $k = 2, \ldots, d, \gamma_k$ maximizes the ratio (3.54) subject to

$$\gamma_k^T S\gamma_h = 0 \qquad (h = 1, \ldots, k - 1). \tag{3.57}$$

Hence the correlation between $\gamma_h^T Y$ and $\gamma_k^T Y$ is zero for $h \neq k = 1, \ldots, d$. The usual normalization of γ_k is

$$\gamma_k^T S \gamma_k = 1 \qquad (k = 1, \ldots, d), \qquad (3.58)$$

which implies that $\gamma_k^T Y$ has unit variance. This normalization along with the constraint (3.57) implies that

$$\boldsymbol{\Gamma}_d S \boldsymbol{\Gamma}_d^T = \boldsymbol{I}_d,$$

where $\boldsymbol{I}_d$ is the $d \times d$ identity matrix. The eigenvalues and eigenvectors of $S^{-1}B$ can be found using a singular-value decomposition algorithm. For example, γ_k is the eigenvector corresponding to the kth largest (nonzero) eigenvalue of $S^{-1}B$.

The sample version of a canonical variate analysis is just the multiple-group generalization of Fisher's (1936) approach to discriminant analysis in the case of $g = 2$ groups for which

$$\gamma_1 \propto S^{-1}(\bar{y}_1 - \bar{y}_2).$$

It will be seen from both allocatory and separatory aspects that the new set of co-ordinates $v_1, \ldots, v_d$ are the complete set of multiple linear discriminant functions. Hence sometimes in the literature they are referred to as discriminant coordinates rather than by the more usual name of canonical variates.

For $p > d$, we let $\gamma_{d+1}, \ldots, \gamma_p$ denote the eigenvectors of $S^{-1}B$ corresponding to its $p - d$ zero eigenvalues, normalized as

$$\gamma_k^T S \gamma_k = 1 \qquad (k = d + 1, \ldots, p).$$

We put

$$\boldsymbol{\Gamma} = (\boldsymbol{\Gamma}_d^T, \boldsymbol{\Gamma}_{p-d}^T)^T, \qquad (3.59)$$

where

$$\boldsymbol{\Gamma}_{p-d} = (\gamma_{d+1}, \ldots, \gamma_p)^T.$$

It follows then that

$$\boldsymbol{\Gamma} Y \sim N(\boldsymbol{\Gamma}\mu_i, \boldsymbol{I}_p) \text{ in } G_i \ (i = 1, \ldots, g) \qquad (3.60)$$

where, corresponding to the partition (3.59) of $\boldsymbol{\Gamma}$,

$$\boldsymbol{\Gamma}\mu_i = \begin{pmatrix} \boldsymbol{\Gamma}_d\mu_i \\ \boldsymbol{\Gamma}_{p-d}\mu_i \end{pmatrix}$$

and

$$\boldsymbol{\Gamma}_{p-d}\mu_i = \boldsymbol{\Gamma}_{p-d}\bar{\mu} \qquad (i = 1, \ldots, g). \qquad (3.61)$$

It is clear from (3.60) and (3.61) that for the purposes of allocation, the last $p - d$ canonical variates can be discarded without an increase in any of the group-specific error rates. This is because $\boldsymbol{\Gamma}_{p-d}Y$ is distributed independently of $\boldsymbol{\Gamma}_d Y$, with the same distribution in each group. Of course, the overall error rate of the Bayes rule will be increased if it is based on a linear projection of y, $C_q y$, where the rank q of C_q is less than d. McLachlan (1992, Section 3.9) has discussed in some detail the allocatory and separatory aspects of discrimination on the basis of $C_q y$. Consider a given $q < d$ in the case of $g > 2$ groups. Then, although an intuitively desirable projection C_q would be one that minimizes the error rate while maximizing the separation between the groups, the optimal choice of C_q depends on whether the error rate or a separatory measure is to be optimized. This point can be illustrated by the example of Habbema and Hermans (1977) in which they considered the univariate projection $C_1^T y$ in the case of $g = 3$ groups with $\nu_1 = \omega_1$, $\nu_2 = \omega_2$, and $\nu_3 = -\omega_1$, where $0 < \omega_2 < \omega_1$ and where $\nu_i = C_1^T \mu_i$ $(i = 1, 2, 3)$. Then the measure of spread (3.54) can be expressed as

$$C_1^T B_o C_1 / C_1^T \Sigma C_1 = \tfrac{1}{3}(\omega_1^2 + \omega_2^2/3), \tag{3.62}$$

while the overall error rate of the Bayes rule is given by

$$eo(C_1) = (2/3)[\Phi\{-\tfrac{1}{2}(\omega_1 + \omega_2)\} + \Phi\{\tfrac{1}{2}(\omega_2 - \omega_1)\}],$$

where $\Phi(y)$ denotes the standard normal (cumulative) distribution. For fixed ω_1, the minimum value of $eo(C_1)$ with respect to ω_2 occurs at $\omega_2 = 0$, at which the measure of spread (3.62) is minimized rather than maximized.

3.21 PARTIAL LEAST SQUARES

For high-dimensional feature vectors, a canonical variate analysis may not be able to be implemented, as the (pooled) within-class covariance matrix will be singular. One way to proceed is to carry out a principal component analysis, ignoring the class memberships of the feature data. But as noted in Section 3.19, a principal component analysis might result in a serious loss in the group structure in the data. As stressed by Antoniadis, Lambert-Lacroix, and Leblanc (2003), a principal component analysis does not use information on the class labels. Thus it would always give the same principal components for two data sets that have the same feature data, but different class labels.

A method that makes use of the information on the class labels of the observations is partial least squares (PLS), a tool often applied in the chemometrics literature after its introduction there by Wold (1966). It constructs weighted linear combinations of the feature variables that have maximal covariance with the outcome (response variable); see (Frank and Friedman, 1993; Garthwaite 1994; Stone and Brooks 1990).

The PLS method resembles that of a principal component analysis (PCA) in that linear combinations of all the feature variables are formed at each stage. Since PLS makes use of the class labels it is more able than principal component analysis to assign patterns of weights that are predictive of the classes.

The PLS solution forces each of the linear combinations of original variables to have a sample correlation of zero, which is likely to be an inappropriate requirement for molecular signatures. More importantly, the main problem of PLS is similar to that of a PCA on the complete set of feature variables: difficulty in the interpretation of components because each component has loadings on all variables.

4

Clustering of Tissue
Samples

4.1 INTRODUCTION

In this chapter, we consider the cluster analysis of tissue samples. There are two distinct clustering problems with microarray data. One problem concerns the clustering of the tissues on the basis of the genes. The clusters of tissues can play a useful role in the discovery and understanding of new subclasses of diseases. Examples of such studies include classifying sixty human cancer cell lines (Ross et al., 2000), distinguishing two different human acute leukemias (Golub et al., 1999), dissecting and classifying breast cancer tumors (Perou et al., 1999), and classifying subtypes of B-cell lymphoma (Alizadeh et al., 2000) and cutaneous malignant melanoma (Bittner et al., 2000). More recent examples include the work of Bullinger et al. (2004) and Valk et al. (2004), who identified prognostic subclasses in acute myeloid leukemia, and of Lapointe et al. (2004), who found tumor subtypes of prostate cancer that may provide a basis for improved prognostication and treatment stratification.

The second problem concerns the clustering of the genes on the basis of the tissues. The clusters of genes obtained can be used to search for genetic pathways or groups of genes that might be regulated together. Also, in the first problem above, we may wish first to summarize the information in the very large number of genes by clustering them into groups, which can be represented by some metagenes. We can then carry out the clustering of the tissues in terms of these metagenes. This second problem of clustering the genes is to be considered in the next chapter.

In microarray studies, the application of clustering techniques is often used to derive meaningful insights into the data. In the past, hierarchical methods have been the primary clustering tool employed to perform this task. The hierarchical algorithms have been mainly applied heuristically to these cluster analysis problems. Further, a major limitation of these methods is their inability to determine the number of clusters. Thus there is a need for a model-based approach to these clustering problems. To this end, McLachlan et al. (2002) developed a mixture model-based algorithm (EMMIX-GENE) for the clustering of tissue samples.

In this chapter, we focus on the EMMIX-GENE procedure developed by McLachlan et al. (2002) for the specific purpose of mixture model-based clustering of tissue samples on the basis of the available genes. It enables elliptical clusters of arbitrary orientation to be imposed on the tissue samples. The method is generic and can be applied to other large data sets that require feature selection. We shall demonstrate the implementation of the EMMIX-GENE procedure by reporting the case studies of McLachlan et al. (2002) and Mar and McLachlan (2003), involving two well-known data sets in the bioinformatics literature. In the latter study, EMMIX-GENE was applied to the breast cancer data of van 't Veer et al. (2002), while in the former, it was applied to the colon data of Alon et al. (1999), which was first introduced in Section 2.6.

Among other work on model-based approaches to the clustering of gene expression data, there are the studies of Yeung et al. (2001, 2003), Ghosh and Chinnaiyan (2002), Medvedovic and Sivaganesan (2002), and Liu et al. (2003). A Bayesian approach is adopted in the latter two papers.

Before we proceed to consider here the clustering of the tissue samples via EMMIX-GENE, we introduce the following notation.

4.2 NOTATION

Although biological experiments vary considerably in their design, the data generated by microarray experiments can be viewed as a matrix of expression levels. For M microarray experiments (corresponding to M tissue samples), where we measure the expression levels of N genes in each experiment, the results can be represented by the $N \times M$ matrix. For each tissue, we can consider the expression levels of the N genes, called its *expression signature*. Conversely, for each gene, we can consider its expression levels across the different tissue samples, called its *expression profile*. The M tissue samples might correspond to each of M different patients or, say, to samples from a single patient taken at M different time points. The expression levels are taken to be the measured (absolute) intensities for oligonucleotide microarrays and the ratios of the intensities for the Cy5-channel (red) images and Cy3-channel (green) images for cDNA microarrays; see, for example, Dudoit et al. (2002b). It is assumed that one starts the clustering process with preprocessed (relative) intensities, such as those produced by RMA (for Affy data), loess-modified log ratios, or differences of logged/generalized-logged data; see, for example, Parmigiani et al. (2003), Huber et al. (2003), Irizarry et al. (2003), Rocke and Durbin (2003), and Speed (2003).

The $N \times M$ matrix is portrayed in Figure 4.1, where each sample represents a separate microarray experiment and generates a set of N expression levels, one for each gene.

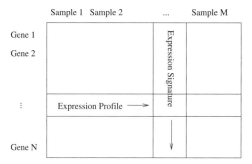

Fig. 4.1 Gene expression data from M microarray experiments represented as a matrix of expression levels with the N rows corresponding to the N genes and the M columns to the M tissue samples.

In the sequel, we shall use the vector $\boldsymbol{y}_j$ to represent the measurement (feature observation) on the jth entity to be clustered. In the context of the classification of the tissues on the basis of the gene expressions, we can represent the $N \times M$ matrix $\boldsymbol{A}$ of gene expressions as

$$\boldsymbol{A} = (\boldsymbol{y}_1, \ldots, \boldsymbol{y}_M),\tag{4.1}$$

where the feature vector $\boldsymbol{y}_j$ (the *expression signature*) contains the expression levels on the N genes in the jth experiment $(j = 1, \ldots, M)$. The latter is a nonstandard problem in parametric cluster analysis because the dimension of the feature space (the number of genes) is typically much greater than the number of observations (the number of tissues).

In the context of the classification of the genes on the basis of the tissues, we can represent the transpose of the matrix $\boldsymbol{A}$ in terms of the feature vectors as

$$\boldsymbol{A}^T = (\boldsymbol{y}_1, \ldots, \boldsymbol{y}_N),\tag{4.2}$$

where the feature vector $\boldsymbol{y}_j$ (the *expression profile*) contains the expression levels on the M tissues on the ith gene $(j = 1, \ldots, N)$ For this clustering problem, the number of observations (the number of genes) is very large relative to the dimension of the feature space (the number of tissues), and so in this sense it falls in the standard framework. However, it is not really a standard problem, as not all the genes are independently distributed.

4.3 TWO CLUSTERING PROBLEMS

In the standard setting of a model-based cluster analysis, the n observations $\boldsymbol{y}_1, \ldots, \boldsymbol{y}_n$ to be clustered are taken to be independent realizations where the sample size n is much larger than the dimension p of each vector $\boldsymbol{y}_j$,

$$n >> p.\tag{4.3}$$

It is also assumed that the sizes of the clusters to be produced are sufficiently large relative to p to avoid computational difficulties with near-singular estimates of the within-cluster covariance matrices.

In the cluster analysis of the M tissue samples on the basis of the N genes, we have $n = M$ and $p = N$. Thus the sample size n will be typically small relative to the dimension p, thus causing estimation problems under the normal mixture model. This is because the g-component normal mixture model (3.22) with unrestricted component-covariance matrices is a highly parameterized model with $\frac{1}{2}p(p+1)$ parameters for each component-covariance matrix Σ_i ($i = 1, \ldots, g$). Some ways to handle this dimensionality problem were outlined in Section 3.2.

An obvious way to handle the very large number of genes is to perform a principal component analysis and carry out the cluster analysis on the basis of the leading components. This approach for the clustering of microarray data has been studied by Alter et al. (2000), Yeung and Ruzzo (2001), and Liu et al. (2003), among many others.

There is also another clustering problem of interest, namely the clustering of the genes on the basis of the tissue samples. For this problem, the sample size n is equal to the number of genes N and the dimension p is equal to the number of tissues M. Thus condition (4.3) for a standard cluster analysis will be satisfied usually. The condition of independent data will not hold given that not all the genes in a given tissue sample are independently distributed. But in practice we can proceed with the standard clustering methodology, ignoring any correlations between genes in the same tissue sample. But tests concerned with the smallest number of components in the mixture model would need to take into account the breakdown in the independence condition. Also, although the dimension p may be very small relative to the n, it can be large in absolute terms. Also, the number of clusters g may be large. Thus one might not be able to fit a normal mixture model directly to the genes. One option would be to use principal components. Another option would be to use mixtures of factor analyzers to effect the clustering of the genes. We consider this in the next chapter.

4.4   PRINCIPAL COMPONENT ANALYSIS

Given the high dimensionality of the feature vector y representing the signature expression of a tissue sample, k-means and hierarchical agglomerative methods of clustering are convenient first choices off the shelf for the scientist. However, as set out in Section 3.7.2, there are advantages to be had by adopting a model-based approach to clustering. Given the high dimensionality of y for the clustering of tissue samples, there is a need to first reduce the dimension of the feature space. As noted in the previous section, an obvious way of dimension reduction in an unsupervised context is to carry out a principal component analysis (PCA) as described in Section 3.19. The shortcomings of a PCA in such a context is that the leading components need not necessarily reflect the direction in the feature space best for revealing the group structure of the tissues. This is because it is concerned with the direction of maximum variance, which is composed of variance within the clusters and variance between the clusters. If the latter are relatively large, then the leading components may not be so useful for the purposes of cluster analysis. But with the analysis of microarray data, this problem is compounded by the very large number of genes and their associated noise. Thus artificial directions can result from noisy genes and highly correlated ones. Consequently, a potential problem with a PCA is the determination of an appropriate number of principal components (PCs) useful for clustering. A common practice is to choose the first few leading components. But as noted in Section 3.2, it may not be clear where to stop and whether some of these components are caused by some artifact or noises in the data. An excellent account of these problems may be found in Liu et al. (2003). They have developed a Bayesian approach to model-based clustering which after an initial PCA simultaneously clusters the observations and selects "informative" variables or components for the cluster analysis.

4.5   THE EMMIX-GENE CLUSTERING PROCEDURE

The EMMIX-GENE procedure handles the problem of a high-dimensional feature vector by using mixtures of factor analyzers whereby the component correlations between the genes are explained by their conditional linear dependence on a small number q of latent or unobservable variables specific to each component. This factor analysis model has been defined explicitly in Section 3.11.

In practice we may wish to work with a subset of the available genes, particularly as the fitting of a mixture of factor analyzers will involve a considerable amount of computation time for an extremely large number of genes. Indeed, the simultaneous use of too many genes in the cluster analysis may serve only to create noise that masks the effect of a smaller number of genes. Also, the intent of the cluster analysis may not be to produce a clustering of the tissues on the basis of all the available genes, but rather to discover and study different clusterings of the tissues corresponding to different subsets of the genes; see the recent paper of Friedman and Meulman (2004) on this point.

Therefore, the EMMIX-GENE procedure has two optional steps before the final step of clustering the tissues. The first step considers the selection of a subset of relevant genes from the available set of genes by screening the genes on an individual basis to eliminate those which are of little use in clustering the tissue samples in terms of the likelihood ratio test statistic. The second step clusters the retained genes n_o into groups on the basis of Euclidean distance so that highly correlated genes are clustered into the same group. The third and final step of the EMMIX-GENE procedure considers the clustering of the tissues by fitting mixtures of factor analyzers. It can be either by considering the groups of genes simultaneously on the basis of their means or by considering the groups individually on the basis of all or a subset of the genes in a given group. We now describe these three steps in more detail.

4.6   STEP 1: SCREENING OF GENES

In step 1 of EMMIX-GENE, we screen the genes by attempting to delete those genes that individually are of little use in clustering the tissue samples into two groups. This screening is undertaken in the absence of tissue samples that are of known classification. The relevance of a gene for clustering the tissue samples can be assessed on the basis of the value of $-2 \log \lambda$, where λ is the likelihood ratio statistic for testing $g = 1$ versus $g = 2$ components in the mixture model. In order to reduce the effect of atypically large observations on the value of λ, we fit mixtures of t components with their degrees of freedom inferred from the data. However, the use of t components in place of normal components still does not eliminate the effect of outliers on inference of the number of groups in the tissue samples. For example, suppose that for a given gene there is no genuine grouping in the tissues, but that there are a small number of gross outliers. Then a significantly large value of λ might be obtained, with one component representing the main body of the data (and providing robust estimates of their underlying distribution) and the other representing the outliers. That is, although the t mixture model may provide robust estimates of the underlying distribution, it does not provide a robust assessment of the number of groups in the data.

Suppose now that for a given gene there are two groups in the tissue samples. If there are no outliers present in the tissue samples, we should obtain a significant value of λ with the two components of the fitted t mixture model corresponding to the two groups. But if there are outliers present, then the two components of the fitted t mixture model may still correspond to the two groups, or it may happen that one component corresponds to the main body of the data and the other component to the outliers. An illustration of the former case is given in Figure 4.2 and of the latter case in Figures 4.3 and 4.4, using the expression levels of two genes over the 62 tissue samples from the colon cancer data of Alon et al. (1999); this data set has been described in detail in Section 2.6 and is to be analyzed further in Section 4.10.

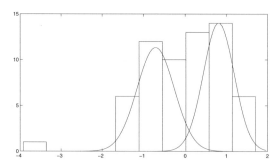

Fig. 4.2   Histogram of gene 1,758 (H20819) with mixture of $g = 2$ fitted t components.

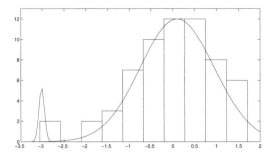

Fig. 4.3 Histogram of gene 474 (T70046) with mixture of $g = 2$ fitted t components.

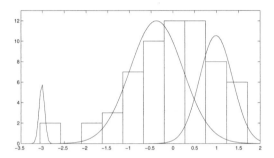

Fig. 4.4 Histogram of gene 474 (T70046) with mixture of $g = 3$ fitted t components.

4.7 STEP 2: CLUSTERING OF GENES: FORMATION OF METAGENES

Concerning the end problem of clustering the tissue samples on the basis of the genes considered simultaneously, we could examine the univariate clusterings provided by each of the selected genes taken individually. But this would be rather tedious when a large number of genes have been selected. Thus with the EMMIX-GENE approach, there is a second (optional) stage for clustering the genes into a user-specified number (N_o) of groups by fitting a mixture in equal proportions of $g = N_o$ normal distributions with covariance matrices restricted to being equal to a multiple of the $(p \times p)$ identity matrix. That is, if the mixing proportions were fixed at 0.5, then it would be equivalent to using a soft version of k-means and grouping the genes in terms of the Euclidean distance between them. One could attempt to make a more objective choice of the number N_o of groups by using, say, the likelihood ratio criterion or BIC. There is an extra complication here since the genes are not independently distributed within a tissue sample.

The groups of genes are ranked in terms of the likelihood ratio statistic calculated on the basis of the fitted mean of a group over the tissues for the test of a single versus two t components. This is provided that the minimum cluster size is greater than a specified threshold. Otherwise, such a group of genes would be put at the end of the list.

A heat map of genes in a group versus the tissues is provided for each of the groups where, in each group, the tissues can be left in their original order or rearranged according to their cluster membership obtained by fitting a univariate t mixture model on the basis of the group mean. Alternatively, one could cluster the tissues by fitting a two-component mixture of factor analyzers on the basis of the genes within the group. Concerning the use of heat maps, they present a grid of colored points where each color represents a gene expression value for a gene in the tissue sample. They are used here primarily to exhibit similarities between groups or clusters of the tissue samples. Thus they are most effective in this role when the tissue samples have been grouped according to their group (cluster) memberships. Of course the heat maps are also useful in revealing similarities between the genes.

We have found in our analyses of microarray data sets that the means of the groups into which the genes have been clustered as above provide a useful representation of the genes in a lower dimensional space (the dimension of this space is equal to the number of groups N_o). If we cluster the tissues on the basis of the group means only, we are ignoring the relative sizes of the groups. This might have some impact on the accuracy of predictions if the aim were to construct a classifier for assigning the tissues to externally existing classes. For instance, one group may contain many genes that are useful in distinguishing between healthy and unhealthy. Thus, if the genes within this group act independently, then there would be a loss in accuracy in using only the mean of this group and not making use of its size. But as the genes have been clustered into groups by working in terms of Euclidean distance (after normalization of the data), the impact of ignoring the size of the groups should be limited. This is because the genes within a group should in the main be at least moderately correlated with each other, as the Euclidean distance between any two genes is equal to $2(p-1)(1-\rho)$, where ρ denotes the sample correlation between them.

Each cluster of genes can be represented by one or more M-dimensional profile vectors over the M tissues. We follow Huang et al. (2003) in referring to these cluster representatives as metagenes. In their work on using classification trees to predict breast cancer, they first clustered the genes into a number of groups via the k-means algorithm. That is, it can be viewed as corresponding to this cluster-genes step of EMMIX-GENE. The only difference is that they do not first eliminate apparently nondifferentially expressed genes as on the select-genes step of EMMIX-GENE. Thus they have to cluster a much larger number of genes and so they summarize them by a larger number of clusters as on the cluster-genes step of EMMIX-GENE. They take the first principal component of a cluster of genes to be the metagene. In EMMIX-GENE, we take the sample mean of the genes within a cluster to be the metagene representing the cluster. This strategy of using a linear combination of the genes within a cluster to represent it and so thereby reducing the dimension of the feature (gene) space also helps smooth out gene-specific noise through the aggregation within a cluster.

In addition to the approach of Huang et al. (2003) in supervised classification, a number of authors in the same context have proposed clustering the genes into groups as a way of reducing the feature space of genes; see Section 7.12.2 Also, in Section 7.12.5, we discuss ways of forming tight clusters of genes or at least having highly correlated genes within the same cluster in the context where the end focus is on the clustering of genes themselves and not the tissues.

In ongoing work on EMMIX-GENE, consideration is being given to the choice of metagene or metagenes for each cluster of genes. It may be that the use of the first principal component might be preferable to the sample mean. The sample mean was chosen for computational convenience. Note that if the genes within a cluster have the same sample correlation, then the first principal component will be proportional to the sample mean since the genes all have the same sample variances from the row (gene) normalization.

We have observed that in some clusters of genes the leading (top) gene has a much larger value of $-2 \log \lambda$ than the remaining genes within the cluster. In this case, it might be worth representing this group of genes by two metagenes, one given by the leading gene within the cluster and the other by the sample mean of the remaining genes within the cluster. These issues are currently being investigated.

4.8 STEP 3: CLUSTERING OF TISSUES

If a clustering is sought on the basis of the totality of the genes, then it can be obtained by fitting a mixture model to these group means. However, it may be that the number of group means N_o is too large to fit a normal mixture model with unrestricted component-covariance matrices. In this circumstance EMMIX-GENE has the option on the third step that allows for the fitting of mixtures of factor analyzers. The use of mixtures of factor analyzers reduces the number of parameters by imposing the assumption that the correlations between the genes can be expressed in a lower space by the dependence of the tissues on q $(q < N)$ unobservable factors. In addition to clustering the tissues on the basis of all of the genes, there may be interest in seeing if the different groups of genes lead to different clustering of the tissues when each is considered separately. For example, only a subset of the genes may be useful in identifying certain subtypes of the cancer being studied.

It can be seen from above that with the EMMIX-GENE procedure, the genes are being treated anonymously. That is, we do not incorporate existing biological information on the function of genes into the selection procedure. Spang (2003) infuses some biological context into an otherwise unsupervised learning task. He structures the feature space by using a functional grid provided by the Gene Ontology annotations.

4.9 EMMIX-GENE SOFTWARE

The EMMIX-GENE program is an interface to EMMIX, which adds functionality to the standard EMMIX through the display of heatmaps of the expression profiles of genes clustered into groups and of metagenes, unsupervised and supervised classification via mixtures of factor analyzers, and many facilities for reordering, sorting, processing, and selecting relevant features of microarray data.

Simple command-line versions of EMMIX-GENE for Linux are available at the website http://www.maths.uq.edu.au/~gjm/emmix-gene/. This webpage also provides updates on the availability and licensing terms for versions with a graphical user interface for Windows.

4.10 EXAMPLE: CLUSTERING OF ALON DATA

As an illustration of the EMMIX-GENE procedure applied to some real data, we report here the results of McLachlan et al. (2002), in their cluster analysis of the colon data in Alon et al. (1999). In particular, we demonstrate the different clusterings that can be obtained by using different subsets of the genes on the tissues as provided by the cluster-genes step of EMMIX-GENE.

In the Alon data set, which was introduced in Section 2.6, prefiltering of the total set of 6,500 genes gave a microarray data matrix A with $N = 2,000$ rows and $M = 62$ columns. The samples comprised 40 tumor and 22 normal colon tissue samples, taken from 40 different patients, with 22 patients supplying both tumor and normal tissue samples.

In Alon et al. (1999), the tissues are not listed consecutively, but here we have rearranged the data so that the tumors are labeled 1 to 40 and the normals 41 to 62. Before we considered the clustering of this set, we processed the data by taking the (natural) logarithm of each expression level in A. Then each column of this matrix was standardized to have mean zero and unit standard deviation. Finally, each row of the consequent matrix was standardized to have mean zero and unit standard deviation.

4.10.1 Clustering on Basis of 446 Genes

On the first screening step of EMMIX-GENE, McLachlan et al. (2002) selected 446 genes as relevant. In practice, clustering tissues on the basis of the entire gene set (2,000 genes) or even this reduced set of 446 genes may not reveal the extent of any group structure in the tissues. For the purpose of this example, McLachlan et al. (2002) fitted a two-component mixture of factor analyzers to the tissues, to cluster on the basis of the 446 selected genes. They fitted mixtures of $g = 2$ factor analyzers for various levels of the number q of factors ranging from $q = 2$ to $q = 8$, but there was little difference between the clustering results. The clustering corresponding to the largest of the local maxima obtained gave the following clustering for $q = 6$ factors,

$$C_1 = \{1\text{--}12, 20, 25, 41\text{--}52\}$$
$$\cup \{13\text{--}39, 21\text{--}24, 26\text{--}40, 53\text{--}62\}. \tag{4.4}$$

Getz et al. (2000) and Getz (2001) reported that there was a change in the protocol during the conduct of the microarray experiments. The 11 tumor tissue samples (labeled 1 to 11 here) and 11 normal tissue samples (41 to 51) were taken from the first 11 patients using a poly detector, while the 29 tumor tissue samples (12 to 40) and normal tissue samples (52 to 62) were taken from the remaining 29 patients using total extraction of RNA. It can be seen from (4.4) that this clustering C_1 almost corresponds to the dichotomy between tissues obtained under the "old" and "new" protocols.

McLachlan et al. (2002) also considered the clustering of the 62 tissue samples on the basis of the top 50 genes in the retained set of 446 genes. Fitting mixtures of factor analyzers with $q = 6$ factors, using 50 random and 50 k-means starts, we obtained the following clustering,

$$C_2 = \{1\text{--}26, 29, 31, 32, 34, 38, 41\text{--}52\}$$
$$\cup \{27\text{--}28, 30, 33, 35\text{--}37, 39, 40, 53\text{--}62\}.$$

This clustering not only splits the tissue samples obtained under "old" and "new" protocols, but it also splits some of the "new" tumor samples and some of the "new" normal tissue samples.

4.10.2 Clustering on Basis of Gene Groups

The tissue samples were also clustered after the retained set of 446 genes had been clustered into $N_o = 20$ groups on the second step of the EMMIX-GENE approach. In Figure 4.5, we have plotted the 18 genes in the first group G_1 for the 62 tissues, with the latter arranged in order of the 40 tumors followed by the 22 normal tissues. In Figure 4.6, we give the corresponding plot of the 24 genes in the second group of genes G_2. A heat map of the genes in a group versus the tissues for all $N_o = 20$ groups may be viewed at `http://www.maths.uq.edu.au/~gjm/emmix-gene/` We have also listed there the heat map of the reduced set of 446 genes to show that this heat map, in contrast to the heat maps for the groups of genes, is not informative visually in revealing group structure in the tissues.

The clustering of the tissues on the basis of the 18 genes in G_1 using $q = 4$ factors in the mixture of factor analyzers model resulted in a partition C_3 of the tissues that is fairly similar to C_2, namely,

$$C_3 = \{1\text{--}26, 29\text{--}32, 41\text{--}52, 55\text{--}56\}$$
$$\cup \{27\text{--}28, 33\text{--}40, 53\text{--}54, 57\text{--}62\}.$$

The clustering of the tissues on the basis of the 24 genes in G_2 resulted in a partition of the tissues in which one cluster contains 37 tumors (1–29, 31–32, 34–35, 37–40)

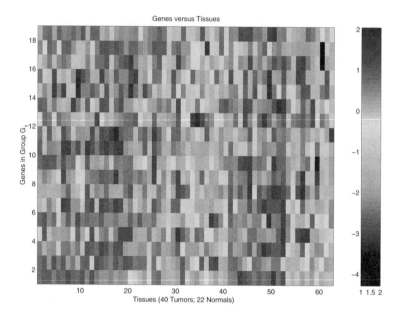

Fig. 4.5 Heat map of 18 genes in group G_1 on 40 tumor and 22 normal tissues in Alon data. See the insert for a color representation.

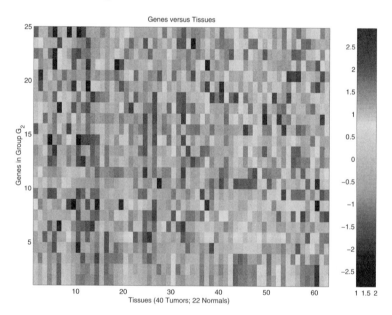

Fig. 4.6 Heat map of 24 genes in group G_2 on the 40 tumor and 22 normal tissues in Alon data. See the insert for a color representation.

and 3 normals (48, 58, 60), and the other cluster contains 3 tumors (30, 33, 36) and 19 normals (41–47, 49–57, 59, 61–62). Calling this clustering C_4, we have that

$$C_4 = \{1\text{–}29, 31\text{–}32, 34\text{–}35, 37\text{–}40, 48, 58, 60\}$$
$$\cup \{30, 33, 36, 41\text{–}47, 49\text{–}57, 59, 61\text{–}62\}.$$

It can be seen from Figure 4.5 that the clustering of the tissues on the basis of the genes in group G_1 gives two clusters with large intercluster differences between the tissues. The clusters are also quite cohesive, but this is accentuated by the fact that we are using genes that were put into the same group by carrying out the grouping effectively in terms of Euclidean distance between genes. Similarly, Figure 4.6 shows that the clustering of the tissues on the basis of the genes in group G_2 gives two cohesive clusters with large intercluster differences. But it appears that the first clustering is stronger in terms of the likelihood ratio statistic λ formed from the individual genes in the groups and on their means. This clustering C_4 produced by the second group of genes G_2 is quite similar to the external classification, as its error rate is only 6.

It can be seen from Figure 4.6 that the genes in group G_2 tend to be more highly expressed in the normal tissues than in the tumors. Alon et al. (1999) and Ben-Dor et al. (2000) noted that the normal colon biopsy also included smooth muscle tissue from the colon walls. As a consequence, smooth muscle-related genes showed high expression levels in the normal tissue samples compared to the tumor samples, which generally had a low muscle content. Ben-Dor et al. (2000) identified a large number of muscle-specific genes as being characteristic of normal colon samples. We note that two of these genes (J02854 and T60155) are in group G_2, while group G_2 also contains two genes (M63391 and X74295) that Ben-Dor et al. (2000) suspected of being expressed in smooth muscle.

The six tissues that are misallocated under this second clustering (tumor tissues 30, 33, and 36 and normal tissues 48, 58, and 60) occur among those tissues that have been misallocated in other cluster and discriminant analyses of this data set. Tissues 30, 33, and 36 are taken from tumor tissue on patients labeled 30, 33, and 36 in Alon et al. (1999), while tissues 48, 58, and 60 are taken from normal tissue on patients 8, 34, and 36. These six tissues have been misallocated in previous analyses even in a discriminant analysis context where use is made of the external classification of these tissues. For example, with the support vector machine classifier formed in Chow et al. (2000) using the known classification of tissues, these six tissues along with tumor tissue 35 were misallocated in the (leave-one-out) cross-validation of this classifier. There is thus some doubt as to the validity of the so-called "true" classification of these six tissues, which was determined by biopsy. An inspection of Figure 4.6 reveals that at least for the 24 genes in this plot, tumor tissues 30, 33 and 36 are very similar to the normal ones, while the normal tissues 48, 58, and 60 are very similar to the tumors. As explained in Chow et al. (2000), misclassification might be due to, say, simple error during sample handling, RNA preparation, data acquisition, and data analysis. They also noted that the normal tissues could have been misclassified because pathologically "normal" regions of the colon could have substantial tumor-like properties from a molecular standpoint.

Applying a hierarchical procedure to cluster the 62 tissues on the basis of the 2,000 genes, Alon et al. (1999) observed that the topmost division in the dendrogram divides the samples into two groups that misallocates three normal and five tumor tissues (tissues 2, 30, 33, 36, 37, 48, 52 and 58). The method used by Alon et al. (1999) can be viewed as fitting a normal mixture model with common spherical component-covariance matrices (although the variance was not estimated from the data; it was varied deterministically during the fitting process). Also, Alon et al. (1999) did not log the data. It is of interest to note that in fitting mixtures of diagonal normal components to the tissues on the basis of all the genes, the only way we could get the algorithm to converge to a local maximum that gave an implied clustering the same as C_4 or a perturbation of it (that is, similar to the external classification) was to use the unlogged data and to impose the condition of common spherical component-covariance matrices. Hence when the data are logged (as is appropriate), or when Euclidean distance is not used as the metric, the smooth muscle-related genes have a diminished capacity in the presence of other genes to distinguish between normal and tumor tissues.

4.10.3   Clustering on Basis of Metagenes

McLachlan et al. (2002) also clustered the 62 tissues on the basis of the $N_o = 20$ fitted group means obtained above by fitting a mixture of $g = 2$ factor analyzers for various levels of the number of factors q. The largest local maximum so located with $q = 8$ factors gives a clustering (C_5) that is similar to C_2 and C_3 with

$$C_5 \;\; = \;\; \{1\text{--}23, 25, 26, 41\text{--}52, 58\}$$
$$\cup \,\{24, 27\text{--}40, 53\text{--}57, 59\text{--}62\}.$$

4.11   EXAMPLE: CLUSTERING OF VAN 'T VEER DATA

As a second example of the cluster analysis of tissue samples, we consider the work of Mar and McLachlan (2003), who applied the EMMIX-GENE procedure to the breast cancer data set of van 't Veer et al. (2002). Here we show how EMMIX-GENE may typically be applied to a data set. We describe the clustering of selected genes into gene groups and discuss how these gene groups compare with those of van 't Veer. Then we show how Mar and McLachlan (2003) cluster the tissues on the basis of the metagenes and discuss the relation of the tissue clusters with other known clinical indicators.

In their original study, van 't Veer et al. (2002) used inkjet synthesized oligonucleotide arrays to measure the expressions of 24,881 genes in 98 primary breast cancers acquired from three groups of patients: 44 representing a good-prognosis group (that is, those who remained metastasis free after a period of more than 5 years), 34 from a poor-prognosis group (those who developed distant metastases within 5

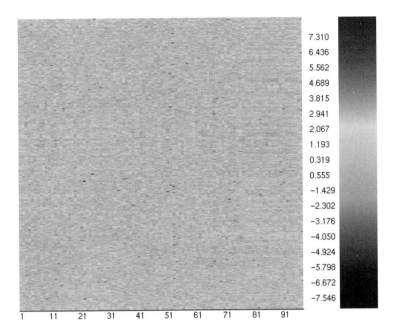

Fig. 4.7 Heat map of the filtered set of 4,869 genes on the 98 tumor tissues in van 't Veer data. See the insert for a color representation.

years), and 20 representing a hereditary form of cancer, due to a BRCA1 (18 tumors) or BRCA2 (2 tumors) germline mutation. The 78 sporadic (non-BRCA) breast cancer patients were chosen specifically on the basis of their clinical outcome. An aim of the study was to identify gene expression profiles which could discriminate between good- and poor-prognosis patients.

van 't Veer et al. (2002) applied a filter in which only genes with both a P-value of less than 0.01 and at least a two fold difference in more than five out of the ninety-eight tissues for the gene were retained. This filter effectively reduced the initial set of genes to 4,869. This gene set was used by Mar and McLachlan (2003) to give a microarray data matrix A with $N = 4,869$ rows and $M = 98$ columns. As can be seen by the heat map displayed in Figure 4.7, discerning an underlying class structure in the data on the basis of this set of 4,869 genes would be extremely difficult without further reduction in the number of genes.

4.11.1 Screening and Clustering of Genes

Mar and McLachlan (2003) used the first step of EMMIX-GENE to select the most relevant genes from this filtered set of 4,869 genes, further reducing the number to 1,867. The 1,867 retained genes were clustered into $N_o = 40$ groups using the second step of the EMMIX-GENE procedure, and the majority of gene groups produced were

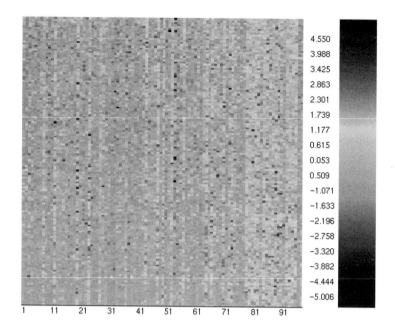

4.550	
3.988	
3.425	
2.863	
2.301	
1.739	
1.177	
0.615	
0.053	
0.509	
-1.071	
-1.633	
-2.196	
-2.758	
-3.320	
-3.882	
-4.444	
-5.006	

Fig. 4.8   Heat map of 146 genes in group G_1 on 98 tumor tissues in van 't Veer data. See the insert for a color representation.

reasonably cohesive and distinct. Based upon these forty group means, the tissue samples were clustered into two and three clusters using a mixture of factor analyzers with $q = 4$ factors.

They found that the heat maps of the genes in a group tend to mainly support the same breakup of the 98 tissues. To illustrate this, we list in Figures 4.8 to 4.10 the heat maps for the top three groups G_1, G_2, and G_3 , which contain 146, 93, and 61 genes, respectively. An important feature to note from these heat maps is that they each indicate a change in gene expression is apparent between the sporadic (first 78 tissue samples) and hereditary (last 20 tissue samples) tumors. For instance, in Figure 4.8, the genes in this cluster are generally down-regulated for the former group of tumors, and up-regulated in the latter. Genes in G_2 were largely constant in expression across the sporadic tumors but notably down-regulated for the hereditary tumors.

Additionally, the final two tissue samples, which represent the two BRCA2 tumors show consistent patterns of expression in each of the clusters that are different from those exhibited by the set of BRCA1 tumors.

It can be seen from these groups that the problem of trying to distinguish between the two classes, patients who were disease-free after 5 years Π_1 and those with metastases within 5 years Π_2, is not straightforward on the basis of the gene expressions.

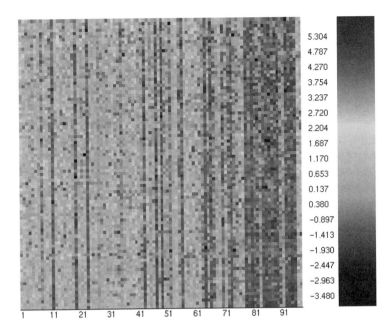

Fig. 4.9 Heat map of 93 genes in group G_2 on 98 tumor tissues in van 't Veer data. See the insert for a color representation.

4.11.2 Usefulness of the Selected Genes

In clustering the genes, van 't Veer et al. (2002) relied upon an agglomerative hierarchical algorithm to organize the genes into dominant genes groups. Two of these clusters were highlighted in the paper and the genes contained in these two groups correspond to biologically significant features. We denote cluster A as the group of genes van 't Veer et al. (2002) have identified as containing genes coregulated with the ER-a gene (ESR1) and cluster B as the group containing coregulated genes that are the molecular reflection of extensive lymphocytic infiltrate, and comprise a set of genes expressed in T and B cells. Both of these clusters contain 40 genes.

Of these 80 genes, the first step of the EMMIX-GENE algorithm select-genes retains only 47 genes (24 from cluster A, 23 from cluster B). When compared to the 40 groups that the cluster-genes step of the EMMIX-GENE algorithm produces, subsets of these 47 genes appeared inside several of these 40 groups (see Table 4.1 below).

The motivation behind select-genes is to isolate the most informative genes to be used for the cluster analysis. For any clustering algorithm, genes that lack distinctive expression pattern changes across different tumor groups only serve to confuse the clustering algorithm and increase the number of misallocation errors made.

The 21 genes that appear in luster A have been grouped in the second cluster constructed by EMMIX-GENE. In Figure 4.11 (below), these genes demonstrate

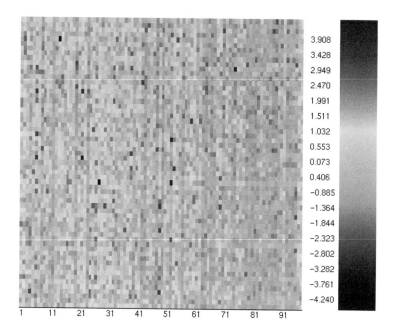

Fig. 4.10 Heat map of 61 genes in group G_3 on 98 tumor tissues in van 't Veer data. See the insert for a color representation.

Table 4.1 Comparing Clusters Constructed by an Hierarchical Algorithm with those Produced by the EMMIX-GENE Algorithm

	Cluster Index (EMMIX-GENE)	Number of Genes Matched	Percentage Matched (%)
Cluster A	2	21	87.5
	3	2	8.33
	14	1	4.17
Cluster B	17	18	78.3
	19	1	4.35
	21	4	17.4

21 Genes were in Cluster 2 and Cluster A

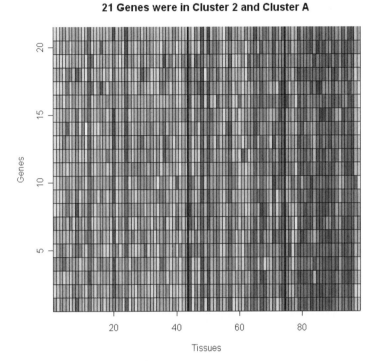

Fig. 4.11 Genes in van 't Veer data retained by EMMIX-GENE appearing in cluster A. See the insert for a color representation.

clear expression changes for the three groups of tumors (indicated by the vertical blue lines).

For the remaining sixteen genes that were rejected by select-genes but belong to cluster A, it is evident from Figure 4.12 that these genes bear very little information in distinguishing between the tumor groups.

The heat maps displayed in Figures 4.13 and 4.14 display the corresponding information for the genes in cluster B. The genes in Figure 4.14 (those retained by EMMIX-GENE) show much variation across the tumor groups. In contrast, the genes in cluster B (those rejected by EMMIX-GENE) show little variation between the tumor groups.

The expression profile of the gene that received the highest $-2 \log \lambda$ value is shown in Figure 4.15. This gene is notably up-regulated for the disease-free tumor group and the metastases tumor group, and down-regulated in the hereditary tumor group.

An expression profile is shown in Figure 4.16 for a gene which appeared in cluster A, but whose $-2 \log \lambda$ value was not high enough for it to be retained by the select-genes step. The overall expression of the gene is essentially unchanging, however,

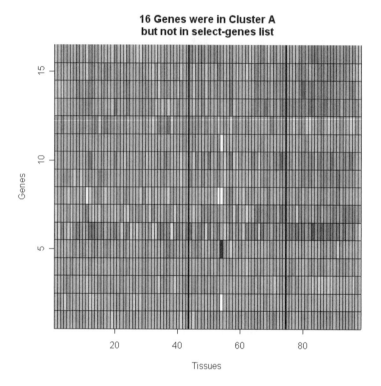

Fig. 4.12 Genes in van 't Veer data rejected by EMMIX-GENE appearing in cluster A. See the insert for a color representation.

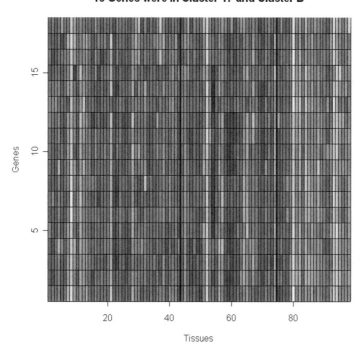

Fig. 4.13 Genes in van 't Veer data retained by EMMIX-GENE appearing in cluster B. See the insert for a color representation.

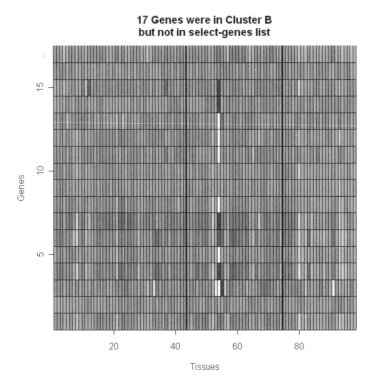

Fig. 4.14 Genes in van 't Veer data rejected by EMMIX-GENE appearing in cluster B. See the insert for a color representation.

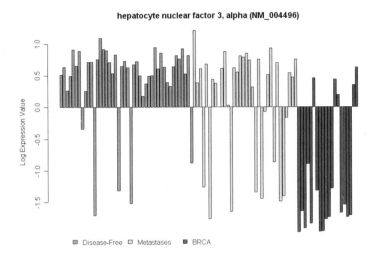

Fig. 4.15 Expression profile for the gene with the highest $-2\log\lambda$ value.

excessively large values for the seventeenth disease-free patient in the first tumor group and the sixth BRCA patient in the third tumor group appear to dominate the expression profile. These outliers seem to account for this gene's inclusion in cluster A.

4.11.3 Clustering of Tissues

We turn next to the problem of clustering tissues on the basis of gene expression. Mar and McLachlan (2003) investigated the clusters constructed by the EMMIX-GENE algorithm in light of the genuine tissue grouping. The tissue samples can be subdivided into two groups corresponding to the 78 sporadic tumors and 20 hereditary tumors. Figure 4.17 shows the two-cluster assignment produced by EMMIX-GENE with respect to this genuine grouping (black denotes the hereditary tumor cluster, white denotes the sporadic tumor cluster; gray distinguishes the genuine grouping).

Clearly, EMMIX-GENE has correctly clustered the majority of the hereditary tumors (misallocation error of 1/20), although 37 of the sporadic tumors were incorrectly assigned to the cluster of hereditary tumors. The set of sporadic tumors has been divided into good- and poor-prognosis groups (that is, 44 patients who continued to be disease-free after 5 years, and 34 patients who developed metastases within 5 years, respectively). Hence they also considered the partitioning of the tissues into three clusters, corresponding to the disease-free, metastases, and hereditary groups. Figure 4.18 shows the tissue samples rearranged according to the three cluster assignments allocated by EMMIX-GENE when a mixture of factor analyzers model with $q = 4$ factors.

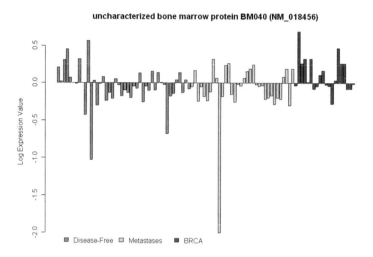

Fig. 4.16   Example of a gene rejected by select-genes but retained by cluster A.

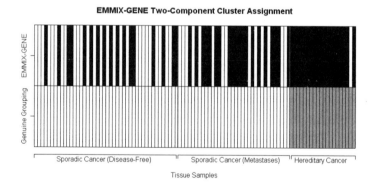

Fig. 4.17   Comparing EMMIX-GENE cluster assignments with the genuine two-group structure (Sporadic; Hereditary).

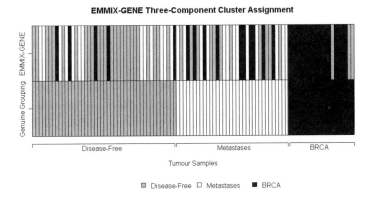

Fig. 4.18 Comparing EMMIX-GENE cluster assignments with a genuine three-group structure (Disease-free; Metastases; BRCA).

Using a mixture of factor analyzers model with $q = 8$ factors, they would misallocate 7 of the 44 members of Π_1 and 24 of the 34 members of Π_2; one member of the 18 BRCA1 samples would be misallocated.

The misallocation rate of 24/34 for the second class Π_2 is not surprising given the gene expressions as summarized in the groups of genes (see Figures 4.8 to 4.10). Also, one has to bear in mind that we are classifying the tissues in an unsupervised manner without using the knowledge of their true classification. But even when such knowledge is used in a supervised classification via an SVM using RFE, the cross-validated error rate (with allowance for the selection bias) is not less than 31% depending on the number of genes used; see Table 7.6. Further analysis of this data set in a supervised context by Tibshirani and Efron (2002) confirms the difficulty in trying to discriminate between the disease-free class Π_1 and the metastases class Π_2.

4.11.4 Use of Underlying Signatures with Clinical Data

For each of the tumor samples in this data set, additional clinical predictors containing information about histological grade, angioinvasion, and lymphocytic infiltrate was included. Mar and McLachlan (2003) investigated whether the three clusters constructed by EMMIX-GENE followed patterns according to these biological indicators. The tumor samples have been ordered in Figure 4.19 according to the three clustered groups.

Tumors assigned to cluster 3 appear to match tumors labeled ER positive, while the majority of tumors in clusters 1 and 2 were ER negative. A close association was also noted between tumors assigned to cluster 1 and a histological grade of 3, while the tumors in clusters 2 and 3 were more likely to have a histological grade of 1 or 2. Some association was visible between clusters 1 and 2 and the lymphocytic infiltrate score, where the majority of tumors in these clusters had scores of 0, while tumors

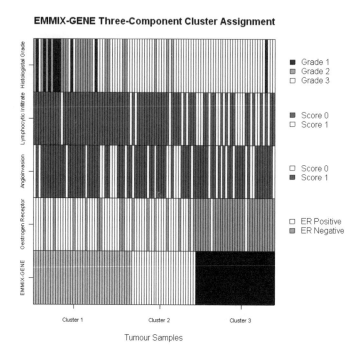

Fig. 4.19 Comparing EMMIX-GENE cluster assignments with other clinical indicators.

in cluster 3 had scores of 1. Indicators related to angioinvasion did not bear a strong association with the EMMIX-GENE clusters. These observations were consistent with those reported by van 't Veer et al (2002).

4.12 CHOOSING THE NUMBER OF CLUSTERS IN MICROARRAY DATA

In this section we demonstrate how the mixture model-based approach can be used for assessing how many clusters of tissues there are in the data. A test for the smallest number of components in the mixture model compatible with the data can be performed on the basis of the likelihood ratio statistic. Firstly, we briefly describe other work done in this area.

4.12.1 Some Previous Attempts

Fraley and Raftery (1998) have provided a summary of methods used to determine the number of clusters in the cluster analysis of gene expression data. Many of the methods rely on graphical display and visual inspection to choose the number of clusters (Eisen et al., 1998; Tamayo et al., 1999). In the cluster affinity search technique

(CAST) method of Ben-Dor, Shamir, and Yakhini (1999) for the clustering of genes, a cluster is incrementally grown by adding one gene after the other. Subsequent cleaning steps enable spurious members of the cluster to be removed. Genes are added and removed until the cluster stabilizes, and then the process starts with a new cluster.

In the hierarchical clustering approach of Hastie et al. (2000), the model size and hence the number of clusters is chosen by cross-validation. In the plaid model approach of Lazzeroni and Owen (2002), layers are added only if they make a significant contribution to the model (in terms of minimizing a sum of squared errors). Thus the number of clusters is effectively determined by the number of layers added to the model. van der Laan and Bryan (2000) have considered the use of the average silhouette width measure in the PAM method of clustering (Kaufman and Rousseeuw, 1990).

On resampling approaches, Tibshirani et al. (2001) proposed a gap statistic as a general method for determining the number of clusters. This method compares an observed internal index, such as the within-cluster sum of squares, to its expectation under a reference null distribution; see Section 3.17.1. More recently, Dudoit and Fridlyand (2002) proposed a prediction-based resampling method to estimate the number of clusters in a data set. This nonparametric resampling method was defined in Section 3.17.2 and compared in some simulations with the parametric approach based on the likelihood ratio statistic (LRT) in Section 3.18.

4.13   LIKELIHOOD RATIO TEST APPLIED TO MICROARRAY DATA

To further demonstrate the effectiveness of the resampling approach based on the LRT for the choice of number of clusters, we now report the results of McLachlan and Khan (2004), who applied it to some microarray data sets as available in the literature and used in the comparisons of Dudoit and Fridlyand (2002).

4.13.1   Golub Data

We firstly consider the clustering of the leukemia data set as in Golub et al. (1999). They used Affymetrix oligonucleotide arrays to measure gene expressions in two types of acute leukemias: acute lymphoblastic leukemia (ALL), and acute myeloid leukemia (AML). The entire data set comprised $M = 72$ tissue samples and $N = 7,129$ genes. The 72 samples were made up of 47 cases of acute lymphoblastic leukemia (ALL) of which there were $n_1 = 38$ ALL B-cell cases and $n_2 = 9$ ALL T-cell cases, along with $n_3 = 25$ cases of acute myeloid leukemia (AML). McLachlan and Khan (2004) followed the processing steps of Dudoit et al. (2002a) of (1) thresholding: floor of 100 and ceiling of 16,000; (2) filtering: exclusion of genes with $\max/\min \leq 5$ and $(\max - \min) \leq 500$, where max and min refer, respectively, to the maximum and minimum expression levels of a particular gene across a tissue sample; (3) taking the natural logarithm of the expression levels. Before they standardized the genes (the rows of the (logged) data matrix A) to have means zero and unit standard deviations over the tissue samples, they first standardized the arrays (the columns of the data matrix A) to have zero means and unit standard deviations. This was done in an attempt to remove systematic sources of variation, as discussed, for example, in Dudoit and Fridlyand (2003). This preprocessing of the genes resulted in 3,731 genes being retained.

Obviously, there were far too many genes relative to the tissue samples to fit a normal mixture model directly to the $n = 72$ samples on the basis of all the genes. Thus they used the EMMIX-GENE program to first remove those genes assessed as having little discriminatory capacity across the $n = 72$ tissue samples by fitting a mixture of t-distributions to each of the 3,731 genes considered separately. This led to 2,069 genes being retained where, for the retention of a gene, the threshold for the increase in twice the log likelihood $(-2 \log \lambda)$ was set to be 8 and the minimum cluster size was set to be 5. They then summarized the retained genes by clustering them into $N_o = 40$ groups on the basis of the 72 tissue samples by fitting in equal proportions a mixture of 40 normal components with a common spherical covariance matrix, $\sigma^2 I_{72}$.

We first clustered the 72 tissue samples by fitting a mixture of $g = 2$ factor analyzers to the means of the 40 gene groups produced by the EMMIX-GENE program. It resulted in two clusters of size 48 and 24, respectively, that correspond almost perfectly to the external classification $n_1 = 47$ ALL tissues and $n_2 = 25$ AML tissues. One AML tissue is put in the first cluster, corresponding to the ALL tissues.

We then fitted a mixture of $g = 3$ factor components with the same number of factors and constraints as for the two-component model. It led to three clusters with one cluster corresponding to the AML cases (with 24 AML tumors and one B-cell ALL); a second cluster with eight of the nine T-cell ALL tumors, along with 15 B-cell ALL tumors; and a third cluster with 22 B-cell ALL tumors, one T-cell ALL tumor, and one AML tumor.

We carried out the LRT test of $g = 2$ versus $g = 3$ component factor analyzers via a resampling approach using $B = 39$ bootstrap samples. As the value of $-2 \log \lambda$ for the original sample is greater than the largest of the 39 bootstrap values of $-2 \log \lambda$, the P-value is estimated to be less than 0.025. This suggests that there is strong support for $g = 3$ clusters in this set.

4.13.2 Alizadeh Data

The second data set to be considered here concerns the case study of Alizadeh et al. (2000), which measured the gene expression levels using a specialized cDNA microarray, the Lymphochip. The data consist of $M = 80$ tissue samples and $N = 4,062$ genes. The former consist of $n_1 = 29$ cases of B-cell chronic lymphocytic leukemia (B-CLL), $n_2 = 9$ cases of follicular lymphoma (FL), and $n_3 = 42$ cases of diffuse large B-cell lymphoma (DLBCL). The missing data were imputed as in Dudoit et al. (2002a).

We first clustered the 80 tissue samples by fitting a mixture of $g = 3$ factor analyzers to the means of the 40 gene groups produced by the EMMIX-GENE program. If we start the iterative fitting process from the aforementioned external classification of the tissues, we obtain a solution (S_1) of the likelihood equation that leads to an outright clustering (C_1) that corresponds perfectly with the external classification. However, if we use 10 random and k-means-based partitions to start the iterative fitting, we obtain a nonspurious solution (S_2) at which the likelihood has greater value than for S_1. The clustering C_2 produced by the solution S_2 has one cluster consisting of the 29 B-CLL cases, another consisting of the 9 FL cases and 7 DLBCL cases, and a third cluster consisting of the remaining 35 DLBCL cases. On the basis of the likelihood ratio test, it was concluded that a mixture model of $g = 3$ factor analyzers (with $q = 4$ factors) is adequate for describing the group structure in the data set $(0.075 < P < 0.1$ using $B = 39$ bootstrap samples).

4.13.3 Bittner Data

The third data set we considered concerns the melanoma data of Bittner et al. (2000). It consists of $M = 31$ tissue samples and $N = 3,613$ genes. We again used EMMIX-GENE (with the same thresholds as for the leukemia data) to reduce the number of genes in this set to 571, which were then clustered into 15 groups. As an inspection of the heat maps in which the genes within a cluster group are displayed for the 31 tissue samples shows that the first cluster group (containing some 49 genes) is one of the more useful groups for revealing the separation of the last 19 tissues from the rest, we worked with this cluster of genes for our subsequent testing for cluster structure among the 31 tissues. As noted in Dudoit and Fridlyand (2002), there are no *a priori* classes known for this data set. The analysis of Bittner et al. (2000) suggests that two classes may be present, as they identified a major cluster of 19 tissues. The Clest procedure (Dudoit and Fridlyand (2002)) also yielded two classes, although their clustering has four tissues joined to the 19-member cluster identified in Bittner et al. (2000).

Application of the likelihood ratio test in conjunction with the fitting of a mixture of g factor analyzers with $q = 4$ clusters gives a significant result (but close to the borderline) at the 5% level $(0.04 < P < 0.05$, using $B = 99$ bootstrap samples) for the test of $g = 2$ versus $g = 3$. The two-cluster solution has the last 19 tissues along with the first, seventh, eighth, and tenth tissues in one cluster with the remaining 9 tissues in the other cluster. If we use the solution obtained from starting with the partition that has the first 12 tissues in one group and the last 19 in another, then we obtain this clustering, but it corresponds to a smaller local maximum.

4.13.4 van 't Veer Data

We finally considered the choice of the number of components g to be used in our normal mixture for the breast cancer data of van 't Veer et al. (2002) discussed previously in Section 4.11. The likelihood ratio statistic was adopted for this purpose, and we used the resampling approach of McLachlan (1987) to assess the P-value. This is because the usual chi-squared approximation to the null distribution of $-2 \log \lambda$ is not valid for this problem, due to the breakdown in regularity conditions. We proceeded sequentially, testing the null hypothesis $H_o : g = g_o$ versus the alternative hypothesis $H_1 : g = g_o + 1$, starting with $g_o = 1$ and continuing until a nonsignificant result was obtained. We concluded from these tests that $g = 3$ components were adequate for this data set.

4.14 EFFECT OF SELECTION BIAS ON THE NUMBER OF CLUSTERS

Note that in the examples above we did not attempt to account for the preprocessing in these applications of the resampling approach. Caution must be exercised in drawing conclusions in analyses based on a subset of the genes selected from a very large number due to the consequent selection bias. This selection bias is to be pursued in some depth in Chapter 7 on unbiased estimation of the error rates of a discriminant rule formed from a subset of a very large number of genes. Hence we need to consider how to allow for this selection in drawing conclusions from tests concerning the smallest number of components in a mixture model fitted to the tissue samples based on a relatively small subset of the genes.

One way of proceeding would be to carry out the likelihood ratio test on the basis of the leading principal components formed from all the genes. That is, we do not carry out any gene selection before performing the principal component analysis. As noted in Section 4.4, an analysis of a full set of genes implies that all aspects of the variation are incorporated into the singular value decomposition, and so the principal components are affected by the noise affecting each of the genes. But this full set analysis should at least provide a lower bound on the smallest number of components needed in a mixture modeling of the distribution on the basis of a selected set of informative genes.

4.15 CLUSTERING ON MICROARRAY AND CLINICAL DATA

In Section 3.14, we considered the case where, in addition to the microarray expression data, there are also available data of a clinical nature on the cases on which the tissue samples have been recorded. Two types of mixture models (unconditional and conditional) are proposed there for the simultaneous use of clinical and microarray data for the clustering of tissue samples.

We illustrate the approaches above on the breast cancer data of van 't Veer et al. (2002), as described in detail in Section 4.11. The EMMIX-GENE algorithm was run as in Section 4.11.1, and the 1,867 genes were clustered into 40 groups over the 98 tissues as before. The forty groups were ranked in decreasing order of the clustering capacity of their means. For the purposes of our illustration here, the vector y of microarray data was taken to be the means of the top 15 ranked groups.

The clinical data vector x consisted of six binary clinical variables, as considered in the Supplementary Information of van 't Veer et al. (2002). The six variables comprised tumor grade ($x_1 = 0$, sizes 0 and 1, and $x_1 = 1$, size 2); estrogen receptor (ER) status ($x_3 = 0, \le 10$, and $x_3 = 1, > 10$); progesterone receptor (PR) status

($x_4 = 0, \leq 20$, and $x_4 = 1, > 20$); age ($x_5 = 0, \leq 40$, and $x_5 = 1, > 40$), and angioinvasion ($x_6 = 0$, no, and $x_6 = 1$, yes).

We clustered the $n = 78$ tissue samples into $g = 2$ clusters on the basis of the full (unconditional) model (3.36) and the conditional model (3.38) fitted to the microarray and clinical data together. We first report the results for the former model, using the NAIVE-independent version of it. That is, the binary variables are taken to be independent within a class, and the microarray-data vector y is taken to be independent of the clinical data (within a class).

The clustering corresponding to the largest of the local maxima located gave the following clustering:

$$C_1 = \{1\text{–}17, 9\text{–}11, 13\text{–}23, 25\text{–}27, 29\text{–}42, 44\text{–}46\}$$
$$\cup \{48\text{–}53, 56, 58\text{–}63, 66, 69\text{–}70, 72, 76\text{–}78\}.$$

$$(4.5)$$

In Table 4.2, we have listed the fitted values of $f_{iv}(x_v = 1)$, which is the probability that the vth clinical (binary) variable is equal to one given its membership of the ith component of the mixture model ($i = 1, 2$). Concerning these clinical variables, it can be seen from Table 4.2 that the estimated probability of a high-grade tumor in the second component is close to one. However, high-grade tumors are not confined to this second component, as the estimated probability of a high-grade tumor in the first component is 0.426.

Table 4.2 Estimates of Component Probabilities for Clinical Binary Variables

v	$f_{1v}(x_v = 1)$	$f_{2v}(x_v = 1)$
1 (grade)	0.426	0.968
2 (ER)	0.894	0.451
3 (PR)	0.787	0.226
4 (size)	0.277	0.710
5 (age)	0.745	0.710
6 (angioinvasion)	0.213	0.387

The first cluster contains 36 of the 44 tumors in the metastasis-free class G_1 of tumors; the second cluster, however, contains only 12 of the 34 tumors in the metastases class G_2. The misallocation rate of 22/34 for the metastases class G_2 is not surprising given the gene expressions as summarized in the groups of genes (see Figures 4.1 and 4.2). It can be seen from these heat maps that there are several tumors in class G_2 that have gene expression patterns similar to those of tumors in class G_1. Thus, it is very difficult to distinguish between the two tissue classes G_1 (metastasis-free) and G_2 (with metastases) on the basis of these gene expressions.

It would therefore be helpful if the clinical data could be used to aid in the cluster analysis of the tissue samples. But using just the microarray data (that is, ignoring the clinical data), we obtained the same clustering as C_1. Thus, it would appear that for the purposes of clustering these tissue samples into two clusters corresponding to the external classes Π_1 and Π_2, the clinical data do not contribute any additional useful information.

We also fitted this mixture model by starting the EM algorithm from the external classification as given by the two classes Π_1 and Π_2. It led to a clustering corresponding to a smaller local maximum than C_1, where the first cluster still contains 36 of the 44 tumors in the metastasis-free class Π_1, but where the second cluster corresponds to 22 of the 34 tumors in the metastases class Π_2. This second clustering clearly corresponds more closely to the external classification Π_1 and Π_2. But we could locate it only by starting the EM algorithm using this classification of the tissues. In any application of the EM algorithm using random starts, we always located the solution that yielded the first clustering C_1. It thus shows that this clustering C_1 corresponds to a local maximum of the likelihood function that dominates the other local maxima. Although C_1 does not correspond directly with the external classification Π_1 and Π_2, it is nonetheless an interesting finding. For example, it sheds light on the group structure that exists among the tissue samples in terms of the available clinical variables and gene expression data. Concerning the use of the conditional model (3.38), we obtained results similar to those reported above for the full mixture model.

4.16 DISCUSSION

A major advantage of the mixture model-based approach to cluster analysis is that it provides a sound mathematical basis for clustering and the subsequent testing for group structure in a data set. In the case of microarray data, the dimension p of the feature vector (the number of genes) is so much greater than the number n of observations (the tissues) to be clustered, that the normal mixture model cannot be fitted directly.

In this chapter, we show how the EMMIX-GENE program handles this dimensionality problem in a three-step approach as follows: (1) eliminate those genes with little variation across the tissue samples; (2) cluster the remaining genes into a manageable number of groups, using essentially Euclidean distance; and (3) cluster the tissue samples either by considering the clusters of genes individually or collectively with each cluster represented by its sample mean (metagene). If the gene count is too large, even after our reduction steps, to fit the normal mixture model of interest directly, then we fit mixtures of factor analyzers.

We illustrate the EMMIX-GENE procedure on two well-known data sets, and highlight different aspects of the cluster analysis. For the Alon data, cluster analyses performed on the basis of various subsets of the genes selected as being relevant by EMMIX-GENE tended to provide strong support for a partitioning of the tissues into two classes that split the tissue samples obtained under "old" and "new" protocols. There is also support for the splitting of some of the "new" tumor samples and some of the "new" normal tissue samples, which can be partly explained by some of these tissues being outliers if the external classification is valid.

In the analysis of the van 't Veer data, Mar and McLachlan (2003) demonstrated how EMMIX-GENE can be applied to cluster a limited number of tissue samples (98 breast cancer tumors) on the basis of subsets of 1,867 genes selected from a filtered set of 4,869 genes. The tissue samples were clustered into two and three clusters using a mixture of factor analyzers model with $q = 4$ factors. Identification of the clusters produced by EMMIX-GENE with the externally existing classes Π_1 (disease-free group), Π_2 (metastases group), and Π_3 (BRCA), gives a large error rate. However, this clustering is consistent with the gene expressions as displayed in the heat maps for the 40 groups of similar genes. For example, in the first three groups given in Figures 4.8 to 4.10, it can be seen that those tissues of class Π_2 that have been misallocated to Π_1 (Π_3) have gene expression patterns similar to those of the majority of the tissues in class Π_1 (Π_3). Similarly, the tissues of class Π_1 that have been misallocated to Π_2 have gene expression patterns similar to those of the majority of the tissues in class Π_2. This comparison provides some insight into why even in a supervised context there is difficulty in trying to discriminate between the disease-free class Π_1 and the metastases class Π_2 .

In the second part of the chapter we have shown how we can use a mixture model-based approach to determine the number of tissue clusters, and we illustrated the application of this method to four microarray cancer data sets. We propose that the choice of the number of clusters be made by testing for the smallest number of components in the mixture model compatible with the data. The test can be carried out on the basis of the likelihood ratio test statistic $-2 \log \lambda$ with its null distribution approximated by resampling. In the examples, this approach was implemented starting with a single component factor analyzer and proceeding to add a component factor analyzer into the mixture model until the test for an additional component is nonsignificant. We also consider how to allow for selection bias in determining the number of components.

5

Screening and Clustering of Genes

In this chapter, we consider initially the detection of differentially expressed genes in known classes of tissue samples. Although the identification of differentially expressed genes is the first objective in an exploratory gene expression study, it is not the only objective. The identification of clusters of genes that are over- or under-expressed in the same class of, say, cancer tissues would be of interest. This is because biological insights into the pathways and pathogenesis of cancer may result. Thus in the latter part of this chapter, we focus on the clustering of the genes on the basis of the tissue samples. Another problem of interest after the identification of genes that are differentially expressed is the selection of small subsets of genes (gene markers) that are useful in being able to discriminate between the classes of tissues. This is known as the feature-selection problem in discriminant analysis and it is to be covered in Chapter 7 on supervised classification of tissue samples.

5.1 DETECTION OF DIFFERENTIALLY EXPRESSED GENES

5.1.1 Introduction

An important and common problem in microarray experiments is the detection of genes that are differentially expressed in a given number of classes $C_1, \ldots, C_m$. The classes may correspond to tissues (cells) that are at different stages in some process, in distinct pathological states, or under different experimental conditions. By comparing gene expression profiles across these classes, researchers gain insight into the roles and reactions of various genes. One can compare, for example, healthy cells to cancerous cells within subjects in order to learn which genes tend to be over-

(or under-) expressed in the diseased cells; regulation of such genes could produce effective cancer treatment and/or prophylaxis. DeRisi et al. (1996) suppressed the tumorigenic properties of human melanoma cells and compared gene expression profiles among "normal" and modified cells. This experiment allowed investigators to study the differential expression that is associated with tumor suppression (van der Laan and Bryan, 2000).

In screening for cancer, there is interest in detecting genes that are differentially expressed in cancer tissue compared with normal tissue. The cancer tissue may not be able to be used directly for population screening. However, if a gene is found that is expressed differentially in cancer tissue, then the corresponding protein product (or an antibody to it) may be detectable in blood or urine, and could be the basis for a population screening test; see Pepe et al. (2001, 2003). In finding differentially expressed genes, in particular overexpressed genes, with the potential for use in cancer screening, a sizeable number of genes need to be obtained. This is because many genes cannot lead to screening markers; for example, genes that relate simply to inflammation or growth are not candidates as these processes also occur naturally in the body. Also, clinical assays for some gene products may be too difficult to develop for technical reasons (Pepe et al., 2003).

We are concerned with the identification of a subset of genes that are differentially expressed between tissue types from a large pool of candidate genes in microarray experiments. The same statistical problem arises in experiments involving other recently developed high-throughput technologies. For example, the methods to be discussed will also be useful in analyzing experiments in protein mass spectrometry.

5.1.2 Fold Change

The simplest method for identifying differentially expressed genes is to evaluate the log ratio between two classes (or averages of the log ratios when there are replicates) and consider all genes that differ by more than an arbitrary cutoff value to be differentially expressed; see Schena et al. (1996) and DeRisi et al. (1997). For example, if a two-fold difference is chosen as the cutoff value, then genes are taken to be differentially expressed if the expression in one class is over two-fold greater or less than in the other class. This test, sometimes called a 'fold' change, is not a statistical test, and there is no associated level of confidence in the designation of a gene as being differentially expressed or not differentially expressed (Cui and Churchill, 2003). Also, this method ignores the variance of the replicates in each class. Thus it is widely recognized now that simply using the fold changes is unreliable and inefficient (Chen et al., 1997).

5.1.3 Multiplicity Problem

The statistical significance of the differential expressions can be tested by performing a test for each gene. When many hypotheses are tested, the probability that a type I error (a false positive error) is committed increases sharply with the number of hypotheses. This multiplicity problem is not unique to microarray analysis, but its magnitude where each experiment may involve many thousands of genes dramatically intensifies the problem. Correlation between the test statistics attributed to gene coregulation and dependency in the measurement errors of the gene expression levels further complicates the problem.

There is now a very large literature on the problem of detecting differentially expressed genes with microarray data. Dudoit et al. (2002) was one of the first studies to recognize the importance of the multiplicity problem as one of the key statistical issues in microarray data analysis. An excellent review of this problem has been given recently by Dudoit, Shaffer, and Boldrick (2003).

The Bonferroni method is perhaps the best known method for dealing with multiple testing. It controls the family-wise error rate (FWER), which is the probability that at least one false positive error will be committed. However, it is known to be conservative. Westfall and Young (1989) proposed adjusted P-values for less conservative multiple testing procedures which take into account the dependence structure between the test statistics. But control of the FWER is only appropriate in situations where the intent is to identify only a small number of genes that are truly different. Otherwise the severe loss of power in controlling the FWER is not justified (Reiner, Yekutieli, and Benjamini, 2003). Instead, it is more appropriate to emphasize the proportion of false positives among the identified differentially expressed genes. The expectation of this proportion is essentially the false discovery rate (FDR) of Benjamini and Hochberg (1995). The FDR will be defined formally in the next section.

The noise introduced by the nondifferentially expressed genes may obscure the signal in the data. Preselection of genes that pass an FDR testing at a moderate level of significance may largely suppress this noise, thereby improving more specific analyses such as discriminant and cluster analyses.

Recently, a number of key papers have been written on controlling the FDR, including the papers by Genovese and Wasserman (2002), Storey (2002), Storey and Tibshirani (2003a,b), and Storey, Taylor, and Siegmund (2004). Hence methods for the detection of differentially expressed genes are still evolving. We now briefly review the current literature.

5.1.4 Overview of Literature

Among nonparametric methods that are directed toward controlling the FDR in detecting differentially expressed genes are the nonparametric empirical Bayes approach of Efron et al. (2001), the significance analysis of microarrays (SAM) method of Tusher, Tibshirani, and Chu (2001), and the mixture model method (MMM) of Pan, Lin, and Le (2001, 2003). A review of these methods may be found in Pan (2002, 2003).

The empirical Bayes approach of Efron et al. (2001) is implemented within the framework of a two-component mixture model whose two components correspond to the set G_0 of genes that are not differentially expressed and the set G_1 (the complement of G_0) that are. Recently, Do, Müeller, and Tang (2003) have proposed a model-based version of the empirical Bayes approach of Efron et al. (2001). Previously, Lönnstedt and Speed (2001) considered a parametric empirical Bayes approach for two-color microarray experiments to derive an expression for the posterior odds of differential expression for a gene. More recently, Smyth (2004) has developed the hierarchical model of Lönnstedt and Speed (2001) with an arbitrary number of classes and RNA samples.

In other work, a parametric frequentist approach to comparing classes via mixture models in terms of gene-specific summary statistics has been given by Lee et al. (2000), Pan, Lin, and Le (2002), and Allison et al. (2002), while a Bayesian approach has been given by Broët, Richardson, and Radvanyi (2002) and Broët et al. (2004). Working in terms of the observed gene expression levels, Newton (2001), Kendziorski et al. (2003), Newton and Kendziorski (2003), and Newton et al. (2004) have adopted parametric empirical Bayes approaches to this problem. In the latter paper, which is on two-class comparisons, a semiparametric approach was also proposed. Ibrahim, Chen, and Gray (2002) and Garrett and Parmigiani (2003) have considered the fitting of mixture models to the gene expression levels by a parametric Bayesian approach.

With a mixture model approach to inference in this problem, the question of whether the genes are differentially expressed can be formulated in a decision-theoretic framework, which is to be considered in more detail in Chapter 7 on supervised classification. A rule for deciding whether gene j is differentially expressed can be formulated in terms of its (estimated) posterior probability $\hat{\tau}_{0j}$ of belonging to component G_0 corresponding to no differential expression $(j = 1, \ldots, N)$. The estimated Bayes rule decides gene j to be differentially expressed if

$$\hat{\tau}_{0j} < c, \tag{5.1}$$

where $(1 - c)/c$ is the relative importance placed on the two errors of the rule. This is to be stated more precisely in Section 5.7.2. The Bayes rule minimizes the linear combination of these two errors so weighted. The link of this rule with those that set out to control the FDR or pFDR was made recently by Storey (2003), who showed that the Bayes error is a linear combination of the pFDR and pFNR. As to be defined formally in Section 5.4.6, pFDR is the conditional expectation of the proportion of false positives among all rejected hypotheses N_r given $N_r > 0$, while pFNR is the conditional expectation of the false negatives among all $N - N_r$ hypotheses that are not rejected given $N - N_r > 0$.

Thus, this mixture model-based rule approach to the multiple testing problem of differential expression among N genes is attractive. It provides gene-specific summaries in terms of the posterior probabilities of differential expression and, considering all multiple comparisons, it minimizes (asymptotically in N) a linear combination of the pFDR and pFNR. By letting the threshold c in (5.1) increase (decrease), we can place less (more) weight on the pFDR relative to the pFNR.

In addition to the papers referenced above, several others have addressed the problem of testing for differential expression, including Ideker et al. (2000), Kerr, Martin, Churchill, (2000), Manduchi et al. (2000), Thomas et al. (2001), van der Laan and Bryan (2001), Westfall, Zaykin, and Young (2001). Wolfinger et al. (2001), Chilingaryan et al. (2002), Troyanskaya et al. (2002), and Tsai, Hsueh, and Chen (2003). There is also the paper of Cui and Churchill (2003), which gives a review of test statistics for differential expression for microarray experiments.

5.2 TEST OF A SINGLE HYPOTHESIS

From the previous section, it has been seen that the biological question of differential expression can be restated as a problem in multiple hypothesis testing. Before we consider the simultaneous testing of all N genes, we consider the test of a single hypothesis.

In the notation of Section 4.2, the M tissue samples on the N available genes are classified with respect to g classes or conditions. These classes might correspond to discrete levels of a covariate, for example, health status (normal versus cancer), subtypes of cancer, or survival outcomes following treatment for a disease. The classes might also correspond to continuous covariates, for example, dose of a drug or time as in time-course experiments. It is assumed that there are n_i tissue samples from each class C_i $(i = 1, \ldots, g)$, where

$$M = n_1 + \cdots + n_g.$$

For simplicity, we shall take $g = 2$, but the methodology to be described applies in the case of multiple classes.

The aim is to detect whether the expression levels of some of the thousands of genes are different in class C_1 than in class C_2. That is, we wish to identify those genes with differential expression. In the context of statistical inference, we can formulate the problem as a multiple hypothesis testing problem. For gene j, we let H_j denote the null hypothesis of no association between its expression level and membership of the response or covariate $(j = 1, \ldots, N)$. Here we consider a response that defines the class membership of the gene. In some cases more specific null hypotheses may be of interest, for example, the null hypothesis of equal mean expressions in the two classes as opposed to identical distributions. We shall say that $H_j = 0$ if the null hypothesis for the jth gene holds and that $H_j = 1$ if it does not hold.

In testing a single hypothesis such as H_j, one can commit two types of errors. One can reject H_j when the null hypothesis H_j holds (a type I error), or one can retain H_j when it does not hold (a type II error). With the traditional approach to the test of a single hypothesis, one aims to maximize the power (one minus the probability of making a Type II error), while keeping the probability of a type I error at or below a specified level.

We let T_j denote a test statistic for testing H_j for the jth gene. A gene-specific summary is given by the observed value t_j of the test statistic T_j or the associated P-value, p_j. Note that t_j should not be confused with the t random variable having a Student's t-distribution with j degrees of freedom. The null hypothesis H_j is rejected if there is enough evidence in favor of the alternative. That is, if T_j falls in some predetermined rejection region Γ_j or if the P-value p_j is less than some conventional value such as $\alpha = 5\%$. For example, the critical region (that is, the rejection region) might have the form $\Gamma_j = \{t_j : |t_j| \geq c_\alpha\}$, where

$$\mathrm{pr}\{|T_j| \geq c_\alpha \mid H_j\} = \alpha.$$

The P-value p_j for an observed value t_j of the test statistic T_j is then given by

$$p_j = \mathrm{pr}\{|T_j| \geq |t_j| \mid H_j\}.$$

In practice, the number of genes N can be very large. Thus if we were to carry out separate tests in the case of $N = 6,000$ genes, the number of false positives could be quite large. For instance, if all $N = 6,000$ genes were not differentially expressed, then the expected number of false positives would be 300. Thus there is a need to control the false positive rate.

Before we proceed to formulate the problem in a multiple hypothesis framework, we consider the calculation of some gene statistics; in particular, some two-sample t-statistics applicable in the case of $g = 2$ classes.

5.3 GENE STATISTICS

5.3.1 Calculation of Interactions via ANOVA Models

We let y_{ijk} denote the (logged) gene expression level for the kth replicate of the jth gene in the ith class $(k = 1, \ldots, n_j; j = 1 \ldots, N; i = 1, \ldots, g)$, where

$$M = n_1 + \cdots + n_g$$

denotes the total number of microarrays. Assuming that the gene expressions have been preprocessed with adjustment for array effects, (for example, by standardization of each tissue sample), we can consider the classical two-way ANOVA model,

$$y_{ijk} = \mu + \alpha_j + \beta_i + \gamma_{ij} + e_{ijk}, \tag{5.2}$$

where μ, α_j, β_i, and γ_{ij} refer to the overall effect, the jth gene-main effect, the ith class-main effect, and the ith class-jth gene interaction effect, respectively, and e_{ijk} is the error term.

In the detection of differential expression, the effect of interest is the interaction term γ_{ij}, as it captures the expression of the jth gene specifically attributable to the ith class $(j = 1, \ldots, N; i = 1, \ldots, g)$. Hence the raw data for the formation of the gene statistics to be considered below are the y_{ijk}^*, where

$$y_{ijk}^* = y_{ijk} - \hat{\mu} - \hat{\alpha}_j - \hat{\beta}_i \tag{5.3}$$

and where $\hat{\mu} = \bar{y}_{...}, \hat{\alpha}_j = \bar{y}_{.j.} - \bar{y}_{...}$, and $\hat{\beta}_i = \bar{y}_{i..} - \bar{y}_{...}$ denote the estimates of the overall, the jth gene- and the ith class-main effects, respectively; see, for example, Lee et al. (2000) and Broët et al. (2004). Note that if we let A denote the $N \times M$ matrix of (logged) expression values, then the process above is achieved by column and row standardization of the elements of the matrix A. This same standardization was carried out before the clustering of the tissue samples as considered in the previous chapter.

As an alternative to (5.2), we could allow for array effects through the fitting of a three-way model with main and interaction terms involving the arrays (Kerr et al., 2002); see, for example, Tsai et al. (2004).

5.3.2 Two-Sample *t*-Statistics

A commonly used statistic for testing for a difference in the means of two classes is the well-known Student's *t*-statistic defined by

$$T_j = \frac{\overline{y}_{1j} - \overline{y}_{2j}}{\sqrt{s_{1j}^2/n_1 + s_{2j}^2/n_2}}, \tag{5.4}$$

where $\overline{y}_{ij}$ and s_{ij}^2 denote the sample mean and variance of the n_i replicates y_{ijk}^* ($k = 1, \ldots, n_i$) for the jth gene in the ith class C_i ($i = 1, 2$). Under the assumption of normality, it can be shown using the Satterthwaite (1946) approximation that T_j has a t-distribution with degrees of freedom

$$\nu = \frac{(s_{i1}^2/n_1 + s_{i2}^2/n_2)^2}{(s_{i1}^2/n_1)^2/(n_1 - 1) + (s_{i2}^2/n_2)^2/(n_2 - 1)}.$$

In the case of an assumed common variance in the two classes, the pooled form of (5.4) has the form

$$T_j = \frac{\overline{y}_{1j} - \overline{y}_{2j}}{s_j \sqrt{1/n_1 + 1/n_2}}, \tag{5.5}$$

where

$$s_j^2 = \{(n_1 - 1)s_{1j}^2 + (n_2 - 1)s_{2j}^2\}/(M - 2)$$

is the pooled within-class sample variance.

Because of the large number of genes in the microarray experiments, there will always be some genes with a very small sum of squares across replicates, so that their (absolute) t-values will be very large whether or not their averages are large. Tusher et al. (2001) have proposed a refinement that avoids this difficulty. They use the modified t-statistic by adding a constant a_0 to the denominator of (5.4) to give

$$T_j = \frac{\overline{y}_{1j} - \overline{y}_{2j}}{s_j \sqrt{1/n_1 + 1/n_2} + a_0}, \tag{5.6}$$

The constant a_0 was chosen to make the coefficient of variation of T_j approximately constant as a function of s_j. This has the added effect of dampening values of T_j that arise from genes whose expression is near to zero. In Efron et al. (2001), a_0 was chosen to be the percentile of the distribution of the s_j that maximizes the posterior probability of a gene being differentially expressed. Baldi and Long (2003) and Broberg (2003) have used t-statistics with an offset.

5.4 MULTIPLE HYPOTHESIS TESTING

As discussed earlier in this chapter, there is a need to consider a measure that is less stringent than the false positive rate. To this end, we now consider some measures of error suitable when multiple hypotheses are being tested simultaneously. The focus

Table 5.1 Possible Outcomes from N Hypothesis Tests

	Accept Null	Reject Null	Total
Null True	N_{00}	N_{01}	N_0
Non-True	N_{10}	N_{11}	N_1
Total	$N - N_r$	N_r	N

will be on the rate of false positives with respect to the number of rejected hypotheses N_r. One may be willing to bear some false positives as long as their number is small in comparison to N_r.

5.4.1 Outcomes with Multiple Hypotheses

The field of multiple hypothesis testing tries to extend the basic paradigm for testing a single hypothesis to the present situation where several hypotheses need to be carried out simultaneously.

Table 5.1 describes the various outcomes when applying some significance test to perform N hypothesis tests. The specific N hypotheses are assumed to be known in advance, but the number N_0 and N_1 of true and false null hypotheses are unknown parameters. The number of rejected null hypotheses N_r is observable, while the number of false positives N_{01}, the number of false negatives N_{10}, the number of true negatives N_{00}, and the number of true positives N_{11} are unobservable random variables.

We have seen that the family wise error rate (FWER) is the probability of yielding one or more false positives out of all hypotheses tested; that is,

$$\text{FWER} = \text{pr}\{N_{01} \geq 1\}. \tag{5.7}$$

5.4.2 Controlling the FWER

There are many adjustment methods for multiple comparisons, including control of the FWER. The most commonly used method for controlling the FWER is the Bonferroni method. The test of each H_j is controlled so that the probability of a Type I error is less than or equal to α/N for some α. This ensures that the overall FWER is less than or equal to α. Closely related to Bonferroni's method is the Šidák (1967) method which calculates the adjusted gene-specific P-value as $p_j^{adj} = 1 - (1 - p_j)^N$, where p_j is the unadjusted P-value for the jth gene $(j = 1, \ldots, N)$. A less conservative adjustment method is the Holm (1979) method that orders the P-values and makes successively smaller adjustments. Let the ordered gene-specific P-values be denoted by $p_{(1)} \leq p_{(2)} \leq \ldots \leq p_{(N)}$. Then the Holm method calculates the adjusted P-values

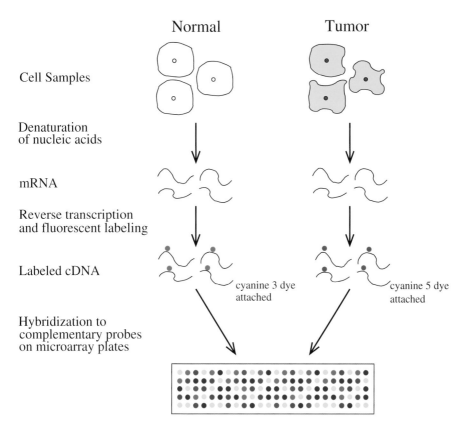

Fig. 1.1 Preparation of target samples: the process from the cell samples to the microarray.

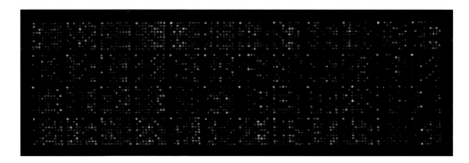

Fig. 1.2 Microarray image showing differentially expressed genes. Red spots: gene transcripts of high expression in target labeled with cyanine 5 dye. Green spots: gene transcripts of high expression in target labeled with cyanine 3 dye. Yellow spots: gene transcripts with similar expression in both target samples.

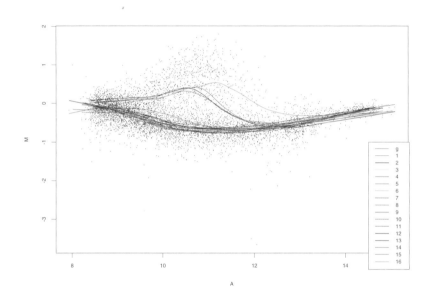

Fig. 2.5 M versus A plot for within-slide normalization. The plot depicts loess curves corresponding to 16 print tips and for the entire data set (labeled g). Data were collected from mice (eight treatment mice with the apo AI gene knocked out, and a control group of eight normal mice). Data details are given in Callow et al. (2000). From Yang et al. (2001a).

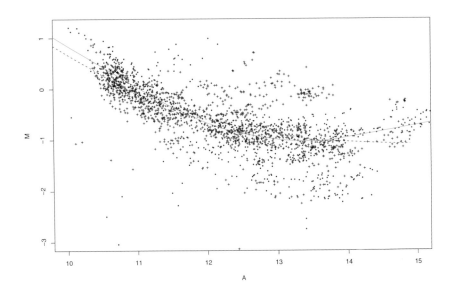

Fig. 2.6 *M* versus *A* plot for paired-slide normalization. The plot depicts loess curves obtained from a follow-up experiment of the apo AI mice study. Details of the experiment are described in Yang et al. (2001a). The dots represent log ratios for slide C3K5 with a corresponding solid loess curve. The crosses represent log ratios for slide C5K3 with a corresponding dotted loess curve. From Yang et al. (2001a).

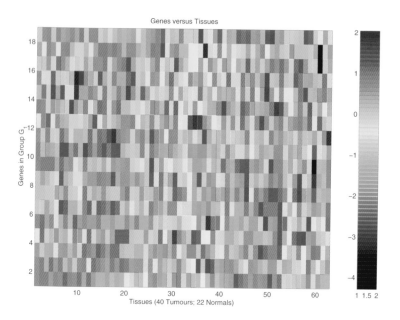

Fig. 4.5 Heat map of 18 genes in group G_1 on 40 tumor and 22 normal tissues in Alon data.

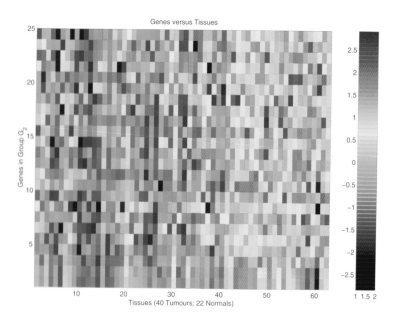

Fig. 4.6 Heat map of 24 genes in group G_2 on 40 tumor and 22 normal tissues in Alon data.

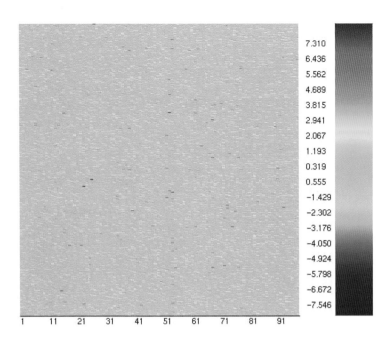

Fig. 4.7 Heat map of the filtered 4,869 genes on 98 tumor tissues in van 't Veer data.

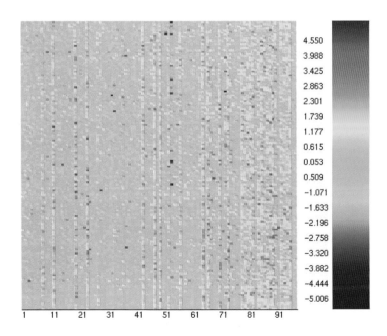

Fig. 4.8 Heat map of 146 genes in group G_1 on 98 tumor tissues in van 't Veer data.

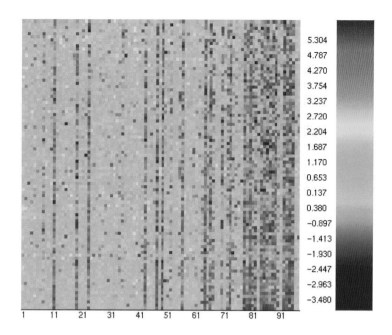

Fig. 4.9 Heat map of 93 genes in group G_2 on 98 tumor tissues in van 't Veer data.

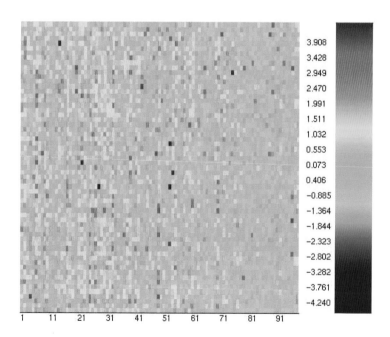

Fig. 4.10 Heat map of 61 genes in group G_3 on 98 tumor tissues in van 't Veer data.

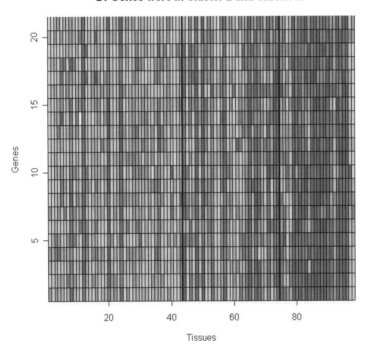

Fig. 4.11   Genes in van 't Veer data retained by EMMIX-GENE appearing in cluster A.

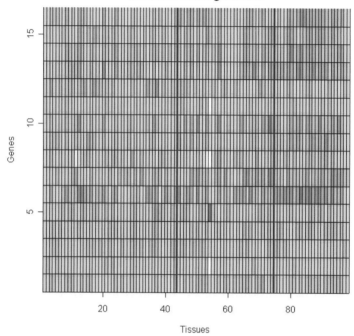

Fig. 4.12 Genes in van 't Veer data rejected by EMMIX-GENE appearing in cluster A.

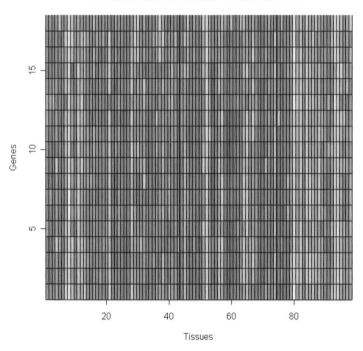

Fig. 4.13 Genes in van 't Veer data retained by EMMIX-GENE appearing in cluster B.

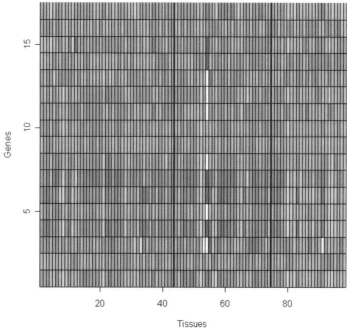

Fig. 4.14 Genes in van 't Veer data rejected by EMMIX-GENE appearing in cluster B.

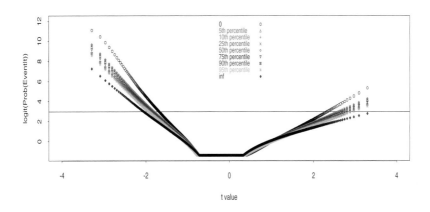

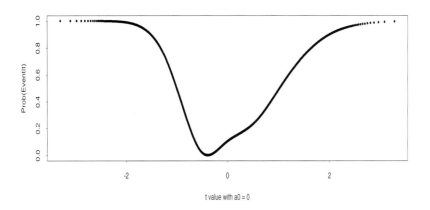

Fig. 5.4 Plots of Prob{Event | t} where 'Event' refers to a gene being differentially expressed; that is, Prob{Event | t} is denoting the (estimated) posterior probability $\hat{\tau}_1(t)$ of differential expression given the value t of the test statistic. Upper panel: Choice of a_0 in the modified t-statistic. Logit of $\tau_{1j}(t)$ was estimated with $\pi_0 = 1$; "inf" is limit as $a_0 \to \infty$. The horizontal line corresponds to the threshold lower bound of 0.9 for $\tau_{1j}(t)$; $a_0 = 0$ is the best choice in terms of maximizing $\hat{\tau}_{1j}(t)$. Lower panel: Empirical Bayes-based curve depicting $\hat{\tau}_{1j}(t)$.

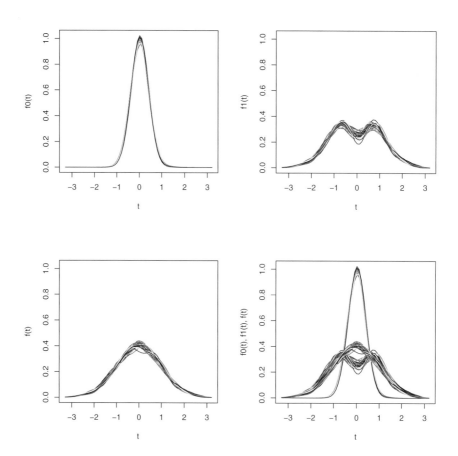

Fig. 5.6 Posterior distributions for the unknown densities (Alon data). The first three panels summarize the posterior information on f_0, f_1 and f, respectively, by showing these densities for 10 draws from the posterior distribution of their parameters. For easier comparison the fourth panel combines the plots from the first three panels into one figure.

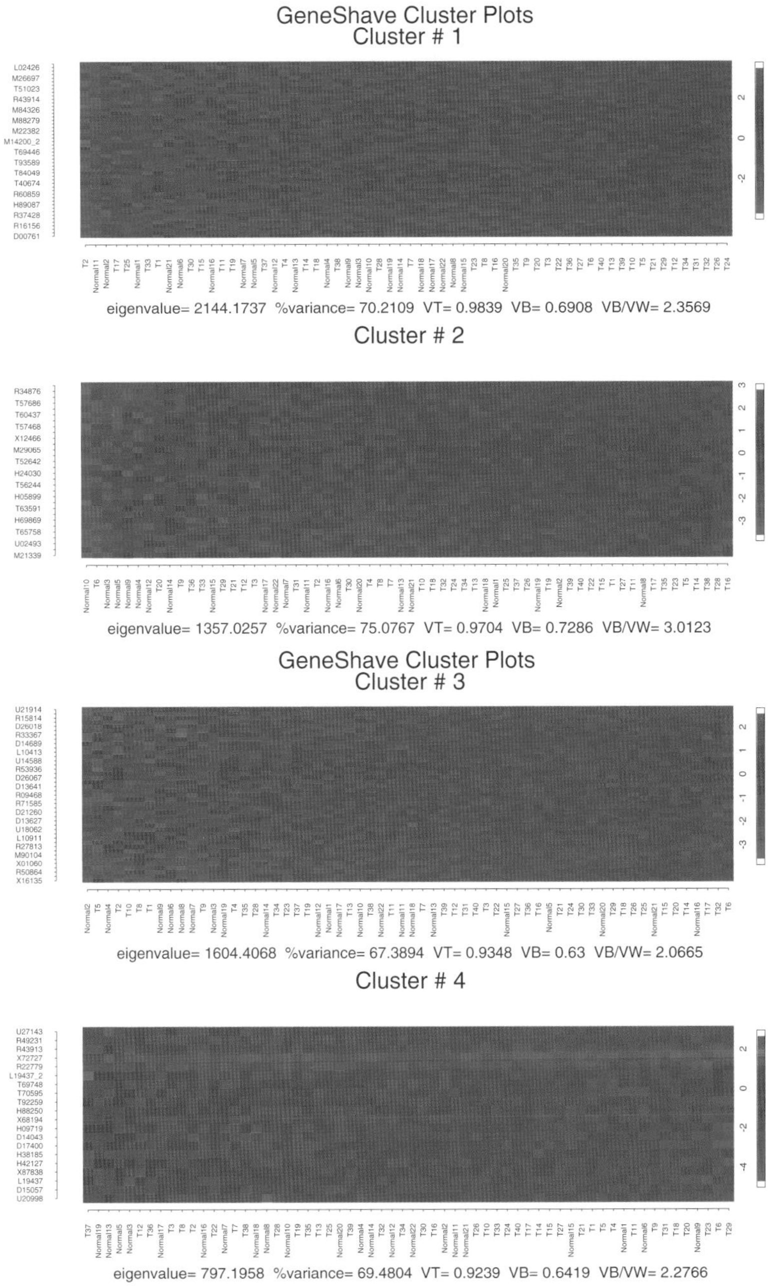

Fig. 5.10 Heat maps of the first four unsupervised gene-shaving clusters for Alon data, sorted by the column-mean gene.

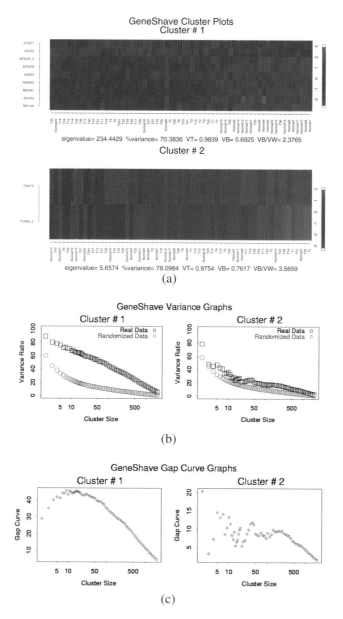

Fig. 5.12 Analysis of Alon data (2,000 genes) under full supervision. (**a**) Heat maps of the first two gene-shaving clusters with samples sorted by the column-mean gene. (**b**) Variance plots for the original and randomized data. Each plot depicts the percent variance explained by each cluster, both for the original data and for an average of over 20 randomized versions for different cluster sizes. (**c**) Gap estimates of cluster size. The gap curve corresponds to the difference between the pair of variance curves.

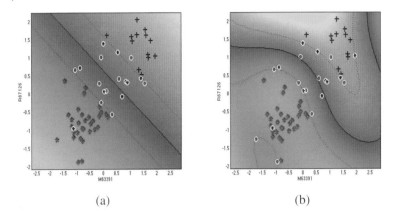

(a)   (b)

Fig. 6.1   Separating surface spawned by SVM with (a) a linear kernel and (b) a Gaussian kernel.

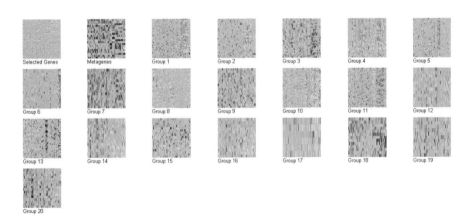

Fig. 8.6   Heat maps for all of the genes, the 20 metagenes, and the 20 gene groups on the 73 tissues in the Stanford data. Tissues are ordered by their histological classification: Adenocarcinoma (1-41), Fetal Lung (42), Large cell (43-47), Normal (48-52), Squamous cell (53-68), Small cell (69-73). From Ben-Tovim Jones et al. (2004b).

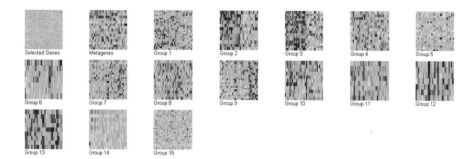

Fig. 8.7 Heat maps for all of the genes, the 15 metagenes, and the 15 gene groups on the 35 tissues in the Stanford data. Tissues are ordered according to the AC groups of Garber et al. (2001): AC group 1 (1-19), AC group 2 (20-26), AC group 3 (27-35). From Ben-Tovim Jones et al. (2004b).

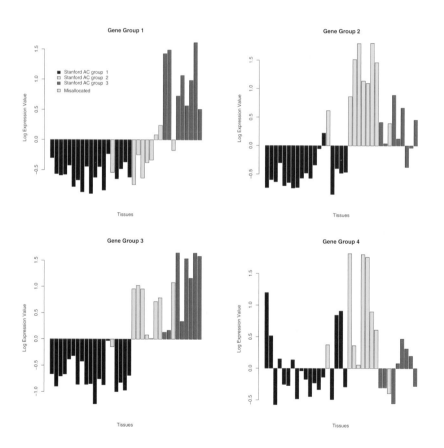

Fig. 8.8 Expression profiles for top metagenes (Stanford 35 AC tissues). From Ben-Tovim Jones et al. (2004b).

by

$$p_{(1)}^{adj} = N \times p_{(1)},$$

$$p_{(j)}^{adj} = \max\{p_{(j-1)}, (N - j + 1) \times p_{(j)}\} \quad (1 < j \leq N).$$

To cater for correlated gene expression levels in certain groups of genes, a more preferable method for bioinformaticians is one that accounts for dependence structure between the genes, such as that proposed by Westfall and Young (1993). This method requires the estimation of the joint null distribution of the unadjusted unknown P-values. Dudoit et al. (2002b) proposed to estimate the joint null distribution of the test statistics T_j for the N genes by permuting the class labels of the M tissue samples. This approach ensures that the dependence structure between genes is preserved, while allowing one to estimate the joint distribution of interest via the empirical distribution generated by the B permutations. Let $t_j^{(b)}$ denote the value of T_j computed for the jth gene from the bth permutation. Then the unadjusted permutation P-values for the t-statistics are

$$p_j = \frac{\sum_{b=1}^{B} I(|t_j^{(b)}| \geq |t_j|)}{B}.$$

Suppose that the observed t-statistics are sorted such that $|t_{(1)}| \geq |t_{(2)}| \geq ... \geq |t_{(p)}|$. Then the adjusted Westfall and Young P-values estimated from the permuted data are given by

$$p_j^{adj} = \frac{\sum_{b=1}^{B} I(|u_j^{(b)}| \geq |t_{(j)}|)}{B},$$

where

$$u_N^{(b)} = |t_{(N)}^{(b)}| \quad \text{and}$$

$$u_j^{(b)} = \max(u_{j+1}^{(b)}, |t_{(j)}^{(b)}|) \quad (1 \leq j < N).$$

5.4.3 False Discovery Rate (FDR)

There is a fundamental issue with approaches that set out to control the FWER. They control the probability of at least one false positive regardless of the number of hypotheses N being tested. When N is very large, they are much too strict, and will lead to many missed findings. The goal in the present situation is therefore to identify as many genes with significant differences as possible, while incurring a relatively low proportion of false positives.

In a seminal paper, Benjamini and Hochberg (1995) introduced a new multiple hypothesis testing error measure called the false discovery rate (FDR), which they define as

$$\text{FDR} = E\{\frac{N_{01}}{N_r \vee 1}\}, \tag{5.8}$$

where $N_r \vee 1 = \max(N_r, 1)$. The effect of $N_r \vee 1$ in the denominator of the expectation in (5.8) is to set $N_{01}/N_r = 0$ when $N_r = 0$. The FDR can be written also as

$$\text{FDR} = E\{N_{01}/N_r \mid N_r > 0\}\,\text{pr}\{N_r > 0\}. \tag{5.9}$$

The FDR can also be written in terms of the proportions $A_{11} = N_{11}/N_1$ and $A_{00} = N_{00}/N_0$, corresponding to the apparent sensitivity and specificity of a prediction rule,

$$\text{FDR} = E\{\frac{N_0(1 - A_{00})}{N_0(1 - A_{00}) + N_1 A_{11}}\}. \tag{5.10}$$

5.4.4 Benjamini-Hochberg Procedure

Controlling the FDR offers a less strict multiple-testing criterion than the FWER. As set out in Storey, Taylor and Siegmund (2004), two approaches to providing conservative FDR procedures are the following. One is to fix the acceptable FDR level beforehand, and to find a data-dependent thresholding rule so that the FDR of this rule is less than or equal to the prechosen level. This is the approach adopted by Benjamini and Hochberg (1995). Another is to fix the thresholding rule, and to form an estimate of the FDR whose expectation is greater than or equal to the FDR rule over the significance region. This was the approach taken by Storey (2002). It avoids having to choose a single acceptable FDR level before any data are seen which, as noted by Storey and Tibshirani (2003a), is often too impractical and restrictive.

In either case, the desire is to be conservative regardless of the value of N_0, which is usually unknown. If this holds, then strong control is provided. Weak control is provided when the procedure is conservative only when $N_0 = N$; in general this is not of interest in the FDR case (Benjamini and Hochberg, 1995).

Benjamini and Hochberg (1995) proved by induction that the following procedure (referred to here as the BH procedure) controls the FDR at level α when the P-values following the null distribution are independent and uniformly distributed.

Step 1. Let $p_{(1)} \leq \cdots \leq p_{(N)}$ be the observed P-values.

Step 2. Calculate

$$\hat{k} = \arg \max_{1 \leq k \leq N} \{k : p_{(k)} \leq \alpha k/N\}. \tag{5.11}$$

Step 3. If $\hat{k}$ exists, then reject null hypotheses corresponding to $p_{(1)} \leq \cdots \leq p_{(\hat{k})}$. Otherwise, reject nothing.

The BH procedure was originally introduced by Simes (1986) to control weakly the FWER when all P-values are independent, but it provides strong control of the FDR as well.

Benjamini and Yekutieli (2001) showed that

$$\text{FDR} \leq \alpha N_0/N$$

for positively dependent test statistics as well. Since the BH procedure controls the FDR at a level too low by a factor of N_0/N, it is natural to try to estimate N_0 and use

$$\alpha^* = \alpha(N/N_0)$$

instead of α to gain more power. Estimating N_0 from a set of P-values goes back to Schweder and Spjøvtoll (1982), which was later formalized by Hochberg and Benjamini (1990). Benjamini and Hochberg (2000) suggested an adaptive procedure that combines the estimation of N_0 with the BH procedure. A similar version has been suggested by Storey (2003), as to be considered further in Section 5.6.

5.4.5 False Nondiscovery Rate (FNR)

Genovese and Wasserman (2002) noticed that one can define a dual quantity to the FDR, which they call the false nondiscovery rate (FNR). It is defined to be the expected proportion of false negatives among all hypotheses that are not rejected, with the ratio being set to zero if all hypotheses are rejected. That is,

$$
\begin{aligned}
\text{FNR} &= E\{\frac{N_{10}}{(N - N_r) \vee 1}\} \\
&= \{E(N_{10}/(N - N_r) \mid (N - N_r) > 0\} \operatorname{pr}\{(N - N_r) > 0\}. \quad (5.12)
\end{aligned}
$$

5.4.6 Positive FDR

The term $\operatorname{pr}\{N_r > 0\}$ in the definition (5.9) of the FDR was included by Benjamini and Hochberg (1995) as a way of allowing the FDR to be bounded in the case when all the null hypotheses are true, $N_0 = N$. But as argued in Storey (2003), if all the null hypotheses are true, one would want the false discovery rate to be one, and one is not interested in cases where no test is significant. These considerations led Storey (2003) to propose the the positive false discovery rate (pFDR), defined by

$$
\text{pFDR} = E(N_{01}/N_r \mid N_r > 0). \quad (5.13)
$$

It is called the pFDR because it is conditioned on the fact that at least one positive outcome has resulted.

5.4.7 Positive FNR

Analogous to the FNR, we can define the positive false nondiscovery rate (pFNR) by

$$
\text{pFNR} = E(N_{10}/(N - N_r) \mid (N - N_r) > 0). \quad (5.14)
$$

5.4.8 Linking False Rates with Posterior Probabilities

Storey (2003) showed under the assumptions of N independent hypotheses with critical region Γ and with $\operatorname{pr}\{H_j = 0\} = \pi_0$ that

$$
\text{pFDR} = \operatorname{pr}\{H_j = 0 \mid T_j \in \Gamma\}, \quad (5.15)
$$

where the right-hand side of (5.15) is the same for each j because of the independent and identically distributed data assumptions. Similarly,

$$
\text{pFNR} = \operatorname{pr}\{H_j = 1 \mid T_j \notin \Gamma\}. \quad (5.16)
$$

5.5   NULL DISTRIBUTION OF TEST STATISTIC

Before we proceed further with procedures that control the FDR or pFDR, we discuss the assessment of the null distribution of the test statistic T_j under the null hypothesis H_j, using the two-sample test statistic (5.4) as an illustration.

5.5.1   Permutation Method

In practice it may not be valid to assume that the null distribution of T_j is a t-distribution, especially as small sample sizes are very common with microarray data. Even if it were, the null distribution of the modified t-statistic (5.6) would still need estimating. We now discuss nonparametric methods for assessing the null distribution of T_j, where T_j is given by (5.4). As pointed out in Pan (1992), the loss in power may not justify the use of the robust Wilcoxon rank sum test, which also requires that the two class distributions have the same shape with the only difference in their location parameters. The null distribution of T_j can be found easily by permuting the class labels or by using bootstrap methodology described in Efron and Tibshirani (1993). As discussed in Storey and Tibshirani (2003b), the permutation method has the strength that if the null distribution is true, then we are able to calculate the null distribution exactly. Storey and Tibshirani (2003b) go on to discuss whether the null distribution of T_j should be calculated on the basis of just the data on the jth gene or by pooling across the N available genes as, for example, in the empirical Bayes approach in Efron et al. (2001). If the null distribution of T_j is calculated on the basis of just the data on the jth gene, then it suffers from a granularity problem. For example, there are only ten ways to divide six microarrays into two equal sized groups; see also Gadbury et al. (2004). The null distribution has a resolution on the order of the number of permutations. If we perform B permutations, then the P-value will be estimated with a resolution of $1/B$. If we assume that each gene has the same null distribution and combine the permutations, then the resolution will be $1/(NB)$ for the pooled null distribution.

Using just the B permutations of the class labels for the gene-specific statistic T_j, the P-value for $T_j = t_j$ is assessed as

$$p_j = \frac{\#\{b : |t_{0j}^{(b)}| \geq |t_j|\}}{B}, \tag{5.17}$$

where $t_{0j}^{(b)}$ is the null version of t_j after the bth permutation of the class labels. If we pool over all N genes, then

$$p_j = \sum_{b=1}^{B} \frac{\#\{j : |t_{0j}^{(b)}| \geq |t_j|, j = 1 \ldots, B\}}{NB} \tag{5.18}$$

The drawback of pooling the null statistics $t_{0j}^{(b)}$ across the genes to assess the null distribution of T_j is that one is using different distributions unless all H_j are true; see Dudoit et al. (2003, Page 84) on this point with respect to the SAM method of Tusher et al. (2001). As explained in Storey and Tibshirani (2003b), this is not a problem in estimating the FDR. The null statistics just have to converge to some overall "null" distribution in N, which will generally hold as N is very large. Storey and Tibshirani (2003b) recommend pooling the null statistics $t_{0j}^{(b)}$ across the genes to make for more accurate and more powerful procedures as well as reducing the computation time. The empirical results of Pan (2003) support this for the SAM method.

5.5.2 Null Replications of the Test Statistic

The empirical Bayes approach to the selection of differentially expressed genes is to be discussed in Section 5.8.1. It will be seen there that with the nonparametric method of Efron et al. (2001), the model-based version proposed by Do et al. (2003), and the MMM method of Pan et al. (2001, 2003), we need to be able to provide suitable estimates of the density of the test statistic T_j, $f_0(t_j)$, in the population G_0 of nondifferentially expressed genes. That is, we need to be able to form replications $t_{0j}^{(b)}$ of t_j that will have a distribution that is as near as possible to the null distribution of T_j, regardless of whether H_j holds or not.

Working, for example, with the t-statistic T_j given by (5.4), Pan, Lin, and Le (2002) suggested null replications $t_{0j}^{(b)}$ of t_j be formed by randomly computing the tissue samples within each class separately in the statistic

$$T_{0j} = a_j / \sqrt{s_{1j}^2/n_1 + s_{2j}^2/n_2}, \tag{5.19}$$

where

$$a_j = \frac{\sum_{j=1}^{n_{11}} y_{1jk}}{n_{11}} - \frac{\sum_{j=n_{11}+1}^{n_1} y_{1jk}}{n_{12}} + \frac{\sum_{j=1}^{n_{21}} y_{2jk}}{n_{21}} - \frac{\sum_{j=n_{21}+1}^{n_2} y_{2jk}}{n_{22}}. \tag{5.20}$$

The term a_j in the numerator of (5.19) is formed by dividing each sample into two almost equally-sized subsamples. For $i = 1, 2$, define $n_{i2} = n_{i1}$ if $n_i = 2n_{i1}$, and $n_{i2} = n_{i1} + 1$ if $n_i = 2n_{i1} + 1$. Also, for use in (5.21) below, we let s_{ij1}^2 and s_{ij2}^2 denote the sample variances of the subsamples y_{ijk} $(k = 1, \ldots, n_{i1})$ and y_{ijk} $(k = n_{i1} + 1, \ldots, n_1)$, respectively $(i = 1, 2)$. The motivation behind the formation of a_j is under the assumption that the errors are distributed symmetrically about the origin, the distribution of T_{0j} should be the same as the null distribution of T_j. However, Pan (2002) noticed that the use of null replications of T_{0j} led to too many genes being declared significant with its use in the MMM method. He pointed out that whereas the numerator and denominator of (5.4) are independently distributed, this is not so for the statistic T_{0j}. He subsequently proposed a modified version,

$$\tilde{T}_{0j} = a_j / \tilde{s}_j, \tag{5.21}$$

where

$$\tilde{s}_j = \sqrt{\frac{s_{11j}^2}{n_{11}} + \frac{s_{12j}^2}{n_{12}} + \frac{s_{21j}^2}{n_{21}} + \frac{s_{22j}^2}{n_{22}}}. \tag{5.22}$$

The degrees of freedom of this statistic is $n_1 + n_2 - 4$. A limitation in its use is that at least four or more tissue samples are required in each class, whereas only two or more are required with T_{0j}.

Under the assumption of normality of the genes within a class, the numerator and denominator in (5.22) are independently distributed. Thus the distribution of $\tilde{T}_{0j}$ should provide a better approximation to the null distribution of the statistic T_j. This has been confirmed in the empirical results of Pan (2003).

Replications $\tilde{t}_{0j}^{(b)}$ ($b = 1, \ldots, B$) of $\tilde{T}_{0j}$ are obtained by randomly computing the tissue samples within each class before the sample from each class is partitioned into two subsamples for the formation of (5.21). For example, if $B = 50$, then the null density $f_0(\tilde{t}_j)$ can be estimated on the basis of the $B = 50$ replications $\tilde{t}_{0j}^{(b)}$ ($j = 1, \ldots, N; b = 1, \ldots, B$).

5.5.3 The SAM Method

The permutation method is used to calculate the null distribution of the modified t-statistic (5.6) in the significance analysis of microarrays (SAM) method of Tusher et al. (2001). But rather than use a symmetric critical region of the form $|t_j| > c$, SAM derives cutoff points c_1 and c_2 so that the critical region has the form $t_j > c_2$ or $t_j < c_1$, and so can lead to a more powerful test in situations where more genes are over expressed than under expressed, or vice-versa. With SAM, the order statistics $t_{(1)}, \ldots, t_{(N)}$ are plotted against their null expectations assessed as

$$\bar{t}_{0(j)} = (1/B) \sum_{b=1}^{B} t_{0(j)}^{(b)} \quad (j = 1, \ldots, N). \tag{5.23}$$

Then as described in Storey and Tibshirani (2003b), one proceeds as follows. Firstly, for a fixed threshold Δ, start at the origin and, moving up to the right, find the first $j = j_1$ such that $t_{(j)} - \bar{t}_{0(j)} \geq \Delta$. All genes past j_1 are called "significant positive." Secondly, start at the origin and, moving down to the left, find the first j_2 such that $\bar{t}_{0(j)} - t_{(j)} \geq \Delta$. All genes past j_2 are called "significant negative." For each Δ, define the upper cutoff point $t_2(\Delta)$ as the smallest t_j among the significant positive genes, and similarly define the lower cutoff point $t_1(\Delta)$. If $t_1(\Delta) > t_2(\Delta)$, then set $t_1(\Delta) = t_2(\Delta) = 0$.

The FDR can be estimated from the same permuted null statistics. In Tusher et al. (2001), the FDR is taken to be the proportion of the $t_{0j}^{(b)}$ that are found to be significant, while in Storey and Tibshirani (2003b), the estimates developed in Storey (2002) are used. We shall discuss these latter estimates of the FDR in Section 5.6.

5.5.4 Application of SAM Method to Alon Data

We applied the SAM method to the cancer colon data of Alon et al. (1999), as considered in Chapter 4. The aim was to identify potentially significant changes in expression of the genes. There are two classes of tissue samples consisting of 2,000 genes with $n_1 = 40$ tumor samples in one class and $n_2 = 22$ samples in the other.

We first applied SAM to the quantile normalized data set above. Figure 5.1, produced by the computer package SAM, presents a scatter plot of the observed relative difference t_j versus the expected relative difference $\bar{t}_{0(j)}$. For the vast majority of genes, $t_{(j)}$ and $\bar{t}_{0(j)}$ are almost equivalent. Some genes are represented by points displaced from the $t_{(j)} = \bar{t}_{0(j)}$ line by a distance greater than a threshold $\Delta = 0.7$, illustrated by the dashed lines in Figure 5.1. The estimated number of falsely significant genes was the average of the number of genes called significant from all the permutations. For $\Delta = 0.7$, the permuted data sets generated an average of 27 falsely significant genes, compared with 540 genes called significant, yielding an estimated FDR of 5% (Table 5.2). As Δ decreased, the number of genes identified as significant by SAM increased along with an increasing FDR. Investigators can use Table 5.2 to calibrate the results that give the best biological interpretation.

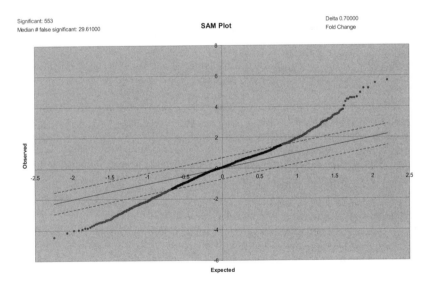

Fig. 5.1 SAM plot for the Alon data set. The dashed lines are drawn at a distance of $\Delta = 0.7$ from the solid line. The upper and lower cut points are the first ones that lie outside the band. All genes to the right of the upper cut point and to the left of the lower cut point are called significant.

Table 5.2   SAM False Positive Results for Different Values of the Threshold Δ

Δ	# False Positive Genes (N_{01})		# Significant Genes (N_r)	FDR (%)	
	Median	90th percentile		Median	90th percentile
0.10	811.88	873.13	1,676	48.44	52.10
0.20	597.56	695.75	1,478	40.43	47.07
0.30	352.78	485.83	1,200	29.40	40.49
0.40	227.57	355.32	1,000	22.76	35.53
0.50	86.01	178.79	720	11.95	24.83
0.60	58.66	135.42	660	8.89	20.52
0.70	27.07	78.96	540	5.01	14.62
0.80	11.28	41.17	430	2.62	9.57
0.90	5.92	25.38	375	1.58	6.77
1.00	2.26	12.46	287	0.79	4.34
1.20	0.56	3.38	204	0.28	1.66
1.30	0.56	1.69	160	0.35	1.06

5.6   RECENT APPROACHES FOR STRONG CONTROL OF THE FDR

We come now to the recent and fruitful results of Storey (2002, 2003), Storey and Tibshirani (2003a), and Storey et al. (2004) on controlling the FDR.

5.6.1   The q-Value

We now consider the pFDR analogue of the P-value, which Storey (2003) called the q-value. An excellent account of the role of the q-value in multiple testing has been given recently by Storey and Tibshirani (2003a). As they note, the P-value can be used to assign to each gene a level of significance in terms of the false positive rate. A P-value threshold of 5% yields a false positive rate of 5% among all null hypotheses being tested. However, it does not provide a measure of the errors among the genes declared to be significant. Information on this is provided by the FDR, which gives a measure of the proportion of false positives among the significant genes. However, as pointed out by Storey and Tibshirani (2003a), one would also like a measure of significance that can be attached to each individual gene. The q-value is a measure designed to reflect this level of attachment.

As to be made more precise in the next section, the q-value for a gene is the expected proportion of false positives incurred when calling that gene significant. Suppose that the genes are ranked in increasing order of their P-values. Then calculating the q-values for each gene in the list and thresholding them at some q-value level α produces a set of genes (each with $q \leq \alpha$) so that a proportion α of them is expected to be false positives. Further, with this thresholding, the q-value can be estimated to control conservatively for large N the FDR at a level $\leq \alpha$ when all N genes are considered simultaneously.

The P-value can be viewed as the probability under the null hypothesis of obtaining a value of the test statistic as or more extreme than its observed value. The q-value for an observed test statistic can be viewed as the expected proportion of false positives among all genes with values of their test statistics as or more extreme than the observed value. This analogy between the P-value and the q-value is considered further in the next section.

5.6.2 Technical Definition of q-Value

Firstly, we consider the test of a single null hypothesis H on the basis of a test statistic T, where the event $\{H = 0\}$ denotes that the null hypothesis holds and $\{H = 1\}$ implies that the alternative hypothesis holds. As explained in Storey (2003), hypothesis tests are usually derived according to a nested set of significance regions. As long as the null and alternative densities $f_0(t)$ and $f_1(t)$ have common support, we can denote this nested set of significance regions without loss of generality by $\{\Gamma\}_{\alpha=0}^1$, where α is such that

$$\alpha = \operatorname{pr}\{T_j \in \Gamma_\alpha \mid H = 0\}. \tag{5.24}$$

Note that $\alpha^* \leq \alpha$ implies that $\Gamma_{\alpha^*} \subseteq \Gamma_\alpha$, giving the nested property. Using this notation, the P-value $p(t)$ of an observed statistic $T = t$ is defined to be

$$p(t) = \inf_{\{\Gamma_\alpha : t \in \Gamma_\alpha\}} \operatorname{pr}\{T_j \in \Gamma_\alpha \mid H = 0\}. \tag{5.25}$$

It is the minimum type I error that can occur when rejecting a statistic value t_j, given the set of nested significance regions. Storey (2003) defined an analogous quantity in terms of the pFDR,

$$q(t) = \inf_{\{\Gamma_\alpha : t \in \Gamma_\alpha\}} \operatorname{pFDR}(\Gamma_\alpha). \tag{5.26}$$

In words, (5.26) says that the q-value is a measure of the strength of an observed statistic with respect to the pFDR; it is the minimum pFDR that can occur when rejecting a statistic with value t for the set of nested significance regions. Under the mixture model (5.36), Storey (2003) showed that

$$q(t) = \inf_{\{\Gamma_\alpha : t \in \Gamma_\alpha\}} \operatorname{pr}\{H_j = 0 \mid T_j \in \Gamma_\alpha\}. \tag{5.27}$$

Storey (2003) called (5.27) a q-value because it is equivalent to the P-value (5.25) with the events $\{T \in \Gamma_\alpha\}$ and $\{H = 0\}$ reversed.

For N identical tests T_j with observed values t_j and P-values p_j, Storey (2003) showed that the q-value can be calculated from either the original statistics or their P-values. Also, when the tests are independent and follow the mixture distribution,

$$q(p_j) = \min_{\gamma \geq p_j} \operatorname{pFDR}(\gamma) \tag{5.28}$$

if and only $K(\alpha)/u$ is a decreasing function of α, and

$$K(\alpha) = \mathrm{pr}\{T_j \in \Gamma_\alpha \,|\, H = 1\}.$$

From (5.28), a q-value of a statistic is the minimum pFDR at which that statistic can be called significant. Note that this definition of the q-value guarantees that the $q(p_j)$ are in the same order as the P-values p_j.

We can estimate $\mathrm{pFDR}(\gamma)$ by the estimator $\widehat{\mathrm{FDR}}_\xi(\gamma)$ to be defined in the next section. Because N is large here,

$$\mathrm{pr}\{N > 0\} \approx 1, \tag{5.29}$$

and so

$$\begin{aligned}
\mathrm{FDR} &\approx \mathrm{pFDR} \\
&\approx \frac{E(N_{01})}{E(N_r)}.
\end{aligned} \tag{5.30}$$

Thus the distinction between pFDR and FDR is not crucial here (Storey and Tibshirani et al., 2003a).

5.6.3 Controlling FDR Strongly

Let $\mathrm{FDR}(\gamma)$ denote the FDR when rejecting all null hypotheses with $p_j \leq \gamma$ for $j = 1, \ldots, N$, and

$$N_r(\gamma) = \#\{p_j : p_j \leq \gamma\}.$$

Storey (2002) proposed $\mathrm{FDR}(\gamma)$ be estimated as

$$\widehat{\mathrm{FDR}}_\xi(\gamma) = \frac{\hat{\pi}_0(\xi)\gamma}{\{N_r(\gamma) \vee 1\}/N}. \tag{5.31}$$

The term $\hat{\pi}_0(\xi)$ is an estimate of $\pi_0 = N_0/N$, the proportion of true null hypotheses. This estimate depends on the tuning parameter ξ and is defined as

$$\hat{\pi}_0(\xi) = \frac{(N - N_r(\xi))}{(1 - \xi)N}. \tag{5.32}$$

As explained in Storey (2002), when the P-values corresponding to the null hypotheses are uniformly distributed, we have

$$(N - N_r(\xi))/N \approx \pi_0(\xi)(1 - \xi) \tag{5.33}$$

for a well-chosen ξ. This is because most of the P-values near 1 should be null if each test has a reasonable power. There is an inherent bias-variance trade-off in the choice of ξ. In most cases as ξ grows smaller, the bias of $\hat{\pi}_0(\xi)$ grows larger, but the variance becomes smaller. Therefore, ξ can be chosen to try to balance this trade-off. A general algorithm for the automatic calculation of ξ is given in Storey and Tibshirani (2003a).

Under the assumption that the P-values are independent and identically distributed under the null hypotheses, Storey et al. (2004) showed that

$$E\{\widehat{\text{FDR}}_\xi(\gamma)\} \geq \text{FDR}(\gamma). \tag{5.34}$$

Storey et al. (2004) showed that $\widehat{\text{FDR}}_\xi(\gamma)$ can be used to provide a step-up rule that controls strongly the FDR. They also showed that this rule has a useful interpretation in the context of the BH procedure. Under the assumption of independent and uniformly distributed null P-values, the BH procedure controls the FDR at exactly level $\pi_0(\xi)\alpha$. Thus, if N is replaced by $\pi_0(\xi)N$ in the BH procedure, then the FDR is controlled exactly at level α; see also Benjamini and Yekutieli (2001). The reader is referred to Storey (2002), Storey (2003), and Storey et al. (2004) for a more detailed and mathematical account of these results.

5.6.4 Selecting Genes via the q-Value

Rather than fixing the threshold α on FDR and estimating k in (5.11) (that is, estimating the critical region), Storey (2002) suggested fixing the critical region Γ. We can list the genes in increasing order of their P-values p_j. If a threshold α is chosen, then we can call all genes with $p_j \leq \alpha$ significant. The q-value of a particular gene j is the expected proportion of false positives incurred when calling that gene significant. Therefore calculating the q-values and then thresholding them at q-value level α produces a set of significant genes so that a proportion α is expected to be false.

As explained in Storey and Tibshirani (2003a), there are two results concerning the accuracy of the estimated q-values that hold for large N under what we call "weak dependence" of the P-values. Weak dependence can loosely be described as any form of dependence whose effect becomes negligible as the number of genes increases to infinity. The first result is that if we call all genes significant with q-values less than or equal to α, then for large N the FDR $\leq \alpha$. The second result is that the estimated q-values are simultaneously conservative for the true q-values. This means that the estimated q-value of each gene is greater than or equal to its true q-value, across all genes at once. Under this result, one can consider each gene's significance simultaneously without worrying about inducing bias. In a sense, the second result implies that one can consider all cutoffs simultaneously, which is a much stronger generalization of the first result. These conservative properties are desirable because one does not want to underestimate the true q-values or the true proportion of false positives. The reader is referred to Storey and Tibshirani (2003a) for further discussion on these results and on the most likely form of dependence between genes in a microarray data set.

5.6.5 Application to Hedenfalk Data

We report here the results of Storey and Tibshirani (2003a) for one of the data sets that they considered to demonstrate their approach to the selection of a list of differentially genes on the basis of their q-values. It concerns the study of Hedenfalk et al. (2001), which consisted of $n_1 = 7$ BRCA1 arrays and $n_2 = 8$ BRCA2 arrays, along with some arrays from sporadic breast cancer, which Storey and Tibshirani (2003a) did not use. One of the goals of this study of Hedenfalk et al. (2001) was to find genes that are differentially expressed between BRCA1- and BRCA2-mutation-positive tumors by obtaining several microarrays from each cell type. In their analysis they computed a modified F-statistic and used it to assign a P-value to each gene. A threshold of $\alpha = 0.001$ was selected to find 51 genes from a total of $N = 3,226$ that show differential gene expression. These authors subsequently used a threshold of $\alpha = 0.0001$ and they concluded that 9-11 genes are differentially expressed.

Storey and Tibshirani (2003a) considered 3,170 genes after eliminating any gene that had one or more expression levels exceeding 20, which was several interquartile ranges away from the interquartile range of all the data. They calculated the P-values for the retained $N = 3,170$ genes using the t-statistic (5.4) with the associated P-values p_j calculated using the permutation-based estimate (5.18) with $B = 100$ permutations. These P-values so obtained were plotted in histogram form which is repeated here in Figure 5.2. In this figure, the dashed line gives the uniform density that would be expected if all the genes were null (not differentially expressed). Using the estimate (5.32) with $\xi = 0.5$, Storey and Tibshirani (2003a) found that at least 33% genes of the 3,226 genes are differentially expressed between BRCA1- and BRCA2-mutation-positive tumors. This compares with only 9-11 genes that Hedenfalk et al. (2001) were comfortable in concluding to be differentially expressed, using traditional P-value thresholds. Storey and and Tibshirani (2003a) noted that thresholding genes with q-values less than or equal to $\alpha = 0.05$ yields 160 genes significant for differential expression. This means that 8 of the 160 genes called significant are expected to be false positives. As stressed by Storey and Tibshirani (2003a), the q-value threshold of $\alpha = 0.05$ is arbitrary, and they do not recommend that this value necessarily be used.

Storey and Tibshirani (2003a) used this data set to demonstrate how one can use various plots to calibrate the q-value cutoff point, notwithstanding that a single cutoff is not always necessary, as each estimated q-value could simply be reported. We report here in Figure 5.3 the plots from Storey and Tibshirani (2003a), displaying (a) a plot of the q-values versus their statistics t_j; (b) a plot of the q-values versus their P-values p_j; (c) a plot of the number of significant genes N_r versus each q-value; (d) a plot of the expected number of false positives versus the number of genes called significant.

From Plot (c) in Figure 5.3, it can be seen that for estimated q-values slightly greater than 0.02, a sharp increase occurs in the number of significant genes over a small increase in the q-value. This allows one to easily see that a slightly larger q-value cutoff results in many more significant genes. From Plot (b) in Figure 5.3, one can see the expected proportion of false positives for different P-value cutoff points. As Storey and Tibshirani (2003a) point out, the last three plots can be used

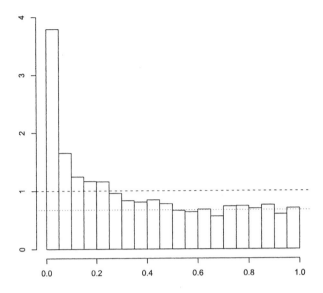

Fig. 5.2 A histogram of the 3,70 P-values from the Hedenfalk data set. The dashed line is the histogram that would be expected if all genes were null (not differentially expressed). The dotted line is at the height of the estimate of the proportion of null P-values. From Storey and Tibshirani (2003a).

concurrently to give the researcher a comprehensive view as to what genes to examine further.

Storey and Tibshirani (2003a) commented on their results for some particular genes in this data set. For example, the MSH2 gene (clone 32,790) is the eighth most significant gene for differential expression with a q-value of 0.013 and a P-value of 5.05×10^5. This gene is overexpressed in the BRCA1-mutation-positive tumors, indicating increased levels of DNA repair. This P-value for the MSH2 genes means that the probability a null (nondifferentially expressed) gene would be as or more extreme than MSH2 is 5.05×10^5. The estimated q-value for this gene is 0.013, meaning that 0.013 of the genes that are as or more extreme than MSH2 are false positives. As cautioned by Storey and Tibshirani (2003a) the q-value is not the probability that a gene is a false positive just as the P-value is also not the probability that a gene is a false positive. For example, the aforementioned value of 0.013 for the q-value for the gene MSH2 does not imply that it is a false positive with probability 0.013. Rather, 0.013 is the expected proportion of false positives incurred if we call MSH2 significant. Because the q-value measure includes genes that are possibly much more significant than MSH2, the probability that MSH2 is itself a false positive may be substantially higher.

As Storey and Tibshirani (2003a) point out, when assigning significance to multiple test statistics, it is necessary to account for the fact that decisions are made for N features simultaneously. The q-value measure of significance accomplishes this by

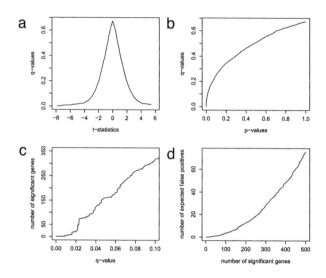

Fig. 5.3 Results for the Hedenfalk data set. (a) The q-values of the genes versus their respective observed test statistics t_j. (b) The q-values versus their respective P-values p_j. (c) The number of genes occurring on the list up through each q-value versus their respective q-value. (d) The expected number of false positive genes versus the total number of significant genes given by the q-values. From Storey and Tibshirani (2003a).

conditioning, based on the fact that every gene with a test statistic as or more extreme will also be called significant. As they comment, the probability that a gene is a false positive does not do this. But ideally one would like to have an estimate of the probability that a significant gene such as the aforementioned MSH2 is actually a false positive. We will see that with a mixture model approach to be presented in the next three sections, we are indeed able to provide such an estimate.

5.7 TWO-COMPONENT MIXTURE MODEL FRAMEWORK

5.7.1 Definition of Model

We can approach the problem of finding genes that are differentially expressed using a prediction rule approach based on a two-component mixture model as formulated in Lee et al. (2000) and Efron et al. (2001). We let G denote the population of genes under consideration. It can be decomposed into G_0 and G_1, where G_0 is the population of genes that are not differentially expressed, and G_1 is the complement of G_0; that is, G_1 contains the genes that are differentially expressed.

We let the random variable Z_{ij} be defined to be one or zero according as the jth gene belongs to G_i or not $(i = 0, 1; j = 1, \ldots, N)$. In Section 5.2, we defined H_j to be zero or one according as to whether the null hypothesis of no differential expression does or does not hold for the jth gene. Thus Z_{1j} is zero or one according as to whether H_j is zero or one.

The prior probability that the jth gene belongs to G_0 is assumed to be π_0 for all j. That is,

$$\pi_0 = \mathrm{pr}\{H_j = 0\}, \tag{5.35}$$

and $\pi_1 = \mathrm{pr}\{H_j = 1\}$. Assuming that the test statistics T_j all have the same distribution in G_i, we let $f_i(t_j)$ denote the density of T_j in G_i $(i = 1, 2)$. The unconditional density $f(t_j)$ of T_j is given by the two-component mixture model

$$f(t_j) = \pi_0\, f_0(t_j) + \pi_1\, f_1(t_j). \tag{5.36}$$

Thus the posterior probability that the jth gene is not differentially expressed (that is, belongs to G_0) is given by

$$\tau_0(t_j) = \pi_0 f_0(t_j)/f(t_j) \quad (j = 1, \dots, N). \tag{5.37}$$

Insofar as the fitted mixture model provides a good fit and dependence among some of the genes can be ignored, the estimates of the gene-specific posterior probabilities $\tau_0(t_j)$ form the basis of optimal statistical inference about differential expression (Newton et al., 2004).

5.7.2 Bayes Rule

Let e_{01} and e_{10} denote the two errors when a rule is used to assign a gene to either G_0 or G_1, where e_{ij} is the probability that a gene from G_i is assigned to G_j $(i, j = 0, 1)$. That is, e_{01} is the probability of a false positive and e_{10} is the probability of a false negative. Then from Section 6.4, the risk is given by

$$\text{Risk} = (1 - c)\pi_0 e_{01} + c\pi_1 e_{10}, \tag{5.38}$$

where $(1 - c)$ is the cost of a false positive. As the risk depends only on the ratio of the costs of misallocation, they have been scaled to add to one without loss of generality.

The Bayes rule, which is the rule that minimizes the risk (5.38), assigns a gene to G_1 if

$$\tau_0(t_j) \le c; \tag{5.39}$$

otherwise, the jth gene is assigned to G_0. In the case of equal costs of misallocation $(c = 0.5)$, the cutoff point for the posterior probability $\tau_0(t_j)$ in (5.39) reduces to 0.5.

5.7.3 Estimated FDR

In practice, we do not know the prior probability π_0 nor the densities $f_0(t_j)$ and $f(t_j)$, which will have to be estimated. If $\hat{\pi}_0$, $\hat{f}_0(t_j)$, and $\hat{f}_1(t_j)$ denote estimates of $\pi_0, f_0(t_j)$, and $f_1(t_j)$, respectively, the gene-specific summaries of differential expression can be expressed in terms of the estimated posterior probabilities $\hat{\tau}_0(t_j)$, where

$$\hat{\tau}_0(t_j) = \hat{\pi}_0 \hat{f}_0(t_j)/\hat{f}(t_j) \quad (j = 1, \dots, N) \tag{5.40}$$

is the estimated posterior probability that the jth gene is not differentially expressed. An optimal ranking of the genes can therefore be obtained by ranking the genes according to the $\hat{\tau}_0(t_j)$ ranked from smallest to largest. A short list of genes can be obtained by including all genes with $\hat{\tau}_0(t_j)$ less than some threshold c_o or by taking the top N_o genes in the ranked list.

Suppose that we select all genes with

$$\hat{\tau}_0(t_j) \leq c_o. \tag{5.41}$$

Then an estimate of the FDR rate is given by

$$\widehat{\text{FDR}} = \sum_{j=1}^{N} \hat{\tau}_0(t_j) \, I_{[0,c_o]}(\hat{\tau}_0(t_j))/N_r, \tag{5.42}$$

where

$$N_r = \sum_{j=1}^{N} I_{[0,c_o]}(\hat{\tau}_0(t_j)) \tag{5.43}$$

is the number of the selected genes in the list.

Thus we can find a data-dependent $c_o \leq 1$ as large as possible such that

$$\widehat{\text{FDR}} \leq \alpha. \tag{5.44}$$

This assumes that there will be some genes with $\hat{\tau}_0(t_j) \leq \alpha$, which will be true in the typical situation in practice. This bound is approximate due to the use of estimates in forming the posterior probabilities of nondifferential expression and so it depends on the fit of the densities $f_0(t_j)$ and $f(t_j)$. We shall discuss various methods that have been used to fit these densities in Sections 5.8 and 5.9.

5.7.4   Bayes Risk in terms of Estimated FDR and FNR

The Bayes prediction rule minimizes the risk of an allocation defined by (5.38). We can estimate the error of a false positive e_{01} and the error of a false negative e_{10} by

$$\hat{e}_{01} = \sum_{j=1}^{N} \hat{\tau}_0(t_j)\hat{z}_{1j} \Big/ \sum_{j=1}^{N} \hat{\tau}_0(t_j) \tag{5.45}$$

and

$$\hat{e}_{10} = \sum_{j=1}^{N} \hat{\tau}_1(t_j)\hat{z}_{0j} \Big/ \sum_{j=1}^{N} \hat{\tau}_1(t_j) \tag{5.46}$$

respectively, where $\hat{z}_{0j}$ is taken to be zero or one according as to whether $\hat{\tau}_0(t_j)$ is less than or greater than c in (5.39), and $\hat{z}_{1j} = 1 - \hat{z}_{0j}$. Also, we can estimate the prior probability π_0 as

$$\hat{\pi}_0 = \sum_{j=1}^{N} \hat{\tau}_0(t_j)/N. \tag{5.47}$$

On substituting these estimates (5.45) to (5.47) into the right-hand side of (5.38), the estimated risk is

$$
\begin{aligned}
\widehat{\text{Risk}} &= (1-c)\sum_{j=1}^{N}\hat{\tau}_0(t_j)\hat{z}_{1j}/N + c\sum_{j=1}^{N}\hat{\tau}_1(t_j)\hat{z}_{0j}/N \\
&= (1-c)\widehat{\text{FDR}}\sum_{j=1}^{N}\hat{z}_{1j}/N + c\widehat{\text{FNR}}\sum_{j=1}^{N}\hat{z}_{0j}/N,
\end{aligned}
\tag{5.48}
$$

where

$$
\widehat{\text{FDR}} = \sum_{j=1}^{N}\hat{\tau}_0(t_j)\hat{z}_{1j}/\sum_{j=1}^{N}\hat{z}_{1j}
\tag{5.49}
$$

and

$$
\widehat{\text{FNR}} = \sum_{j=1}^{N}\hat{\tau}_1(t_j)\hat{z}_{0j}/\sum_{j=1}^{N}\hat{z}_{0j}
\tag{5.50}
$$

are estimates of the FDR and FNR, respectively. Thus from (5.48), it can be seen that if we choose to select a gene as being differentially expressed if its estimated posterior probability $\hat{\tau}_0(t_j)$ is less than c, then we are approximately minimizing the risk, which can be estimated as

$$
\widehat{\text{Risk}} = (1-c)\hat{\omega}\widehat{\text{FDR}} + c(1-\hat{\omega})\widehat{\text{FNR}},
\tag{5.51}
$$

where

$$
\begin{aligned}
\hat{\omega} &= \sum_{j=1}^{N}\hat{z}_{1j}/N \\
&= N_r/N
\end{aligned}
\tag{5.52}
$$

is an estimate of the probability that a gene is selected; that is, belongs to the critical region Γ of the rule.

Thus unlike the tests or rules that are designed to control just the FDR, the Bayes rule approach in its selection of the genes can be viewed as controlling a linear combination of the FDR and FNR. The balance between the FDR and the FNR is controlled by the threshold c. An early reference on the Bayes rule in the context of hypothesis testing is Lehmann (1959).

Recently, Storey et al. (2002) established formally a result similar to (5.51) in which the weight $\hat{\omega}$ was replaced by the probability that the test statistic T_j falls in the critical region Γ and the estimated FDR and FNR are replaced by the pFDR and pFNR. To see this, we have from the results (5.15) and (5.16) for the pFDR and pFNR, respectively,

$$
\begin{aligned}
\text{pFDR} &= \text{pr}\{H_j = 0 \,|\, T_j \in \Gamma\} \\
&= \pi_0 e_{01}/\omega
\end{aligned}
\tag{5.53}
$$

and

$$
\begin{aligned}
\text{pFNR} \quad &= \quad \text{pr}\{H_j = 1 \mid T_j \notin \Gamma\} \\
&= \quad \pi_1 e_{10}/(1-\omega)
\end{aligned}
\tag{5.54}
$$

where

$$
\omega = \text{pr}\{T_j \in \Gamma\}
$$

is the probability that a gene is selected (that is, T_j falls in the critical region Γ). On using (5.53) and (5.54) in (5.38), we have that

$$
\text{Risk} = (1-c)\omega\, \text{pFDR} + c(1-\omega)\, \text{pFNR}.
\tag{5.55}
$$

5.8 NONPARAMETRIC EMPIRICAL BAYES APPROACH

5.8.1 Method of Efron et al. (2001)

To apply the nonparametric empirical Bayes approach of Efron et al. (2001) in the present context, the null density $f_0(t_j)$ and the ratio $f_0(t_j)/f(t_j)$ are estimated using the empirical distributions for the $t_{0j}^{(b)}$ and the t_j. More specifically, Efron et al. (2001) used logistic regression to estimate the ratio $f_0(t)/f(t)$. The N values of t_j and the NB null values $t_{0j}^{(b)}$ are plotted on a line, with values of t_j considered as "successes" and values of $t_{0j}^{(b)}$ as "failures". The probability of a success is given in terms of the ratio $f_0(t)/f(t)$ as

$$
1/[1 + B\{f_0(t)/f(t)\}].
$$

The probability π_0, which is not estimable in a nonparametric setting, is estimated using the following inequality

$$
\pi_0 \le \min_t \{f(t)/f_0(t)\},
\tag{5.56}
$$

or more stable upper bounds can be constructed by integrating over an interval $\mathcal{A}$ near $t = 0$, that is,

$$
\pi_0 \le \frac{\int_{\mathcal{A}} \{f(t)/f_0(t)\} f_0(t)\, dt}{\int_{\mathcal{A}} f_0(t)\, dt} = \frac{\int_{\mathcal{A}} f(t)\, dt}{\int_{\mathcal{A}} f_0(t)\, dt}.
\tag{5.57}
$$

The upper bound in (5.57) is directly estimated by the proportion of t_j in $\mathcal{A}$ divided by the proportion of t_{0j} in $\mathcal{A}$.

5.8.2 Mixture Model Method (MMM)

Following the idea in Efron et al. (2001) and Tusher et al. (2001) of creating replications of a null version of the test statistic, Pan (2002, 2003) and Zhao and Pan (2003) considered a nonparametric approach, which they called the mixture model method (MMM). They advocated modeling the densities $f_0(t_j)$ and $f(t_j)$ in the two-component mixture model (5.36) by normal mixtures. With this mixture model method approach, the likelihood ratio test statistic

$$\lambda(T_j) = f_0(T_j)/f(T_j)$$

can be used to test the null hypothesis that the jth gene is not differentially expressed. The critical region Γ thus has the form

$$\Gamma_j = \{t_j : \lambda(t_j) < k\},$$

where for a test of size α, the constant k can be obtained by solving the equation

$$\alpha = \int_{\Gamma_j} f_0(t)\, dt, \tag{5.58}$$

using the bisection method. Pan (2002) suggests a simplification by working with the critical region $\Gamma_j = \{t_j : |t_j| > k^*\}$ and solving (5.58). An advantage of this mixture method approach is that it is amenable to power calculations (Pan, Lin, and Le et al., 2002) for determining how many replicates are needed.

Following the spirit of SAM (Tusher et al., 2001), we can estimate the FDR as

$$\widehat{\text{FDR}} = \frac{1}{B} \sum_{b=1}^{B} \#\{j : |t_{0j}^{(b)}| > k^*\}/N_r, \tag{5.59}$$

where

$$N_r = \#\{j : |t_j| > k^*\}. \tag{5.60}$$

Guo et al. (2003) have extended the MMM and SAM methods to their application to longitudinal gene expression data from time-course microarray experiments, where each class corresponds to a time point. They developed a test statistic that properly accounts for the within-subject correlation in longitudinal microarray data.

5.8.3 Nonparametric Bayesian Approach

Do et al. (2003) proposed an extension to the nonparametric approach of Efron et al. (2001) by adopting a fully model-based approach. While Efron's method proceeds by plugging in point estimates, the fully model-based approach constructs a probability model for the unknown mixture, allowing investigators to deduce the desired inference about differential expression as posterior inference in that probability model. Dirichlet process mixture models are chosen to represent the probability model for the unknown distributions. Markov chain Monte Carlo (MCMC) posterior simulation was developed to generate samples from the relevant posterior and posterior predictive distributions.

5.8.4 Application of Empirical Bayes Methods to Alon Data

We return to the colon cancer data set of Alon et al. (1999) as analyzed in Section 5.5.4 to demonstrate how the nonparametric empirical Bayes method of Efron's can be applied to this data set to identify a subset of genes that differentially express between the colon tumors and the colon normal tissues. Concerning the use of the modified t-statistic (5.6), Figure 5.4 (top panel) shows that the best choice for a_0 is zero. The lower panel of Figure 5.4 gives the empirical Bayes estimate of the posterior probability of differential expression, $\hat{\tau}_1(t)$, which is seen to increase with $|t|$. The positive end of the t-axis corresponds to genes that overexpress for tumor samples, while negative values of t indicate underexpression when compared to normal samples. Out of the 2,000 genes 227 genes had $\hat{\tau}_1(t)$ exceeding .90, more on the negative than positive end of the t scales. The estimate of π_0, which is derived as a bound on possible values for π_0, is given by $\hat{\pi}_0 = 0.39$ for this data set. In contrast, Figure 5.5 shows the marginal posterior distribution of π_0. The bound $\hat{\pi}_0$ is far out in the tail of the posterior distribution, indicating that $\hat{\pi}_0$ might lead to very conservative estimates for π_0 and π_1 (by underestimating π_1). Figure 5.6 summarizes the posterior distributions for $f_0(t)$, $f_1(t)$, and $f(t)$ obtained from applying the nonparametric Bayesian mixture model approach of Do et al. (2003).

5.9 PARAMETRIC MIXTURE MODELS FOR DIFFERENTIAL GENE EXPRESSION

In the previous sections, we have considered approaches to the detection of genes that are differentially expressed via the two-component mixture model (5.36). Nonparametric methods have been described for estimating the distribution of the statistic T_j given its membership of the null component of the mixture corresponding to nondifferential expression and to its unconditional distribution represented by the two-component mixture model (5.36). In this section, we consider methods that use parametric mixture models to estimate these distributions.

5.9.1 Parametric Empirical Bayes Methods

Newton (2001), Kendziorski et al. (2003), Newton and Kendziorski (2003), and Newton et al. (2004) have adopted parametric empirical Bayes approaches to the problem of the detection of differential expression. They have used hierarchical mixture models to form estimates of the density $f(\boldsymbol{y}_{ij})$, where

$$\boldsymbol{y}_{ij} = (y_{ij1}, \ldots, y_{ijn_i})^T,$$

and y_{ijk} denotes the expression level for the kth replicate on the jth gene in the ith class $(k = 1, \ldots n_j; j = 1, \ldots, N; i = 1, \ldots, g)$. These models are hierarchical in the sense that at the lower level, the component distribution in the mixture model describes the conditional variation in the expression profiles given their expected means while, at the upper level, there is a distribution that describes the variation in the expected means. As explained by Newton et al. (2004), such hierarchical modeling enables the sharing of information among the genes; genes become linked by

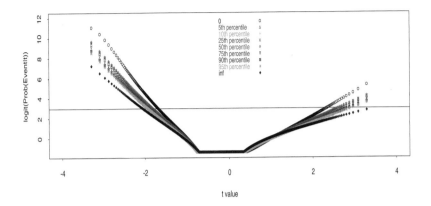

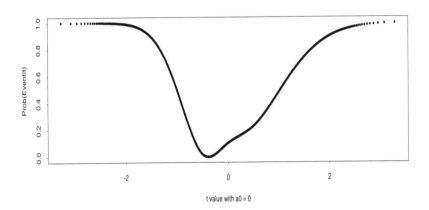

Fig. 5.4 Plots of Prob{Event|t} where 'Event' refers to a gene being differentially expressed; that is, Prob{Event|t} is denoting the (estimated) posterior probability $\hat{\tau}_1(t)$ of differential expression given the value t of the test statistic. Upper panel: Choice of a_0 in the modified t-statistic. Logit of $\tau_{1j}(t)$ was estimated with $\pi_0 = 1$; "inf" is limit as $a_0 \to \infty$. The horizontal line corresponds to the threshold lower bound of 0.9 for $\tau_{1j}(t)$; $a_0 = 0$ is the best choice in terms of maximizing $\hat{\tau}_{1j}(t)$. Lower panel: Empirical Bayes-based curve depicting $\hat{\tau}_{1j}(t)$. See the insert for a color representation.

virtue of having expected expression values drawn from a common, albeit unknown, probability distribution.

For multiple classes, Kendziorski et al. (2003) considered mixtures of gamma and log normal component distributions with an inverse gamma distribution for the

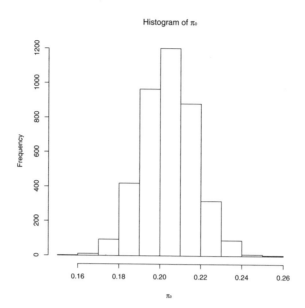

Fig. 5.5   Analysis of Alon data set. The histogram depicts the marginal posterior distribution of π_0 from the nonparametric Bayesian model. Compare with the point estimate $\hat{\pi}_0 = 0.39$ under the nonparametric empirical Bayes method.

component means. In the case of $g = 2$ classes, Newton et al. (2004) modeled the joint distribution of Y_{1jk} and Y_{2jk} by a mixture of $h = 3$ components with restrictions on the means of Y_{1jk} and Y_{2jk} in the two components corresponding to the cases of under- and over-expression under the alternative hypothesis. These means were set equal to each other in the other component corresponding to the null hypothesis. Newton et al. (2004) also considered a nonparametric approach by postulating a nonparametric prior distribution for the component means.

As noted in Newton et al. (2004), their approach which is based on models for the actual gene expression levels should be more sensitive than those that consider models for a test statistic T_j (or its P-value p_j) that has been formed for the test of no differential expression between the classes for the jth gene. This is because one-dimensional gene-specific summary statistics are usually isolated from each other in the sense that evaluation of the test statistic for one gene does not use data from any other gene. By contrast, information sharing can be beneficial because it can counteract the effects of low sample size. However, the approach using one-dimensional gene-specific summary statistics is more straightforward to carry out, as it involves modeling only a one-dimensional random variable. We henceforth focus on approaches that are based on mixture models for the test statistics or their P-values.

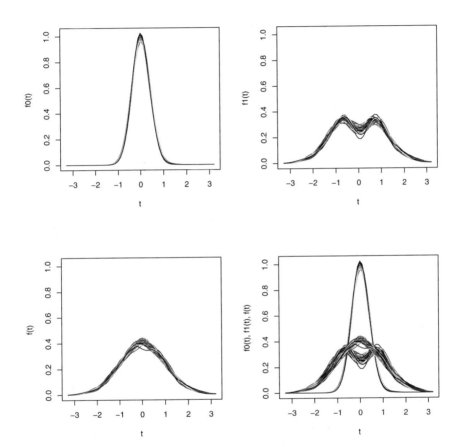

Fig. 5.6 Posterior distributions for the unknown densities (Alon data set). The first three panels summarize the posterior information on f_0, f_1 and f, respectively, by showing these densities for 10 draws from the posterior distribution of their parameters. For easier comparison the fourth panel combines the plots from the first three panels into one figure. See the insert for a color representation.

5.9.2 Finding Clusters of Differentially Expressed Genes

For a continuous test statistic, the normal mixture model will provide an arbitrarily accurate estimate of its density under both the null and alternative hypotheses. Thus we can model the density of the test statistic T_j for the jth gene by a finite normal mixture model,

$$f(t_j; \boldsymbol{\Psi}) = \sum_{i=1}^{h} \pi_i \phi(t_j; \mu_i, \sigma_i^2), \qquad (5.61)$$

where $\boldsymbol{\Psi}$ is the parameter vector containing the mixing proportions π_i, the means μ_i, and the variances σ_i^2 $(i = 1, \ldots, h)$.

Ignoring the correlations between the test statistics T_j, we can fit this mixture model (5.61) of h univariate normal components to the observed values t_j for the T_j $(j = 1, \ldots, N)$. The genes can then be put into h clusters on the basis of the fitted posterior probabilities of component membership.

The clusters corresponding to the components in the fitted normal mixture model with means close to zero can be considered as containing those genes that are not differentially expressed.

This was the approach adopted by Pan, Lin and Le (2002a) in the case of $g = 2$ classes. We shall give an example from their paper in the next section. A Bayesian approach to this problem has been considered by Broët et al. (2002).

In the case of more than two classes, we can proceed above with the T_j-statistic (5.5) replaced by, say, an F-statistic F_j

$$F_j = (\boldsymbol{B})_{jj}/(\boldsymbol{S})_{jj}, \qquad (5.62)$$

where the between-class sum of squares is $(\boldsymbol{B})_{jj}$ obtained by a one-way analysis of variance of the expression values of the N genes over the M tissues classified into g classes. In (5.62), $\boldsymbol{B}$ and $\boldsymbol{S}$ denote the between- and within-class sums of squares and products matrices (on their degrees of freedom), as given by (3.51) and (3.55), respectively. Their diagonal elements $(\boldsymbol{B})_{jj}$ and $(\boldsymbol{S})_{jj}$ give the relevant sums of squares for the jth gene. Under the null hypothesis that the jth gene is not differentially expressed (that is, given that the jth gene belongs to population G_0), F_j has a F-distribution with $(g - 1)$ and $(M - g)$ degrees of freedom.

In a recent paper, Broët et al. (2004) suggested transforming this F_j statistic as

$$T_j = \frac{\left(1 - \dfrac{2}{9(M - g)}\right) F_j^{\frac{1}{3}} - \left(1 - \dfrac{2}{9(g - 1)}\right)}{\sqrt{\dfrac{2}{9(M - g)} F_j^{\frac{2}{3}} + \dfrac{2}{9(g - 1)}}} \qquad (5.63)$$

The distribution of the transformed statistic T_j is approximately a standard normal under the null hypothesis that the jth gene is not differentially expressed (that is, given its membership of population G_0). As noted in Broët et al. (2004), it is remarkably accurate for $(M - g) \geq 10$ (Johnson and Kotz, 1970).

In their normal mixture model for T_j, Broët et al. (2004) specified the mean of one component to be zero. This component corresponds to the group of genes that are not differentially expressed. The order restriction,

$$\mu_h > \mu_{h-1} > \cdots > \mu_2 > \mu_1 = 0, \tag{5.64}$$

is placed on the means of the other h components in the mixture model corresponding to the genes that are differentially expressed. The assumption that the means of the h nonnull components should be positive is based on the fact that F_j has a central F-distribution under the null hypothesis and a noncentral F-distribution under the alternative hypothesis.

In the mixture model of Broët et al. (2004), just one component (the null component) has mean specified to be zero. This assumes that all genes that are not differentially expressed have the same variance. The case where this is not so can be handled by allowing more than one component in the mixture model to have mean zero. The specification of a component (or components) with mean zero in the mixture model avoids the post-fitting decision that has to be made when the mixture model is fitted without any restrictions on the component means. This decision concerns how small in magnitude the fitted component means should be before they are interpreted as corresponding to genes that are not differentially expressed. In some instances as in the example in the next section, this interpretation is relatively straightforward for the biologist.

5.9.3 Example: Fitting Normal Mixtures to t-Statistic Values

We consider an example from Pan et al. (2002a), who analyzed data from a study carried out at the University of Minnesota. In the original study, they used radioactively labeled cDNA microarrays to measure changes in gene expression in response to pneumococcal middle-ear infection. They extracted mRNA from the middle-ear mucosa of rats infected with pneumococcus, and also from a non-infected control group. They measured the expression levels for 1,176 genes in each of six microarray experiments; two arrays were run with pooled control mRNA and four with mRNA pooled from infected rats. After logging the data, a measure of possible differential expression t_j was calculated for each gene j, using the t-statistic (5.5) $(j = 1, \ldots, 1,176)$.

Pan et al. (2002a) fitted a mixture of h normal components to the 1,176 values of t_j, using the EMMIX program (McLachlan et al., 1999). They investigated the number of components h by using several criteria, including AIC, BIC, and the resampling approach of McLachlan (1987), as described in Section 3.16. Their results are given in Table 5.3, where it can be seen that the null hypothesis of $h = 3$ components would be retained over the alternative hypothesis of $h = 4$ components according to both BIC and the resampling approach (the P-value for the latter is 0.18). The fitted model is given in Table 5.4, where it can be seen that more than 95% of the genes fall into two clusters with either no or little change in their expression levels. On the other hand, 30 genes in the first cluster seem to have a change in expression levels equal to 6.74 on average. To demonstrate this, Pan et al. (2002a) plotted the profiles of gene

Table 5.3 Clustering Results for Various Numbers of Components h

h	BIC	$\log L$	P-value
1	5,877.26	−2,931.56	−
2	5,282.85	−2,623.75	0.01
3	5,248.80	−2,596.12	0.18
4	5,263.06	−2,592.64	−
5	5,280.94	−2,590.98	−

Source: Adapted from Pan et al. (2002a).

Table 5.4 Fitted Normal Mixture Model for $h = 3$ Components

i	π_i	μ_i	σ_i^2
1	0.042	6.74	77.07
2	0.510	0.88	5.56
3	0.448	−0.31	1.15

Source: From Pan et al. (2002a).

expression levels across all six experiments for each cluster in the case of $h = 4$. These plots are displayed here in Figure 5.7.

5.10 USE OF THE P-VALUE AS A SUMMARY STATISTIC

An alternative summary statistic to (5.5) is to use the value p_j, where p_j is the P-value associated with t_j in the test of the null hypothesis that there is no difference in expression between the two classes. This is the approach adopted by Allison et al. (2002). The distribution of p_j has support on the unit interval, and so its distribution can be represented by a mixture of beta distributions of the first kind (Diaconis and Ylvisaker, 1985). Under the null hypothesis of no differential expression for the jth gene, p_j will have a uniform distribution on the unit interval; that is the $\beta_{1,1}$ distribution. The $\beta_{\alpha_1,\alpha_2}$ density is given by

$$f(u; \alpha_1, \alpha_2) = \{u^{\alpha_1-1}(1-u)^{\alpha_2-1}/B(\alpha_1, \alpha_2)\}I_{(0,1)}(t), \qquad (5.65)$$

where

$$B(\alpha_1, \alpha_2) = \Gamma(\alpha_1)\Gamma(\alpha_2)/\Gamma(\alpha_1 + \alpha_2).$$

Allison et al. (2002) discusses the fitting of mixtures of $\beta_{\alpha_1,\alpha_2}$ components to the values of p_j for the N genes, including the caution that needs to be exercised in interpreting the existence of modes in the fitted mixture density as a consequence of the correlation between some of the p_j values.

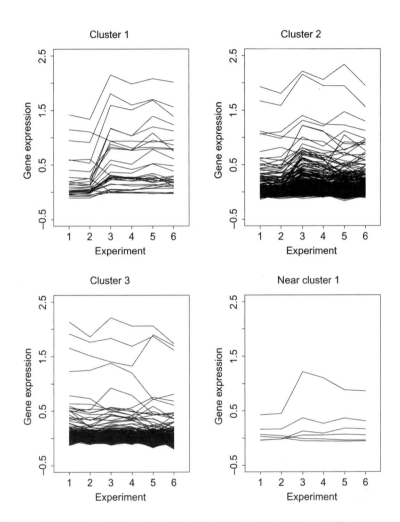

Fig. 5.7 Gene expression profiles of the four clusters. From Pan et al. (2002a).

5.10.1   Beta Mixture for Distribution of P-Values

Instead of working directly with the test statistics for the test of no difference in the expression of the genes over the tissues, we can work in terms of the associated P-values p_j of the test, as discussed above. With this approach of Allison et al. (2002), the distribution of the P-value is modeled by the h-component mixture model

$$f(p_j) = \sum_{i=1}^{h} \pi_i f(p_j; \alpha_{i1}, \alpha_{i2}), \tag{5.66}$$

where $\alpha_{11} = \alpha_{12} = 1$. If it is concluded that an additional component beyond the first component (the uniform) is needed in (5.66) to model the P-values p_j $(j = 1, \ldots, N)$ adequately, then the global null hypothesis of no change in the mean expression level for all the genes can be rejected. Allison et al. (2002) considered carrying out tests on the number of components g in (5.66) using the resampling approach as described above. They used the nonparametric bootstrap (that is, sampling with replacement from the p_j values) to provide standard errors of the fitted parameters. An estimate of the number of genes from the total of N for which there is a true difference in gene expression is given by $N(1 - \hat{\pi}_1)$. They discuss situations where this estimate is not applicable due to the estimate of π_1 being not interpretable. They also discuss potential complications with their approach in small samples. The effect of the assumption of independently distributed expression levels for the genes was investigated by a simulation experiment.

In this experiment, M vectors $\boldsymbol{y}_j$ of dimension N were generated randomly from a multivariate normal distribution with covariance matrix specified to be

$$\boldsymbol{\Sigma} = \sigma^2 \boldsymbol{B} \otimes \boldsymbol{I}_6 \tag{5.67}$$

and

$$\boldsymbol{B} = \boldsymbol{1}_{500} \boldsymbol{1}_{500}^{\mathrm{T}} + (1 - \rho) \boldsymbol{1}_{500}.$$

Here $\boldsymbol{1}_{500}$ denotes the unit vector of length 500 and $\boldsymbol{I}_m$ is the $m \times m$ identity matrix.

For the simulations the common variance was $\sigma^2 = 4$, while the correlation ρ varied over three values of 0 (independence), 0.4 (moderate dependence), and 0.8 (strong dependence). They noted that this covariance structure seems plausible since groups of genes are likely to be coexpressed, but it is unlikely that a particular gene is correlated with all other genes. For 20% of the genes (600 randomly selected), a true mean difference in expression between the two classes of mice was incorporated by adding d to the gene measurements $\boldsymbol{y}_j$ from $j = \frac{1}{2}M + 1$ through to M.

Allison et al. (2002) observed in their simulations that a second beta component was only significant in the mixture model under a certain type of gene expression data. This occurred in the case of high correlation ($\rho = 0.9$), and when $d > 0$. This high dependency created a bimodal distribution of P-values. The first beta distribution modeled the peak near zero, and the second component beta distribution modeled the second peak that occurred, typically, between 0.5 and 1. This second peak corresponded to 80% of the genes that were generated with no difference in mean expression between the two classes. Hence Allison et al. (2002) cautioned that care needs to be exercised in interpreting the mixture model in such situations.

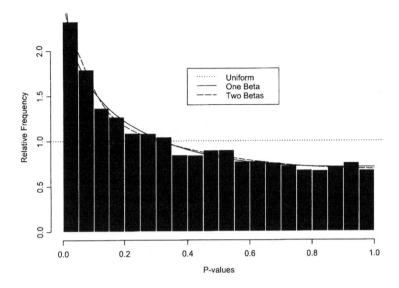

Fig. 5.8 Histogram of 6,347 P-values with beta mixture models for 2 and 3 components. From Allison et al. (2002).

5.10.2 Example: Fitting Beta Mixtures to P-Values

Allison et al. (2002) gave an illustrative example in which they analyzed data described by Lee et al. (2000). In the original study, Lee et al. (2000) used Affymetrix microarrays to measure changes in gene expression related to the aging brain in mice. In particular, they were interested in the effects of caloric restriction, which is known to slow the aging process in mammals. To this end they considered two groups of mice; one group of control mice on a normal diet, while the second group were fed a calorie restricted diet (CR mice). There were three mice in each group, and the expression levels of 6,347 genes were measured from mRNA extracted from the brain tissue of each mouse. For each gene j, the P-value p_j was calculated for the t-statistic as defined by (5.4). In Figure 5.8, we display the histogram of the 6,347 P-values so obtained, along with the fitted beta mixture models for $h = 2$ and $h = 3$ components. It can be seen that there is little difference between the fitted densities for $h = 2$ and $h = 3$. These two mixture distributions model the peak near zero very well.

The P-value for the test of $h = 1$ (uniform) component was estimated by resampling to be less than 0.005. For the $h = 2$ model, the fitted value of π_1 was 0.712, while the parameters in the second beta component were $\hat{\alpha}_{21} = 0.775$ and $\hat{\alpha}_{22} = 3.682$. Given these estimates, the number of genes that have a real difference can be estimated as $6347(1 - 0.712) = 1828$.

In Figure 5.9, we give the plot that Allison et al. (2002) obtained for the posterior probability $\hat{\tau}_2(p_j)$ that the gene j belongs to the second component of the beta mixture model given the P-value p_j $(j = 1, \ldots, 6347)$. This plot shows that as long as the

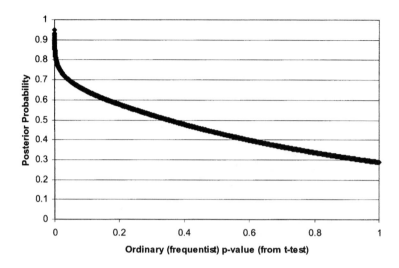

Fig. 5.9 Posterior probability given the P-value. From Allison et al. (2002).

p_j is less than about 0.35, there is more than a 50% chance that the gene is a gene for which there is a real difference in expression. Thus, as pointed out by Allison et al. (2002), if one were to use the unconventionally large α level of 0.35 as indicative of genes that are differentially expressed, then there is at least 50% chance of being correct.

Allison et al. (2002) also make some interesting observations about three specific genes, which we shall repeat here. Firstly, the gene identified by accession number L06451 was more highly expressed among calorie-restricted animals and had a P-value of 0.06 and a posterior probability $\hat{\tau}_2(p_j)$ of 0.68. Thus, although this gene would not be significant at the conventional 5% α level, there is is still a 68% chance that it is a gene with a real difference in expression induced by caloric restriction (CR). This gene encodes a protein that is homologous to agouti signaling protein. As this gene product is believed to be involved in body weight and appetite regulation, it is therefore quite plausible that it is affected by CR. Secondly, the gene identified by accession number M74180 had, by Affymetrix's definition, a "fold change" of 2.7 (increased expression among CR animals) which, while large, was not the largest of those reported by Lee et al. (2000). Nevertheless, this gene had the highest value (0.95) of the posterior probability $\hat{\tau}_2(p_j)$. This gene encodes a protein that is homologous to mouse hepatocyte growth factor-like or macrophage stimulating protein (MSP). Thirdly, they consider the gene identified by accession number W75705 which encodes a protein that is over 80% homologous to mouse cyclophilin. This gene had a fold change of 3.3, which many investigators would consider clearly significant. Nevertheless, its estimated posterior probability $\hat{\tau}_2(p_j)$ was only 0.44, indicating that it has more chance of not being differentially expressed as a function of CR. The reason why the estimated posterior probabilities do not 'agree' (that is, have a one-to-one correspondence) with the fold-change metric is that the latter does not take the within-class variability in gene expression into account. As Allison et al. (2002) point out, the fold-change is a measure of magnitude effect (and not necessarily an optimal one), not a direct measure of strength of evidence for an effect. Their work illustrates how the mixture model can guide the interpretation of the overall suite of class gene expression differences as well as differences in individual gene expression levels.

5.11 CLUSTERING OF GENES

In the previous work in this chapter, we have been concerned with the detection of genes that are differentially expressed. Having drawn up a list of genes that are considered to be differentially expressed, there is the problem of clustering the genes into groups in which they have similar expression profiles. Although genes might be assigned to the same component in the mixture model (5.66) fitted to the test statistics T_j for the N genes, it does not follow that genes with similar values will have similar expression profiles except in simple cases like $g = 2$ classes in which the genes are assumed to have the same variances. But clearly in the case of $g > 2$ classes, genes with similar values of the F_j statistic given by (5.62) need not have similar expression profiles. Thus there is a need to undertake the clustering of the genes that have been identified as being differentially expressed among the classes of tissues.

In the remainder of this chapter, we consider the clustering of the genes on the basis of the tissues, that is, on the basis of the expression-profile vector.

In Chapter 4, we were concerned with the clustering of the tissue samples on the basis of the genes within each sample (the expression-signature vector). Here the problem is to cluster the genes on the basis of the tissues, that is, on the basis of the expression-profile vector.

In the EMMIX-GENE procedure considered in Chapter 4 for the clustering of the tissue samples, there is a step in which the genes are clustered into a number of groups. This clustering of the genes in terms of Euclidean distance is carried out solely as a means of data reduction. Each group of genes is represented by a single vector (a metagene) for the subsequent clustering of the tissue samples. Also, as exhibited in some of the examples in which EMMIX-GENE was applied, the use of different clusters of genes can lead to different clusterings of the tissues. Thus the clustering of genes has an important role to play in the clustering of the tissues. We will also see in Chapter 7 on supervised classification of the tissue samples that the clustering of genes can also be usefully employed to provide summary statistics on which to form parametric or nonparametric discriminant rules (classifiers) for the class prediction of a new tissue sample.

In addition to the above need for a clustering of the genes into a smaller number of sets, the clustering of the genes can be of interest in its own right; that is, independent of any subsequent classification of the tissue samples. Examples where this is the case concern the clustering of the genes in order to discover genes that belong to the same molecular pathway (Segal et al., 2003a). Another aim of clustering the genes might be to find clusters of genes that are potentially coregulated in order to search for common motifs in upstream regions of the genes in each cluster (Segal et al., 2003b). In the work of Eisen et al. (1998), cluster analysis was used to identify genes that showed similar expression patterns over a wide range of experimental conditions in yeast. Such genes are typically involved in related functions and are frequently coregulated (as demonstrated by other evidence, such as shared promoter sequences and experimental verification).

In considering the clustering of the tissue samples, the rows of the microarray matrix A represented by (4.1) were standardized. This can be done without affecting the subsequent clustering of the tissue samples if a location and scale invariant method of clustering is used, such as a normal mixture model-based method. In clustering the genes, we shall also assume that the rows have been normalized. Although a normal mixture model-based clustering of the genes is not invariant under row normalization, it is usually carried out, as researchers are interested only in the relative behavior of the genes across the tissues, rather than in the absolute expression values of the genes. This same standardization was carried out in Section 5.3 prior to implementing procedures for the detection of differentially expressed genes. With this column and row standardization, we are effectively working with the residuals after the fitting of a two-way ANOVA model to the expression levels in A; see, for example, Ghosh and Chinnaiyan (2002). Also, in the sequel we consider the clustering of the genes on the basis of the M tissue samples without consideration of their class memberships. This is because we are primarily concerned with the pattern of the genes over the individual M tissue samples rather than their collective behavior as given by the patterns in the class means. Of course if we were interested in the latter, we could cluster the genes on the basis of the class means, although this approach would not be allowing for differences in the variability of the tissues within each class.

The clustering of the genes on the basis of their expression-profile vectors differs from the standard clustering problem in that the observations to be clustered (the genes) are not all independently distributed. There is interest in finding

(1) Groups of genes that are significantly correlated with each other.

(2) Groups of genes that share similar expressions across the tissues.

Concerning (1) and (2), there is much interest in genes whose expression levels tend to vary together (that is, genes that are coexpressed), because such genes might be part of the same pathway or the same causal mechanism. Clusters of genes that have similar expression levels across biological samples can be investigated for the presence of shared regulatory motifs among the genes (Tavazoie et al., 1999). This may lead to the identification of genes that are not only coexpressed but are also under similar regulatory control. Joint analysis of transcript level and sequence data should lead to greater biological insight into molecular characterization of tumors (Dudoit and Fridlyand, 2002).

The aims (1) and (2) above are closely related in that the Euclidean distance between two (normalized) gene profiles is proportional to one minus the correlation between them. Thus two genes with very similar profiles across the tissues will be highly (positively) correlated.

In the final sections of this chapter, we describe the gene-shaving method proposed by Hastie et al. (2000). Their method has the specific goal of identifying small, homogeneous subsets of genes that have maximal variance across the tissues.

5.12 FINDING CORRELATED GENES

As mentioned above, researchers are interested in finding highly correlated genes. But there is also an interest in finding highly correlated genes to assist in the data analysis of the gene expressions; cluster analysis methods can be applied as in the case of independently distributed observations, but the usual approaches for assessing the number of clusters in the data may be affected by the fact that some of the genes are highly correlated. One way to avoid this problem would be to find highly correlated genes and remove some of them from the data set before undertaking the cluster analysis. For instance, two highly (positively) correlated genes could be replaced by one of them or their mean (Hardin, Rocke, and Woodruff, 2000).

Jaeger, Sengupta, and Ruzzo (2003) have discussed the problem of handling correlated genes in the context of improved feature selection. In feature selection, if two genes are highly correlated, then there is no need to include both of them in the test statistic for detecting differential expression between the classes of tissues or, say, in the discriminant rule for distinguishing between the classes. To this end, they consider some methods that prefilter the genes and then delete those genes that are very similar. The genes are ranked in order of importance according to the P-value of some test statistic for the detection of differential expression. To find the correlated genes, they employ a greedy algorithm whereby genes that have a pairwise correlation below a certain threshold are selected and ranked according to some test statistic. The kth ranked gene selected is the gene with the highest P-value among all the genes whose correlation with each of the first $(k - 1)$ genes is below the specified threshold.

They also consider a method that clusters the genes and then selects one or two representative genes from each cluster. This method is in the spirit of the cluster-genes step of the EMMIX-GENE procedure of McLachlan et al. (2002), as discussed in Section 4.5. With this EMMIX-GENE step, the genes are clustered on the basis of a mixture of normal components with common spherical covariance matrices with the aim that genes that are close in Euclidean distance (that is, highly positively correlated) are put in the same cluster.

5.13 CLUSTERING OF GENES VIA FULL EXPRESSION PROFILES

We now consider the clustering of the genes on the basis of the individual tissues, assuming that a decision has been taken as to which genes are to be used in the cluster analysis. For instance, some genes may have been eliminated because they were highly correlated with others, as discussed in the previous section, or they may have been eliminated because they were judged not to be differentially expressed across the tissue classes as discussed in Section 5.1.

If there is only a small number M of tissue samples, then a mixture of M-variate normal components can be fitted directly to the N expression-profile vectors. Otherwise, one has to give consideration to fitting mixtures of factor analyzers in a manner similar to that of the clustering of the tissue samples considered in the previous chapter. Alternatively, one may choose to work with the principal components of the expression profiles.

In the case where each tissue sample has been obtained independently (say from M different patients), it may be thought that the component-covariance matrices should be taken to be diagonal to reflect the independent sampling of the tissue samples. However, the correlations between some of the genes within a tissue sample can result in the genes belonging to elliptically shaped clusters, and so nondiagonal component-covariance matrices may need to be specified. The latter obviously need to be specified in the case where the tissue samples are part of a time-course experiment (Park et al., 2003).

Examples of mixture model-based clustering of genes on the basis of the expression profiles may be found in Yeung et al. (2001a); see Holmes and Bruno (2000) and Barash and Friedman (2001) for some previous work on mixture models. Hardin, Rocke, and Woodruff (2000) have considered robust model-based clustering of genes in microarray data.

5.14 CLUSTERING OF GENES VIA PCA OF EXPRESSION PROFILES

We now consider an example in which the clustering of genes was undertaken on the basis of the leading principal components (PCs) of the expression profiles. Muro et al. (2003) used adapter-tagged competitive PCR to measure the expression levels of 1,536 genes in 100 colorectal cancer and 11 normal tissues. Although this does not, strictly speaking, represent a microarray experiment, it is a high-throughput technique for which cluster analysis methods are applicable, and we include it here as a nice example of gene clustering. Muro et al. (2003) concluded that the multivariate nature of the gene expression vectors could be represented by the first three PCs. They then fitted a mixture of h trivariate normal distributions to the first three PCs. They concluded that there were $h = 3$ clusters of expressed genes. Two clusters contained large numbers of genes, one of which correlated well with both the differences between tumor and normal tissues and the presence or absence of distant metastasis. The other was found to correlate only with the tumor/normal difference. The third cluster contained only a small number of genes. Approximately half showed an identical expression pattern, and cancer tissues were classified into two groups by their expression levels. The high-expression group had a strong correlation with distant metastasis and a poorer survival rate than the low-expression group, indicating possible clinical applications of these genes.

5.15 CLUSTERING OF GENES WITH REPEATED MEASUREMENTS

In recent times, microarray experiments are being carried out with replication. The importance of replication has been demonstrated by Lee et al. (2002). More recently, Pavlidis, Li, and Noble (2003) and Novak, Sladek, and Hudson (2003) have examined the effect of replication.

The simplest way to proceed with a cluster analysis of tissues with replication is to average the repeated measurements and work with the mean expression for each tissue. However, with this approach, the information on the variability between replicates is discarded and only the information about the mean expression level utilized.

With a model-based approach to clustering, the model is able to be adjusted to allow for repeated measurements. Work on this problem has been considered by Medvedovic and Sivaganesan (2002), Yeung et al. (2003), and Celeux, Martin, and Lavergne (2004).

5.15.1 A Mixture Model for Technical Replicates

We now consider the clustering of the genes where there are available technical replicates for each tissue; that is, replicates obtained from the same biological source. More specifically, we let there be n_v measurements for the vth tissue (patient) for $v = 1, \ldots, n$, where n denotes the number of distinct tissues (patients). The feature vector on the jth gene is therefore given by

$$\boldsymbol{y}_j = (\boldsymbol{y}_{1j}^T, \ldots, \boldsymbol{y}_{nj}^T)^T, \tag{5.68}$$

where

$$\boldsymbol{y}_{vj} = (y_{v1j}, \ldots, y_{vn_vj})^T \quad (v = 1, \ldots, n) \tag{5.69}$$

contains the n_v replications on the jth gene from the vth tissue.

It is assumed that conditional on its membership of the hth component of the normal mixture model to be fitted, the wth replicate on the jth gene from the vth tissue, Y_{vjw}, can be written as

$$Y_{vjw} = \mu_{vi} + b_{vj} + e_{vjw} \quad (i = 1, \ldots, h) \tag{5.70}$$

for $w = 1, \ldots, n_v; v = 1, \ldots, n; \ j = 1, \ldots, N$, where μ_{vi} denotes the ith component mean of Y_{vjw} and the b_{vj} denote the (unobservable) random effects for the biological sources, which are independent of the measurement errors e_{vjw}. Conditional on their membership of the ith component of the mixture, the random effects b_{vj} are taken to be jointly normal with mean zero and variance σ_{Rvi}^2. They are independent of the measurement errors e_{vjw}, which are also taken to be jointly normal with mean zero; their common variance is σ_{vi}^2.

We now illustrate the fitting of a normal mixture model to correlated replicate data. We consider the straightforward case where the random effects b_{vjw} are all uncorrelated and where the measurement errors e_{vjw} are also all uncorrelated. The latter will not hold in practice for all pairs of genes as not all the genes are independently distributed. Aslo, the former assumption would not hold in the case of a time-course study, as Y_{vjw} and $Y_{v'jw}$ would be correlated then; see Celeux et al. (2004).

Under these assumptions, we have that $\boldsymbol{y}_{1j}, \ldots, \boldsymbol{y}_{nj}$ are independently distributed with an hth component-conditional distribution given as

$$\boldsymbol{Y}_{vj} \sim N(\boldsymbol{\mu}_{vi}, \boldsymbol{\Sigma}_{vi}) \quad (v = 1, \ldots, n; j = 1, \ldots, N), \tag{5.71}$$

where $\boldsymbol{\mu}_{vi}$ is a n_v-dimensional vector with common element μ_{vi} and $\boldsymbol{\Sigma}_{vi}$ is $n_v \times n_v$ matrix with

$$
\begin{aligned}
(\boldsymbol{\Sigma}_{vi})_{ww'} &= \sigma_{vi}^2 + \sigma_{Rvi}^2 \quad (w = w'), \tag{5.72}\\
&= \sigma_{Rvi}^2 \quad\quad\;\; (w \neq w'). \tag{5.73}
\end{aligned}
$$

Note that in the case of a time-course study, Y_{vjw} and $Y_{v'jw}$ would not be uncorrelated as assumed here; see Celeux et al. (2004).

We let $\boldsymbol{\Psi} = (\boldsymbol{\omega}^T, \pi_1, \ldots, \pi_{h-1})^T$ be the vector of unknown parameters, where $\boldsymbol{\omega} = (\boldsymbol{\omega}_1^T, \ldots, \boldsymbol{\omega}_h^T)^T$ and where $\boldsymbol{\omega}_i$ is the vector containing the unknown parameters μ_{vi}, σ_{vi}^2, and the σ_{Rvi}^2 $(v = 1, \ldots, n)$ for $i = 1, \ldots, h$.

5.15.2   Application of EM Algorithm

McLachlan and Krishnan (1997, Section 5.9) have demonstrated the use of the EM algorithm to fit a single component linear mixed model. The unobservable random effects (the b_{vjw} in the present context) are treated as missing data in the EM framework. This approach can be extended to the present context where a mixture of h linear mixed models with normal errors is to be fitted. The unobservable component-indicator variables z_{ij}, as defined in Section 3.7.1, are introduced and treated as missing data in addition to the b_{vj}.

5.15.3   M-Step

Following McLachlan (2004), we have that the estimate of the vector of unknown parameters $\boldsymbol{\Psi}$ can be updated on the $(k + 1)$th iteration of the M-step as follows,

$$\pi_i^{(k+1)} = \sum_{j=1}^{N} \tau_{ij}^{(k)} / N, \tag{5.74}$$

$$\mu_{vi}^{(k+1)} = \sum_{j=1}^{N} \sum_{w=1}^{n_v} \tau_{ij}^{(k)} (y_{vjw} - b_{vj}^{(k)}) / n_v \sum_{j=1}^{N} \tau_{ij}^{(k)}, \tag{5.75}$$

$$\sigma_{vi}^{(k+1)^2} = \sum_{j=1}^{N} \tau_{ij}^{(k)} A_{vj} / n_v \sum_{j=1}^{N} \tau_{ij}^{(k)}, \tag{5.76}$$

$$\sigma_{Rvi}^{(k+1)^2} = \sum_{j=1}^{N} \tau_{ij}^{(k)} B_{vj} / \sum_{j=1}^{N} \tau_{ij}^{(k)}, \tag{5.77}$$

where

$$b_{vj}^{(k)} = \sum_{w=1}^{n_v} (y_{vjw} - \mu_{vi}^{(k)})/(n_v + \sigma_{vi}^{(k)^2}), \tag{5.78}$$

$$A_{vj}^{(k)} = \sum_{w=1}^{n_v} (y_{vjw} - \mu_{vi}^{(k)} - b_{vjw}^{(k)})^2 + n_v \sigma_{vi}^{(k)^2} (n_v + \sigma_{vi}^{(k)^2})^{-1}, \tag{5.79}$$

and

$$B_{vj}^{(k)} = b_{vj}^{(k)^2} + \sigma_{vi}^{(k+1)^4} (n_v + \sigma_{vi}^{(k)^2})^{-1}. \tag{5.80}$$

In (5.74) to (5.80), $\tau_{ij}^{(k)}$ denotes the value of the posterior probability that the jth gene belongs to the ith component of the mixture model using $\Psi^{(k)}$ for Ψ in

$$\tau_i(y_j; \Psi) = \pi_i f_i(y_j; \omega_i)/ \sum_{l=1}^{h} \pi_l f_l(y_j; \omega_l), \tag{5.81}$$

where

$$f_i(y_j; \omega_i) = \Pi_{v=1}^{n} \phi(y_{vj}; \mu_{vi}, \Sigma_{vi}) \tag{5.82}$$

for $i = 1, \ldots, h; j = 1, \ldots, N$.

Let $\hat{\Psi}$ be the estimate of Ψ obtained by applying the EM algorithm as above. Then the genes can be clustered on the basis of the fitted posterior probabilities $\tau_i(y_j; \hat{\Psi})$.

5.16 GENE SHAVING

5.16.1 Introduction

The development of the gene-shaving methodology, as proposed by Hastie et al. (2000), was motivated by research to identify distinct sets of genes for which variation in expression could be related to a biological property of the tissue samples. In subsequent sections, we shall discuss *gene shaving* in the context of (i) *unsupervised*, where the genes and samples are treated as unlabeled and the main goal is to identify any coherent patterns; or (ii) *supervised*, either fully or partially, by using known properties of the genes or samples to assist in finding meaningful clusters. For example, if the goal is to identify the subset of genes that can discriminate between cancer classes or different clinical response groups, then the supervised method would be the appropriate choice where the clustering process of the genes (rows) does incorporate prior information about the samples (columns).

5.16.2 Methodology and implementation

Let $A = a_{ij}$ be a row-centered $N \times M$ matrix of real-valued measurements representing the gene expression matrix, assuming no missing values. The rows are genes, the columns are tissue samples or cell lines, and a_{ij} is the measured (log) expression relative to a baseline. Suppose that we are interested in finding g distinct gene clusters that express similar patterns across the samples, with g chosen *a priori*. Gene shaving is an iterative algorithm based on the singular value decomposition (SVD) of the data matrix. It starts with the entire microarray gene expression matrix A and seeks a function of the genes in the direction of maximal variation across the tissue samples. The simplest form of this function is a normalized linear combination of the genes weighted by its largest principal component loadings, referred to as the super gene. The genes may be sorted according to the principal component weights. A fraction α of the genes having lowest correlation (essentially the absolute inner product) with the super gene is then *shaved* off (discarded) from the original data matrix. The process of calculating the leading principal component and shaving off some genes is iterated on the reduced data matrix until only two genes remain. This iterative top-down process produces a sequence of nested gene blocks of sizes ranging from the full set of n genes down to the final block, consisting of just two genes. Let S denote the total number of shaving iterations, and let B_s denote the particular gene block after the sth shave and denote the sequence of gene blocks by $A = B_0 \supset B_1 \supset B_2 \supset \cdots \supset B_S$. An optimal gene block B_{opt}, or cluster C_1, is isolated from this nested sequence via the *Maximum Gap statistic criterion* by comparing the columnwise variance for each block to that obtained by applying the procedure to permuted data.

The next step is to remove the effect of genes in the optimal cluster C_1 from the original matrix A. By computing the average gene or the vector of column averages for C_1, denoted by $\bar{C}_1$, we can remove the component that is correlated with this average. This is equivalent to regressing each row of A on the average gene row $\bar{C}_1$, and replacing the former with the regression residuals. Such a process is referred to as orthogonalization by Hastie et al. (2000), from which a modified data matrix A_{ortho} is produced. With A_{ortho}, the whole process is repeated, including the calculation of the leading principal component, producing another nested sequence of shaved gene blocks, applying the Gap statistic to obtain the next optimal cluster C_2, and orthogonalizing the current data matrix. This sequence of operations is iterated until g gene clusters $C_1, \ldots, C_g$ are found, which can be displayed graphically for visual inspection.

The shaving process requires repeated computation of the largest principal component of a particular data matrix A or its subset (after at least one step of shaving). This process is easily implemented using the singular value decomposition.

5.16.3 Optimal cluster size via the Gap statistic

One important goal for any clustering technique is the ability to assess whether the extracted cluster is real; that is, we should be able to distinguish real patterns from random small clusters. Even for a totally random $N \times M$ matrix where the rows and columns are independent of each other, a naïve application of any clustering technique can still result in the identification of spurious clusters by pure random chance. Therefore, the Gap statistic was devised by Tibshirani et al. (2001) to select a reasonable cluster size from the sequence of nested clusters. It is an adaptation of the usual permutation test based on randomization and an appropriate definition of a quality measure, or test statistic, for each cluster.

5.16.4 Supervised Gene Shaving

Supervised PC shaving can be implemented in the context of class discrimination with a slight modification to the unsupervised algorithm. It also can be generalized to incorporate prior information on the tissues in the form of continuous attributes and also survival times with possible censoring.

Hastie et al. (2000) noted that the fully supervised approach in a general regression setting reduces to finding clusters from simply ranking the genes in order of the regression model score test.

5.16.5 Real Data Example

In this chapter, we consider the application of both supervised and unsupervised gene shaving to the well-known data set, the colon data of Alon et al. (1999), as considered in Chapter 4 and previously in this chapter. Heat maps of the first four gene-shaving clusters using 10% shaving and 20 permutations are presented in Figure 5.10. Figure 5.11 shows the percent-variance curves for both the original and randomized data as a function of cluster size, as well as the gap curves used to select the specific cluster sizes. Visual examination of the first four unsupervised gene-shaving clusters reveal some interesting patterns. The first cluster of 50 genes groups 25 of the tumors to the right, indicating that these specific genes are highly expressed in tumors. The second cluster of 40 genes can be interpreted similarly, although the pattern of high expression is different from that of the first cluster. The third cluster of 41 genes corresponds to the clustering of the "old" versus "new" protocols where most samples (tumor and normal) from the 11 patients using a poly detector are mostly underexpressed for these genes and are grouped toward the left-hand side of the heat map. Subsequent clusters display coherent patterns of expression with high values of the ratio of the between-cluster to the within-cluster sums of squares, but do not suggest any clear clusterings that resemble either the external classification or the change in protocol paradigms identified by Getz et al. (2000).

We also reanalyzed the full Alon data set with different levels of supervision, ranging from 10 to 100% supervision, using the external classification of tumor versus normal. The first cluster (samples are not reordered) shows 50 genes (including the two smooth-muscle genes J02854 and T60155), representing two distinct groups of negatively correlated genes that correspond well to the external classification. The third cluster of five genes (sorted by the column means of the cluster) groups the tissues according to the old versus new protocols. When 100% supervision is used (Figure 5.12), the most coherent cluster that corresponds to the external classification consists of nine genes and classifies the tumors and normals with an error rate of 6, as found by other methods. These nine genes also correspond to those with the top TNoM scores used by Ben-Dor et al. (2000). Inspection of the variance and Gap plots under the fully supervised scenario indicates that only the first cluster captures the full external classification.

We note that the Gap curve of the gene-shaving clusters may be flat near the maximum, or may not be unimodal. This implies that there are larger cluster sizes that may include additional genes highly correlated with the cluster super gene and possessing a Gap statistic almost as large as the Maximum Gap value. An automatic implementation of choosing the cluster size according to the Maximum Gap statistic criterion would usually end up with smaller cluster sizes than those from other methods, but with much higher coherence in the cluster. We relaxed the Maximum Gap statistic criterion to allow for the choice of picking the largest cluster size within 5% of the Maximum Gap value. Under 50% supervision, the relaxed gene-shaving method picks out the first cluster with 77 genes, thus capturing the normal-versus-tumor structure and encompassing all of the six smooth muscle genes (J02854, T60155, M63391, D31885, X74295, X12369), as well as two ribosomal genes (T95018, T62947).

5.16.6   Computer Software

We have implemented the gene-shaving method, for both unsupervised and supervised analyses, in an S-language package that we call **GeneShaving**. The source code is available from the StatLib S-archive collection at

$$\texttt{http://lib.stat.cmu.edu/S/.}$$

A faster implementation, the **GeneClust** software package, described in Do, Broom, and Wen (2003), has the following properties. **GeneClust** has a graphical user interface (GUI) written in JAVA. The GUI allows users to:

- Perform a simple one-way hierarchical clustering by genes or by samples,

- Perform unsupervised, fully supervised, or partially supervised gene shaving; if the latter is chosen the user can also select the amount of partial supervision,

- Specify the number of clusters to extract, the percent to shave for each iteration, the number of permutations used to calculate the Gap statistic, and the level of gap tolerance .

GeneClust may be used either to analyze a real data set by selecting the **Raw Data** mode, or to investigate the performance of gene shaving for simulated data sets, by selecting the **Demo** mode. When the user starts the gene-shaving procedure, the JAVA GUI invokes the back-end statistical analysis process. This is an S-PLUS (or R) application with which the GUI communicates using a pseudo terminal. The computationally intensive gene-shaving algorithm is implemented using C, and is dynamically loaded into S-PLUS (or R) to perform the analysis. After the clusters have been extracted, the S-PLUS (or R) application presents graphically presents the results of the analysis.

The **GeneClust** software has been implemented for the Solaris and Linux operating systems, and for the S-PLUS and R statistical programming environments, discussed in details by Do et al. (2003). For exact details and continuous updates, check the website

$$\texttt{http://odin.mdacc.tmc.edu/\textasciitilde kim/geneclust.}$$

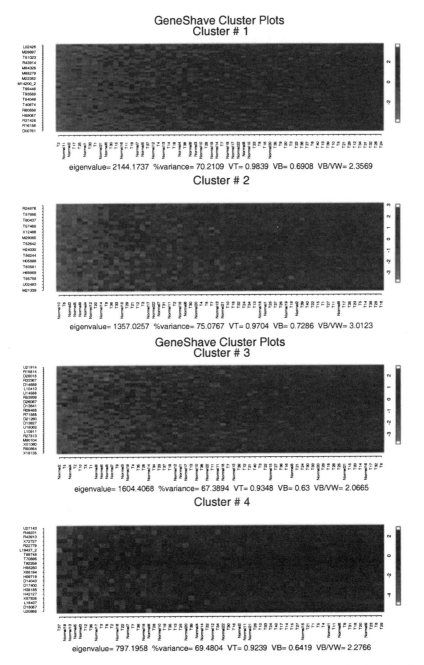

Fig. 5.10 Heat maps of the first four unsupervised gene-shaving clusters for the colon data, sorted by the column-mean gene. See the insert for a color representation.

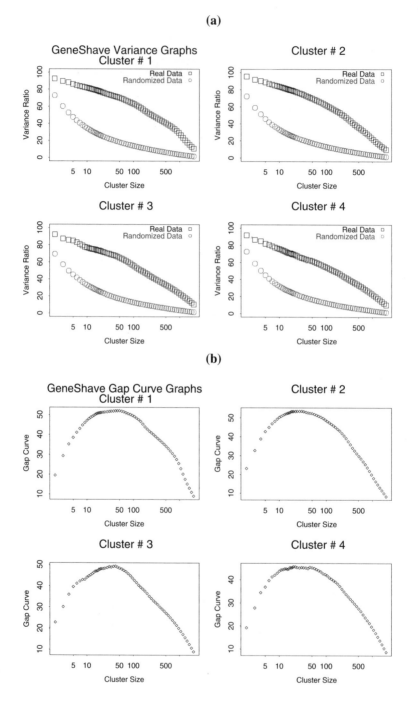

Fig. 5.11 (a) Variance plots for the original and randomized data. Each plot depicts the percent variance explained by each cluster, both for the original data, and for an average of over twenty randomized versions for different cluster sizes. (b) Gap estimates of cluster size. The Gap curve corresponds to the difference between the pair of variance curves.

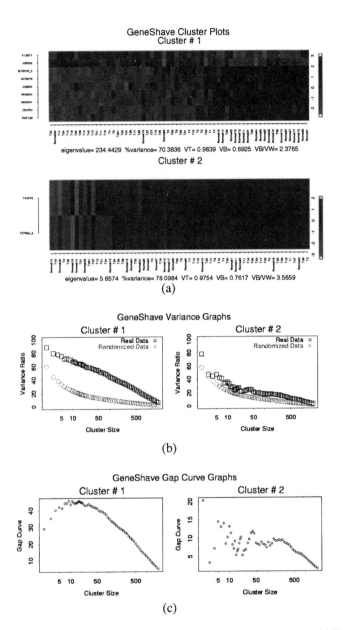

Fig. 5.12 Analysis of the Alon data set (2,000 genes) under full supervision. (**a**) Heat maps of the first two gene-shaving clusters with samples sorted by the column-mean gene (**b**) Variance plots for the original and randomized data. Each plot depicts the percent variance explained by each cluster, both for the original data, and for an average of over twenty randomized versions for different cluster sizes. (**c**) Gap estimates of cluster size. The Gap curve corresponds to the difference between the pair of variance curves. See the insert for a color representation.

6

Discriminant Analysis

6.1 INTRODUCTION

In this chapter, we consider some methods in discriminant analysis or supervised classification in a general context. In the next chapter, these methods are to be applied to the classification of the tumors on the basis of the genes. There are many discriminant rules in the statistical literature, but in this chapter we consider only those rules that we consider sufficient for the class prediction of tissues. For a full account of discriminant analysis, the reader is referred to one of the books on or related to the topic; for example, Hand, Mannila and Smyth (2001), Hastie, Tibshirani, and Friedman (2001), McLachlan (1992), and Ripley (1996).

6.2 BASIC NOTATION

We let Y denote the p-dimensional random feature vector corresponding to the realization y as measured on the entity under consideration. The class of origin of y (really the entity) can be denoted by a g-dimensional vector z of zero-one indicator variables. The ith component of z is defined to be one or zero according as y belongs or does not belong to the ith class C_i $(i = 1, \ldots, g)$; that is,

$$
\begin{aligned}
(z)_i &= 1, & y \in C_i, \\
&= 0, & y \notin C_i,
\end{aligned}
$$

for $i = 1, \ldots, g$.

In discriminant analysis, the aim is to construct a discriminant or prediction rule $r(\boldsymbol{y})$ for assigning $\boldsymbol{y}$ to one of the g classes; that is, the aim is to predict the class of origin of $\boldsymbol{y}$. If $r(\boldsymbol{y}) = i$, then it implies that $\boldsymbol{y}$ should be assigned to the ith class C_i $(i = 1, \ldots, g)$. Also, there may be interest in seeing what features in $\boldsymbol{y}$ are most useful in making this prediction. In the situation where the intention is limited to making an outright assignment of the entity to one of the possible classes, it is perhaps more appropriate to use the term prediction rather than discriminant to describe the rule. However, we shall use either nomenclature regardless of the underlying situation. In the pattern recognition jargon, such a rule is referred to as a classifier.

The probability density function (p.d.f.) of $\boldsymbol{Y}$ in class C_i is denoted by $f_i(\boldsymbol{y})$ for $i = 1, \ldots, g$. These class-conditional densities may be probability functions for a feature vector of discrete variables. It is typically assumed that the observation has been drawn from a mixture of the g classes $C_1, \ldots, C_g$ in some proportions $\pi_1, \ldots, \pi_g$ respectively, where

$$\sum_{i=1}^{g} \pi_i = 1 \quad \text{and} \quad \pi_i \geq 0 \ (i = 1, \ldots, g).$$

The p.d.f. of $\boldsymbol{Y}$ can therefore be represented in the finite mixture form

$$f(\boldsymbol{y}) = \sum_{i=1}^{g} \pi_i f_i(\boldsymbol{y}). \tag{6.1}$$

An equivalent assumption is that the random vector $\boldsymbol{Z}$ of zero-one class indicator variables with $\boldsymbol{z}$ as its realization is distributed according to a multinomial distribution consisting of one draw on g categories with probabilities $\pi_1, \ldots, \pi_g$ respectively; that is,

$$\mathrm{pr}\{\boldsymbol{Z} = \boldsymbol{z}\} = \pi_1^{z_1} \pi_2^{z_2} \cdots \pi_g^{z_g}. \tag{6.2}$$

We write

$$\boldsymbol{Z} \sim \mathrm{Mult}_g(1, \boldsymbol{\pi}), \tag{6.3}$$

where $\boldsymbol{\pi} = (\pi_1, \ldots, \pi_g)^T$. The ith mixing proportion π_i can be viewed as the prior probability that the entity belongs to C_i $(i = 1, \ldots, g)$.

With $\boldsymbol{Y}$ having been observed as $\boldsymbol{y}$, the posterior probability that the observation $\boldsymbol{y}$ belongs to C_i is given by

$$\begin{aligned} \tau_i(\boldsymbol{y}) &= \mathrm{pr}\{\boldsymbol{Y} \in C_i \,|\, \boldsymbol{y}\} \\ &= \mathrm{pr}\{Z_i = 1 \,|\, \boldsymbol{y}\} \\ &= \pi_i f_i(\boldsymbol{y}) / f(\boldsymbol{y}) \qquad\qquad (i = 1, \ldots, g). \end{aligned} \tag{6.4}$$

We are to consider the formation of an optimal discriminant rule (known as the Bayes rule) in terms of these posterior probabilities of class membership $\tau_i(\boldsymbol{y})$. But firstly, we shall define the error rates of a prediction rule.

6.3 ERROR RATES

The error rates associated with the discriminant rule $r(\boldsymbol{y})$ are denoted by $e_{ij}(r)$, where

$$e_{ij}(r) = \text{pr}\{r(\boldsymbol{Y}) = j \mid \boldsymbol{Y} \in C_i\}$$

is the probability that a randomly chosen entity from C_i is allocated to C_j $(i, j = 1, \ldots, g)$. The probability that a randomly chosen member of C_i is misallocated can be expressed as

$$e_i(r) = \sum_{j \neq i}^{g} e_{ij}(r) \tag{6.5}$$

For a diagnostic test using the rule $r(\boldsymbol{y})$ in the context where C_1 denotes the absence of a disease and C_2 its presence, the error rate $e_{12}(r)$ corresponds to the probability of a false positive, while $e_{21}(r)$ is the probability of a false negative. The correct allocation rates

$$e_{22}(r) = 1 - e_{21}(r) \quad \text{and} \quad e_{11}(r) = 1 - e_{12}(r)$$

are known as the sensitivity and specificity, respectively, of the diagnostic test.

6.4 DECISION-THEORETIC APPROACH

Decision theory provides a convenient framework for the construction of prediction rules in the situation where an outright allocation of an unclassified entity is required. The case where the prior probabilities of the classes and the class-conditional densities are taken to be known is relatively straightforward.

Let c_{ij} denote the cost of allocation when an entity from C_i is allocated to class C_j, where $c_{ij} = 0$ for $i = j = 1, \ldots, g$; that is, there is zero cost for a correct allocation. We assume for the present that the costs of misallocation are all the same. We can then take the common value of the c_{ij} $(i \neq j)$ to be unity, since it is only their ratios that are important.

For given $\boldsymbol{y}$, the loss for allocation performed on the basis of the rule $r(\boldsymbol{y})$ is

$$l\{\boldsymbol{z}, r(\boldsymbol{y})\} = \sum_{i=1}^{g} z_i Q[i, r(\boldsymbol{y})] \tag{6.6}$$

where, for any u and v,

$$\begin{aligned} Q[u, v] &= 0, & (u = v), \\ &= 1, & (u \neq v). \end{aligned} \tag{6.7}$$

The expected loss or risk, conditional on $\boldsymbol{y}$, is given by

$$E[l\{\boldsymbol{Z}, r(\boldsymbol{y})\} \mid \boldsymbol{y}] = \sum_{i=1}^{g} \tau_i(\boldsymbol{y}) Q[i, r(\boldsymbol{y})], \tag{6.8}$$

since from (6.4),

$$E(Z_i \mid \boldsymbol{y}) = \tau_i(\boldsymbol{y}).$$

An optimal rule of allocation can be defined by taking it to be the one that minimizes the conditional risk (6.4) at each value $\boldsymbol{y}$ of the feature vector. In decision-theory language, any rule that so minimizes (6.4) for some $\pi_1, \ldots, \pi_g$ is said to be a Bayes rule. It can be seen from (6.6), that the conditional risk is a linear combination of the posterior probabilities, where all coefficients are zero except for one, which is unity. Hence it is minimized by taking $r(\boldsymbol{y})$ to be the label of the class to which the entity has the highest posterior probability of belonging. Note that this is the "intuitive solution" to the allocation problem.

If we let $r_o(\boldsymbol{y})$ denote this optimal rule of allocation, then

$$r_o(\boldsymbol{y}) = i \quad \text{if} \quad \tau_i(\boldsymbol{y}) \geq \tau_j(\boldsymbol{y}) \quad (j = 1, \ldots, g; \ j \neq i). \tag{6.9}$$

We can write this as

$$r_o(\boldsymbol{y}) = \arg\max_i \tau_i(\boldsymbol{y}). \tag{6.10}$$

The rule $r_o(\boldsymbol{y})$ is not uniquely defined at $\boldsymbol{y}$ if the maximum of the posterior probabilities of class membership is achieved with respect to more than one class. In this case the entity can be assigned arbitrarily to one of the classes for which the corresponding posterior probabilities are equal to the maximum value.

As the posterior probabilities of class membership $\tau_i(\boldsymbol{y})$ have the same common denominator $f(\boldsymbol{y})$, $r_o(\boldsymbol{y})$ can be defined in terms of the relative sizes of the class-conditional densities weighted according to the class-prior probabilities; that is,

$$r_o(\boldsymbol{y}) = \arg\max_i \pi_i f_i(\boldsymbol{y}). \tag{6.11}$$

Note that as the optimal or Bayes rule of allocation minimizes the conditional risk (6.8) over all rules r, it also minimizes the unconditional risk

$$
\begin{aligned}
e(r) &= \sum_{i=1}^{g} E\{\tau_i(\boldsymbol{Y})Q[i, r(\boldsymbol{Y})]\} \\
&= \sum_{i=1}^{g} \pi_i e_i(r),
\end{aligned}
$$

which is the overall error rate associated with r.

Up to now we have taken the costs of misallocation to be the same. The reader is referred to McLachlan (1992, Section 1.4) for the form of the Bayes rule in the case of unequal costs of misallocation c_{ij}. For $g = 2$ classes, it reduces to the definition (6.9) for $r_o(\boldsymbol{y})$ in the case of equal costs of misallocation, except that π_1 is replaced now by $\pi_1 c_{12}$ and π_2 by $\pi_2 c_{21}$, where c_{ij} is the cost incurred when an entity from class C_i is assigned incorrectly to class C_j $(i \neq j)$. As it is only the ratio of c_{12} and c_{21} that is relevant to the definition of the Bayes rule, these costs can be scaled so that

$$\pi_1 c_{12} + \pi_2 c_{21} = 1.$$

Hence we can assume without loss of generality that $c_{12} = c_{21} = 1$, provided π_1 and π_2 are now interpreted as the class-prior probabilities adjusted by the relative importance of the costs of misallocation. Due to the rather arbitrary nature of assigning costs of misallocation in practice, they are often taken to be the same in real problems. Further, the class-prior probabilities are often specified as equal. This is not as arbitrary as it may appear at first sight. For example, consider the two-class situation where C_1 denotes a class of individuals with a rare disease and C_2 those without it. Then although π_1 and π_2 are disparate, the cost of misallocating an individual with this rare disease may well be much greater than the cost of misallocating a healthy individual. If this is so, then $\pi_1 c_{12}$ and $\pi_2 c_{21}$ may be comparable in magnitude and, as a consequence, the assumption of equal class-prior probabilities with unit costs of misallocation in the formation of the Bayes rule $r_o(\boldsymbol{y})$ is apt. Also, it would avoid in this example the occurrence of highly unbalanced class-specific error rates. The latter are obtained if $r_o(\boldsymbol{y})$ is formed with extremely disparate prior probabilities π_i and equal costs of misallocation. This imbalance between the class-specific error rates is a consequence of $r_o(\boldsymbol{y})$ being the rule that minimizes the overall error rate. The construction of rules that are optimal with respect to other criteria is discussed in McLachlan (1992, Section 1.5. In particular, it is shown there that by specifying the prior probabilities π_i in $r_o(\boldsymbol{y})$ so that its consequent error rates are equal, we obtain the rule that minimizes the maximum of the class-specific error rates.

6.5 TRAINING DATA

A basic assumption in discriminant analysis is that in order to estimate the unknown class-conditional densities there are observations of known origin on which the feature vector $\boldsymbol{Y}$ has been recorded for each. We let $\boldsymbol{y}_1, \ldots, \boldsymbol{y}_n$ denote these recorded feature vectors and $\boldsymbol{z}_1, \ldots, \boldsymbol{z}_n$ the corresponding vectors of zero-one indicator variables defining the known class of origin of each. The collection of the data

$$\boldsymbol{t} = (\boldsymbol{y}_1^T, \boldsymbol{z}_1^T, \ldots, \boldsymbol{y}_n^T, \boldsymbol{z}_n^T)^T \tag{6.12}$$

is referred to in the literature as either the initial, reference, design, training, or learning data. The last two have arisen from their extensive use in the context of pattern recognition. Also in the latter field, the formation of a prediction rule from training data of known origin is referred to as supervised learning.

There are two major sampling designs under which the training data $\boldsymbol{t}$ may be realized, joint or mixture sampling and $\boldsymbol{z}$-conditional or separate sampling. They correspond, respectively, to sampling from the joint distribution of $\boldsymbol{Y}$ and $\boldsymbol{Z}$ and to sampling from the distribution of $\boldsymbol{Y}$ conditional on $\boldsymbol{z}$. The first design applies to the situation where the feature vector and class of origin are recorded on each of n entities drawn from a mixture of the possible classes. Mixture sampling is common in prospective studies and diagnostic situations. In a prospective study design, a sample of individuals is followed and their responses recorded.

With most applications in discriminant analysis, it is assumed that the training data are independently distributed. For a mixture sampling scheme with this assumption,

$y_1, \ldots, y_n$ are the realized values of n independent and identically distributed (i.i.d.) random variables $Y_1, \ldots, Y_n$ with common distribution function $F(y)$. We write

$$Y_1, \ldots, Y_n \overset{\text{i.i.d.}}{\sim} F.$$

The associated class-indicator vectors $z_1, \ldots, z_n$ are the realized values of the random variables $Z_1, \ldots, Z_n$ distributed unconditionally as

$$Z_1, \ldots, Z_n \overset{\text{i.i.d.}}{\sim} \text{Mult}_g(1, \boldsymbol{\pi}). \tag{6.13}$$

The prior probabilities π_i can be estimated by

$$\hat{\pi}_i = \sum_{j=1}^{n} z_{ij}/n \quad (i = 1, \ldots, g). \tag{6.14}$$

With separate sampling in practice, the feature vectors are observed for a sample of n_i observations taken separately from each class C_i $(i = 1, \ldots, g)$. Hence it is appropriate for retrospective studies, which are common in epidemiological investigations. As many diseases are rare and even a large prospective study may produce few diseased individuals, retrospective sampling can result in important economies in cost and study duration. Note that as separate sampling corresponds to sampling from the distribution of Y conditional on z, it does not provide estimates of the prior probabilities π_i for the classes.

6.6   DIFFERENT TYPES OF ERROR RATES

For a given realization t of the training data T, it is the conditional or actual allocation rates of a sample prediction rule $r(y; t)$ that are of central interest. The conditional allocation rates are given by

$$ec_{ij}(r) = \text{pr}\{r(Y; t) = j \mid Y \in C_i, t\} \qquad (i, j = 1, \ldots, g). \tag{6.15}$$

That is, $ec_{ij}(r)$ is the probability, conditional on t, that a randomly chosen observation from C_i is assigned to C_j by $r(y; t)$.

The unconditional or expected allocation rates of $r(y; t)$ are given by

$$\begin{aligned}
eu_{ij}(r) &= \text{pr}\{r(Y; T) = j \mid Y \in C_i\} \\
&= E\{ec_{ij}(r)\} \qquad (i, j = 1, \ldots, g).
\end{aligned}$$

The unconditional rates are useful in providing a guide to the performance of the rule before it is actually formed from the training data.

Concerning the error rates specific to a class, the conditional probability of misallocating a randomly chosen member from C_i is

$$ec_i(r) = \sum_{j \neq i}^{g} ec_{ij}(r) \qquad (i = 1, \ldots, g).$$

The overall conditional error rate for an entity drawn randomly from a mixture G of $C_1, \ldots, C_g$ in proportions $\pi_1, \ldots, \pi_g$, respectively, is

$$ec(r) = \sum_{i=1}^{g} \pi_i ec_i(r).$$

The individual class and overall unconditional error rates, eu_i and eu, are defined similarly.

If $r(y; t)$ is constructed from t in a consistent manner with respect to the Bayes rule $r_o(y)$, then

$$\lim_{n \to \infty} eu(r) = e(r_o),$$

where $e(r_o)$ denotes the optimal error rate. Interest in the optimal error rate in practice is limited to the extent that it represents the error of the best obtainable version of the sample-based rule $r(y; t)$.

6.7 SAMPLE-BASED DISCRIMINANT RULES

We now consider the construction of a prediction rule from available training data in the situation where the class-conditional densities and perhaps also the class-prior probabilities are unknown. The initial approach to this problem, and indeed to discriminant analysis in its modern guise, was by Fisher (1936). In the context of $g = 2$ classes, he proposed that an entity with feature vector y be assigned on the basis of the linear discriminant function $a^T y$, where a maximizes an index of separation between the two classes. The index was defined to be the magnitude of the difference between the class-sample means of $a^T y$ normalized by the pooled sample estimate of its assumed common variance within a class. The derivation of Fisher's (1936) linear discriminant function is to be discussed further in Section 6.10, where it is contrasted with normal theory-based discriminant rules.

With the development of discriminant analysis through to the decision-theoretic stage, an obvious way of forming a sample-based prediction rule $r(y; t)$ is to take it to be an estimated version of the Bayes rule $r_o(y)$ where, in (6.10), the posterior probabilities of class membership $\tau_i(y)$ are replaced by some estimates $\hat{\tau}_i(y; t)$ formed from the training data t. One approach to the estimation of the posterior probabilities of class membership is to model the $\tau_i(y)$ directly, as with the logistic model. Dawid (1976) calls this approach the diagnostic paradigm.

A more common approach, called the sampling approach by Dawid (1976), is to use the Bayes formula (6.4) to formulate the $\tau_i(y)$ through the class-conditional densities $f_i(y)$. With this approach the Bayes rule is estimated by the so-called plug-in rule, whereby the class-conditional densities are replaced by estimates of the posterior probabilities $\tau_i(y)$ in the form (6.10) of the Bayes rule $r_o(y)$.

6.8 PARAMETRIC DISCRIMINANT RULES

Under the parametric approach to the estimation of the class-conditional distributions, and hence of the Bayes rule, the class-conditional distributions are taken to be known up to a manageable member of parameters. More specifically, the ith class-conditional density is assumed to belong to a family of densities

$$\{f_i(\boldsymbol{y}; \boldsymbol{\theta}_i) \; : \; \boldsymbol{\theta}_i \in \boldsymbol{\Theta}_i\}, \tag{6.16}$$

where $\boldsymbol{\theta}_i$ is an unknown parameter vector belonging to some parameter space $\boldsymbol{\Theta}_i$ $(i = 1, \ldots, g)$. Often the class-conditional densities are taken to belong to the same parametric family, for example, the normal. We can now write the Bayes rule $r_o(\boldsymbol{y})$ as $r_o(\boldsymbol{y}; \boldsymbol{\Psi})$, where

$$\boldsymbol{\Psi} = (\pi_1, \ldots, \pi_{g-1}, \boldsymbol{\omega}^T)^T, \tag{6.17}$$

where $\boldsymbol{\omega}$ is the vector consisting of the elements of $\boldsymbol{\theta}_1, \ldots, \boldsymbol{\theta}_g$ known *a priori* to be distinct. For example, if the class-conditional distributions are assumed to be multivariate normal with means $\boldsymbol{\mu}_1, \ldots, \boldsymbol{\mu}_g$ and covariance matrices $\boldsymbol{\Sigma}_1, \ldots, \boldsymbol{\Sigma}_g$, then $\boldsymbol{\theta}_i$ consists of the elements of $\boldsymbol{\mu}_i$ and of the distinct elements of $\boldsymbol{\Sigma}_i$, while $\boldsymbol{\omega}$ consists of the elements of $\boldsymbol{\mu}_1, \ldots, \boldsymbol{\mu}_g$ and of the distinct elements of $\boldsymbol{\Sigma}_1, \ldots, \boldsymbol{\Sigma}_g$. Note that since the mixing proportions π_i sum to one, one of them is redundant. Here we have (arbitrarily) not included π_g in (6.17).

With the so-called plug-in approach to the choice of a sample-based prediction rule, unknown parameters in the adopted parametric forms for the class-conditional densities are replaced by appropriate estimates obtained from the training data $\boldsymbol{t}$. Hence if $r_o(\boldsymbol{y}; \boldsymbol{\Psi})$ now denotes the optimal rule, then with this approach $r(\boldsymbol{y}; \boldsymbol{t}) = r_o(\boldsymbol{y}; \hat{\boldsymbol{\Psi}})$, where $\hat{\boldsymbol{\Psi}}$ is an estimate of $\boldsymbol{\Psi}$ formed from $\boldsymbol{t}$. Provided $\hat{\theta}_i$ is a consistent estimator of $\boldsymbol{\theta}_i$ and $f_i(\boldsymbol{y}; \boldsymbol{\theta}_i)$ is continuous in $\boldsymbol{\theta}_i$ $(i = 1, \ldots, g)$, then $r_o(\boldsymbol{y}; \hat{\boldsymbol{\Psi}})$ is a Bayes risk-consistent rule in the sense that its risk, conditional on $\hat{\boldsymbol{\Psi}}$, converges in probability to that of the Bayes rule, as n approaches infinity. This is assuming that the postulated model (6.16) is indeed valid and that the class-prior probabilities are estimated consistently, as possible for instance, with mixture sampling of the training data.

Given the widespread use of maximum likelihood as a statistical estimation technique, the plug-in rule $r_o(\boldsymbol{y}; \hat{\boldsymbol{\Psi}})$ is usually formed with $\hat{\boldsymbol{\Psi}}$, or at least $\hat{\boldsymbol{\omega}}$, taken to be the maximum likelihood (ML) estimate. Since their initial use by Wald (1944), Rao (1948, 1954), and Anderson (1951) among others, plug-in rules formed by ML estimation under the assumption of normality have been extensively applied in practice.

Another way of proceeding with the estimation of the class-conditional densities, and hence of $r_o(\boldsymbol{y}; \boldsymbol{\Psi})$, is to adopt a Bayesian approach; see, for example, McLachlan (1992, Section 2.2).

For high-dimensional data, the total sample size n is too small relative to the number p of feature variables in $\boldsymbol{y}$ for a reliable estimate of $\boldsymbol{\theta}$ to be obtained from the full set $\boldsymbol{t}$ of training data. This is referred to as "the curse of dimensionality", a phrase due to Bellman (1961). Consideration then has to be given to which variables in $\boldsymbol{y}$ should be deleted in the estimation of $\boldsymbol{\theta}$ and the consequent allocation rule.

Even if a satisfactory discriminant rule can be formed using all the available feature variables, consideration may still be given to the deletion of some of the variables in y. This is because the performance of a rule fails to keep improving and starts to fall away once the number of feature variables has reached a certain threshold. It is an important problem in its own right in discriminant analysis; as with many applications the primary or sole aim is not one of allocation, but rather to infer which feature variables of an entity are most useful in explaining the differences between the classes.

6.9   DISCRIMINATION VIA NORMAL MODELS

We have seen in Section 6.4 that discriminant analysis is relatively straightforward for known class-conditional densities. Generally in practice, the latter are either partially or completely unknown, and so there is the problem of their estimation from data t on training entities, as defined by (6.12). As with other multivariate statistical techniques, the assumption of multivariate normality provides a convenient way of specifying a parametric structure. Hence normal models for the class-conditional densities provide the basis for a good deal of the theoretical results and practical applications in discriminant analysis. In this section we therefore focus on discrimination via normal-based models.

6.9.1   Heteroscedastic Normal Model

Under a heteroscedastic normal model for the class-conditional distributions of the feature vector Y on an entity, it is assumed that

$$Y \sim N(\boldsymbol{\mu}_i, \boldsymbol{\Sigma}_i) \quad \text{in} \quad C_i \ (i = 1, \ldots, g), \tag{6.18}$$

where $\boldsymbol{\mu}_1, \ldots, \boldsymbol{\mu}_g$ denote the class means and $\boldsymbol{\Sigma}_i, \ldots, \boldsymbol{\Sigma}_g$ the class-covariance matrices. Corresponding to (6.18), the ith class-conditional density $f_i(y; \boldsymbol{\theta}_i)$ is given by

$$\begin{aligned}
f_i(\boldsymbol{y}; \boldsymbol{\theta}_i) &= \phi(\boldsymbol{y}; \boldsymbol{\mu}_i, \boldsymbol{\Sigma}_i) \\
&= (2\pi)^{-\frac{p}{2}} |\boldsymbol{\Sigma}_i|^{-\frac{1}{2}} \exp\{-\tfrac{1}{2}(\boldsymbol{y} - \boldsymbol{\mu}_i)^T \boldsymbol{\Sigma}_i^{-1}(\boldsymbol{y} - \boldsymbol{\mu}_i)\},
\end{aligned}$$

where $\boldsymbol{\theta}_i$ consists of the elements of $\boldsymbol{\mu}_i$ and the $\frac{1}{2}p(p+1)$ distinct elements of $\boldsymbol{\Sigma}_i$ $(i = 1, \ldots, g)$. It is assumed that each $\boldsymbol{\Sigma}_i$ is nonsingular. There is no loss of generality in so doing, since singular class-covariance matrices can always be made nonsingular by an appropriate reduction of dimension.

If $\pi_1, \ldots, \pi_g$ denote the prior probabilities for the classes $C_i, \ldots, C_g$, then we let

$$\boldsymbol{\Psi} = (\pi_1, \ldots, \pi_{g-1}, \boldsymbol{\omega}^T)^T, \tag{6.19}$$

where $\boldsymbol{\theta}$ consists of the elements of $\boldsymbol{\mu}_1, \ldots, \boldsymbol{\mu}_g$ and the distinct elements of $\boldsymbol{\Sigma}_1, \ldots, \boldsymbol{\Sigma}_g$.

The posterior probability that an entity with feature vector $\boldsymbol{y}$ belongs to class C_i is denoted by $\tau_i(\boldsymbol{y}; \boldsymbol{\Psi})$ for $i = 1, \ldots, g$. In estimating these posterior probabilities, it is more convenient to work in terms of their log ratios. Accordingly, we let

$$
\begin{aligned}
\eta_{ig}(\boldsymbol{y}; \boldsymbol{\Psi}) &= \log\{\tau_i(\boldsymbol{y}; \boldsymbol{\Psi})/\tau_g(\boldsymbol{y}; \boldsymbol{\Psi})\} \\
&= \log(\pi_i/\pi_g) + \zeta_{ig}(\boldsymbol{y}; \boldsymbol{\omega}),
\end{aligned}
\tag{6.20}
$$

where

$$
\zeta_{ig}(\boldsymbol{y}; \boldsymbol{\omega}) = \log\{f_i(\boldsymbol{y}; \boldsymbol{\theta}_i)/f_g(\boldsymbol{y}; \boldsymbol{\theta}_g)\} \qquad (i = 1, \ldots, g - 1).
$$

The definition (6.20) corresponds to the arbitrary choice of C_g as the base class.
    Under the heteroscedastic normal model (6.18),

$$
\begin{aligned}
\zeta_{ig}(\boldsymbol{y}; \boldsymbol{\omega}) &= -\tfrac{1}{2}\{\delta(\boldsymbol{y}, \boldsymbol{\mu}_i; \boldsymbol{\Sigma}_i) - \delta(\boldsymbol{y}, \boldsymbol{\mu}_g; \boldsymbol{\Sigma}_g)\} \\
&\quad - \tfrac{1}{2}\{\log|\boldsymbol{\Sigma}_i|/|\boldsymbol{\Sigma}_g|\} \qquad (i = 1, \ldots, g - 1), \quad (6.21)
\end{aligned}
$$

where

$$
\delta(\boldsymbol{y}, \boldsymbol{\mu}_i; \boldsymbol{\Sigma}_i) = (\boldsymbol{y} - \boldsymbol{\mu}_i)^T \boldsymbol{\Sigma}_i^{-1}(\boldsymbol{y} - \boldsymbol{\mu}_i)
$$

is the squared Mahalanobis distance between $\boldsymbol{y}$ and $\boldsymbol{\mu}_i$ with respect to $\boldsymbol{\Sigma}_i$ $(i = 1, \ldots, g)$. The notation $\delta(\boldsymbol{a}, \boldsymbol{b}; \boldsymbol{C})$ for the squared Mahalanobis distance

$$
(\boldsymbol{a} - \boldsymbol{b})^T \boldsymbol{C}^{-1}(\boldsymbol{a} - \boldsymbol{b})
$$

between two vectors $\boldsymbol{a}$ and $\boldsymbol{b}$ with respect to some positive-definite symmetric matrix $\boldsymbol{C}$ applies throughout this book. For typographical brevity, we henceforth abbreviate $\delta(\boldsymbol{y}, \boldsymbol{\mu}_i; \boldsymbol{\Sigma}_i)$ to $\delta_i(\boldsymbol{y})$ for $i = 1, \ldots, g$.
    In this setting, the optimal or Bayes rule $r_o(\boldsymbol{y}; \boldsymbol{\Psi})$ assigns an entity with feature vector $\boldsymbol{y}$ to C_g if

$$
\eta_{ig}(\boldsymbol{y}; \boldsymbol{\Psi}) \leq 0 \qquad (i = 1, \ldots, g - 1)
$$

is satisfied. Otherwise, the entity is assigned to C_h if

$$
\eta_{ig}(\boldsymbol{y}; \boldsymbol{\Psi}) \leq \eta_{hg}(\boldsymbol{y}; \boldsymbol{\Psi}) \qquad (i = 1, \ldots, g - 1; \; i \neq h)
$$

holds. In the subsequent work, we shall refer to $r_o(\boldsymbol{y}; \boldsymbol{\Psi})$ as the normal-based quadratic discriminant rule (NQDR).

6.9.2 Plug-in Sample NQDR

In practice, $\boldsymbol{\omega}$ is generally taken to be unknown and so must be estimated from the available training data $\boldsymbol{t}$, as given by (6.12). With the estimative approach to discriminant analysis, the posterior probabilities of class membership $\tau_i(\boldsymbol{y}; \boldsymbol{\Psi})$ and the consequent Bayes rule $r_o(\boldsymbol{y}; \boldsymbol{\Psi})$ are estimated simply by plugging in some estimate $\hat{\boldsymbol{\omega}}$, such as the ML estimate, for $\boldsymbol{\omega}$ in the class-conditional densities.

The ML estimates of μ_i and Σ_i computed under (6.18) from the training data t are given by the sample mean $\bar{y}_i$ and the sample covariance matrix $\hat{\Sigma}_i$, respectively, where

$$\bar{y}_i = \sum_{j=1}^{n} z_{ij} y_j / n_i$$

and

$$\hat{\Sigma}_i = \sum_{j=1}^{n} z_{ij} (y_j - \bar{y}_i)(y_j - \bar{y}_i)^T / n_i$$

for $i = 1, \ldots, g$. Consistent with our previous notation,

$$n_i = \sum_{j=1}^{n} z_{ij}$$

denotes the number of entities from class C_i in the training data t ($i = 1, \ldots, g$). It is assumed here that $n_i > p$, so that $\hat{\Sigma}_i$ is nonsingular ($i = 1, \ldots, g$). In practice, the unbiased estimator of Σ_i,

$$S_i = n_i \hat{\Sigma}_i / (n_i - 1) \quad (i = 1, \ldots, g),$$

tends to be used instead of the ML estimate $\hat{\Sigma}_i$.

With ω estimated as above,

$$\zeta_{ig}(y; \hat{\omega}) = -\tfrac{1}{2}\{\hat{\delta}_i(y) - \hat{\delta}_g(y)\} - \tfrac{1}{2}\log\{|S_i|/|S_g|\} \quad (i = 1, \ldots, g-1) \tag{6.22}$$

where

$$\begin{aligned}
\hat{\delta}_i(y) &= \delta(y, \bar{y}_i; S_i) \\
&= (y - \bar{y}_i)^T S_i^{-1}(y - \bar{y}_i) \quad (i = 1, \ldots, g). \tag{6.23}
\end{aligned}$$

6.9.3 Homoscedastic Normal Model

Under a homoscedastic normal model for the class-conditional distributions of the feature vector Y on an entity, it is assumed that

$$Y \sim N(\mu_i, \Sigma) \quad \text{in} \quad C_i \ (i = 1, \ldots, g); \tag{6.24}$$

that is,

$$\Sigma_i = \Sigma \ (i = 1, \ldots, g). \tag{6.25}$$

It can be seen from (6.22) that a substantial simplification occurs in the form for the posterior probabilities of class membership and the consequent Bayes rule if the class-covariance matrices $\Sigma_1, \ldots, \Sigma_g$ are all the same. This is because the quadratic term in y, $y^T \Sigma_i^{-1} y$, in the exponent of the ith class-conditional density is now the same for all classes, and so vanishes in the pairwise ratios of these densities as specified by (6.22).

Corresponding to (6.22), we have under the homoscedastic normal model (6.24) that

$$
\begin{aligned}
\eta_{ig}(\boldsymbol{y}; \boldsymbol{\Psi}) &= \log\{\tau_i(\boldsymbol{y}; \boldsymbol{\Psi})/\tau_g(\boldsymbol{y}; \boldsymbol{\Psi})\} \\
&= \log(\pi_i/\pi_g) + \zeta_{ig}(\boldsymbol{y}; \boldsymbol{\omega}),
\end{aligned}
\tag{6.26}
$$

where

$$
\begin{aligned}
\zeta_{ig}(\boldsymbol{y}; \boldsymbol{\omega}) &= -\tfrac{1}{2}\{\delta_i(\boldsymbol{y}) - \delta_g(\boldsymbol{y})\} \\
&= \{\boldsymbol{y} - \tfrac{1}{2}(\boldsymbol{\mu}_i + \boldsymbol{\mu}_g)\}^T \boldsymbol{\Sigma}^{-1}(\boldsymbol{\mu}_i - \boldsymbol{\mu}_g) \qquad (i = 1, \ldots, g-1)
\end{aligned}
\tag{6.27}
$$

and where

$$
\delta_i(\boldsymbol{y}) = \delta(\boldsymbol{y}, \boldsymbol{\mu}_i; \boldsymbol{\Sigma}) \qquad (i = 1, \ldots, g).
$$

The optimal or Bayes rule $r_o(\boldsymbol{y}; \boldsymbol{\Psi})$ assigns an entity with feature vector $\boldsymbol{y}$ to C_g if

$$
\eta_{ig}(\boldsymbol{y}; \boldsymbol{\Psi}) \le 0 \qquad (i = 1, \ldots, g-1)
$$

is satisfied. Otherwise, the entity is assigned to C_h if

$$
\eta_{ig}(\boldsymbol{y}; \boldsymbol{\Psi}) \le \eta_{hg}(\boldsymbol{y}; \boldsymbol{\Psi}) \qquad (i = 1, \ldots, g-1; i \ne h)
$$

holds. We shall refer to $r_o(\boldsymbol{y}; \boldsymbol{\Psi})$ as the normal-based linear discriminant rule (NLDR).

In the case of $g = 2$ classes, we write $\eta_{12}(\boldsymbol{y}; \boldsymbol{\Psi})$ and $\zeta_{12}(\boldsymbol{y}; \boldsymbol{\omega})$ as $\eta(\boldsymbol{y}; \boldsymbol{\Psi})$ and $\zeta(\boldsymbol{y}; \boldsymbol{\omega})$, respectively. For future reference, we express the normal-based linear discriminant function (NLDF) $\zeta(\boldsymbol{y}; \boldsymbol{\omega})$ in the form

$$
\zeta(\boldsymbol{y}; \boldsymbol{\omega}) = \beta_0 + \boldsymbol{\beta}^T \boldsymbol{y},
\tag{6.28}
$$

where

$$
\beta_0 = -\tfrac{1}{2}(\boldsymbol{\mu}_1 + \boldsymbol{\mu}_2)^T \boldsymbol{\Sigma}^{-1}(\boldsymbol{\mu}_1 - \boldsymbol{\mu}_2)
$$

and

$$
\boldsymbol{\beta} = \boldsymbol{\Sigma}^{-1}(\boldsymbol{\mu}_1 - \boldsymbol{\mu}_2).
$$

Thus we can write $\eta(\boldsymbol{y}; \boldsymbol{\Psi})$ as

$$
\begin{aligned}
\eta(\boldsymbol{y}; \boldsymbol{\Psi}) &= \log(\pi_1/\pi_2) + \beta_0 + \boldsymbol{\beta}^T \boldsymbol{y} \\
&= \beta_0^* + \boldsymbol{\beta}^T \boldsymbol{y}
\end{aligned}
$$

where

$$
\beta_0^* = \log(\pi_1/\pi_2) + \beta_0.
$$

6.9.4 Optimal Error Rates

It can be seen from (6.28) that the NLDR $r_o(y; \Psi)$ is linear in y. One consequence of this is that it is straightforward to obtain closed expressions for the optimal error rates, at least in the case of $g = 2$ classes. For it can be seen from (6.26) and (6.27) that in this case, $r_o(y; \Psi)$ is based on the single linear discriminant function $\zeta(y; \omega)$ with cutoff point

$$k = \log(\pi_2/\pi_1).$$

That is, an entity with feature vector y is assigned to C_1 or C_2 according to whether $\zeta(y; \omega)$ is greater or less than k.

The optimal error rate specific to C_1 is therefore given by

$$
\begin{aligned}
eo_{12}(\Psi) &= \text{pr}\{\zeta(Y; \omega) < k \mid Y \in C_1\} \\
&= \Phi\{(k - \tfrac{1}{2}\Delta^2)/\Delta\},
\end{aligned}
$$

where

$$\Delta^2 = (\mu_1 - \mu_2)^T \Sigma^{-1}(\mu_1 - \mu_2)$$

is the squared Mahalanobis distance between C_1 and C_2 and $\Phi(\cdot)$ denotes the standard normal distribution function. Similarly,

$$eo_{21}(\Psi) = \Phi\{-(k + \tfrac{1}{2}\Delta^2)/\Delta\}.$$

In the above, we have followed the notation for the allocation rates as introduced in Section 6.3, so that $eo_{ij}(\Psi)$ denotes the probability that a randomly chosen entity from class C_i is allocated to C_j on the basis of the Bayes rule $r_o(y; \Psi)$ $(i, j = 1, \ldots, g)$.

For a zero cutoff point $(k = 0)$, the class-specific error rates $eo_{12}(\Psi)$ and $eo_{21}(\Psi)$ are equal with the common value of $\Phi(-\tfrac{1}{2}\Delta)$. Often in practice, k is taken to be zero. Besides corresponding to the case of equal prior probabilities, it also yields the minimax rule, as defined in Section 6.4.

6.9.5 Plug-in Sample NLDR

For unknown ω in the case of an arbitrary number g of classes, the maximum likelihood estimate of μ_i is given as under heteroscedasticity by the sample mean $\overline{y}_i$ of the feature observations from C_i in the training data t $(i = 1, \ldots, g)$. The ML estimate $\hat{\Sigma}$ of the common class-covariance matrix Σ is the pooled (within-class) sample covariance matrix. That is,

$$
\begin{aligned}
\hat{\Sigma} &= \sum_{i=1}^{g}(n_i/n)\hat{\Sigma}_i \\
&= \sum_{i=1}^{g}\sum_{j=1}^{n} z_{ij}(y_j - \overline{y}_i)(y_j - \overline{y}_i)^T/n.
\end{aligned}
$$

In using these estimates to form the plug-in sample versions $\zeta_{ig}(y; \hat{\omega})$ and $\tau_i(y; \hat{\Psi})$ of the log likelihood ratios and the class-posterior probabilities, we follow here the usual practice of first correcting $\hat{\Sigma}$ for bias, so that

$$S = n\hat{\Sigma}/(n-g)$$

is used instead of $\hat{\Sigma}$. With ω estimated as above,

$$\zeta_{ig}(y; \hat{\omega}) = \{y - \tfrac{1}{2}(\overline{y}_i + \overline{y}_g)\}^T S^{-1}(\overline{y}_i - \overline{y}_g) \quad (i = 1, \ldots, g-1). \quad (6.29)$$

It can be seen from (6.29) that for $g = 2$ classes, the plug-in sample version $r_o(y; \hat{\Psi})$ of the normal-based linear discriminant rule (NLDR) $r_o(y; \Psi)$ is based on the sample version

$$\zeta(y; \hat{\omega}) = \{y - \tfrac{1}{2}(\overline{y}_1 + \overline{y}_2)\}S^{-1}(\overline{y}_1 - \overline{y}_2), \quad (6.30)$$

of the NLDF $\zeta(y; \omega)$, as defined by (6.28). The sample NLDF $\zeta(y; \hat{\omega})$ is often referred to in the literature as the W classification statistic (Wald, 1944).

6.9.6 Normal Mixture Model

In some instances, the class-conditional distributions of the feature vector Y is unable to be modeled adequately by a single normal distribution. Hence we may consider modeling the ith class-conditional density $f_i(y)$ as a g_i-normal mixture density,

$$f_i(y; \Psi_i) = \sum_{h=1}^{g_i} \pi_{ih}\phi(y; \mu_{ih}, \Sigma_{ih}) \quad (i = 1 \ldots, g_i), \quad (6.31)$$

where Ψ_i contains the mixing proportions π_{ih}, the elements of the μ_{ih} and the distinct elements of the Σ_{ih} ($h = 1, \ldots, g_i$). For continuous data feature data, the normal mixture model (6.31) will provide an arbitrarily accurate estimate of the ith class-conditional density $f_i(y)$; see, for example, Li and Barron (2000). Hastie and Tibshirani (1996) have proposed the use of (6.31) with common component-covariance matrices,

$$\Sigma_{ih} = \Sigma \quad (h = 1, \ldots, g_i; i = 1 \ldots, g). \quad (6.32)$$

Of course the normal mixture model (6.31) for the ith class-conditional density contains many parameters and so the ith class sample size n_i has to be very large. As this will not be the case with high-dimensional data such as in applications to microarray data, this model may not be able to be fitted directly. One way to circumvent this problem is to reduce the dimension of the feature vector (that is, the number of genes with microarray data) by the dimension-reduction options of the EMMIX-GENE procedure, as described in Chapter 4. One can then fit a normal mixture model or mixtures of factor analyzers to the reduced features. This is to be considered further in the context of microarray data in the next chapter.

Although consideration is being given to ways of forming estimates of the posterior probabilities of class membership for a given observation with nonparametric rules like the SVM (Lee and Lee, 2003), these estimates can be formed directly from (6.4) with parametric rules as provided by normal mixture models.

6.10 FISHER'S LINEAR DISCRIMINANT FUNCTION

6.10.1 Separation Approach

The sample NLDF $\zeta(\boldsymbol{y}; \hat{\boldsymbol{\omega}})$ is essentially the same as Fisher's (1936) linear discriminant function derived without the explicit adoption of a normal model. As noted in Section 6.7, Fisher's linear discriminant function is given by $\boldsymbol{a}^T \boldsymbol{y}$, where $\boldsymbol{a}$ maximizes the quantity

$$\{\boldsymbol{a}^T(\overline{\boldsymbol{y}}_1 - \overline{\boldsymbol{y}}_2)\}^2 / (\boldsymbol{a}^T \boldsymbol{S}^T \boldsymbol{a}), \tag{6.33}$$

which is the square of the difference between the class-sample means of $\boldsymbol{a}^T \boldsymbol{y}$ scaled by the (bias-corrected) pooled sample variance of $\boldsymbol{a}^T \boldsymbol{y}$. The maximization of (6.33) is achieved by taking $\boldsymbol{a}$ proportional to $\hat{\boldsymbol{\beta}}$, where

$$\hat{\boldsymbol{\beta}} = \boldsymbol{S}^{-1}(\overline{\boldsymbol{y}}_1 - \overline{\boldsymbol{y}}_2).$$

This leads to the sample linear discriminant function

$$\boldsymbol{y}^T \hat{\boldsymbol{\beta}} = \boldsymbol{y}^T \boldsymbol{S}^{-1}(\overline{\boldsymbol{y}}_1 - \overline{\boldsymbol{y}}_2). \tag{6.34}$$

Since

$$(\overline{\boldsymbol{y}}_1 - \overline{\boldsymbol{y}}_2)^T \boldsymbol{S}^{-1}(\overline{\boldsymbol{y}}_1 - \overline{\boldsymbol{y}}_2) \geq 0,$$

the sample mean of $\boldsymbol{y}^T \hat{\boldsymbol{\beta}}$ in class C_1 is not less than what it is in C_2. Hence an entity with feature vector $\boldsymbol{y}$ can be assigned to C_1 for large values of $\boldsymbol{y}^T \hat{\boldsymbol{\beta}}$ and to C_2 for small values. If the cutoff point for the linear discriminant function (6.34) is taken to be equidistant between its class-sample means, then it is equivalent to the sample NLDF $\zeta(\boldsymbol{y}; \hat{\boldsymbol{\omega}})$ applied with a zero cutoff point.

6.10.2 Regression Approach

Fisher (1936) also derived the sample linear discriminant function (6.34) using a linear regression approach. The discrimination problem can be viewed as a special case of regression, where the regressor variables are given by the feature vector $\boldsymbol{y}$ and the dependent variables by the vector $\boldsymbol{z}$ of class-indicator variables. In the case of $g = 2$ classes, the dependent variable associated with the entity having feature vector $\boldsymbol{y}_j$ can be taken to be $z_{1j} = (\boldsymbol{z}_j)_1$ where, as defined previously, z_{1j} is one or zero according as $\boldsymbol{y}_j$ belongs to C_1 or C_2. Then for a linear relationship between the dependent and regressor variables, we have

$$z_{1j} = a_0 + \boldsymbol{a}^T \boldsymbol{y}_j + \epsilon_j \qquad (j = 1, \ldots, n), \tag{6.35}$$

where $\epsilon_1, \ldots, \epsilon_n$ are the errors. The two values taken by the dependent variable in (6.35) are irrelevant, provided they are distinct for each class. Fisher (1936) actually took

$$z_{1j} = (-1)^i n_i / n \quad \text{if } \boldsymbol{y}_j \in C_i \ (i = 1, 2).$$

The least-squares estimate of a satisfies

$$Va = \sum_{j=1}^{n}(z_{1j} - \bar{z}_1)(y_j - \bar{y}),$$ (6.36)

where

$$V = \sum_{j=1}^{n}(y_j - \bar{y})(y_j - \bar{y})^T$$

denotes the total sums of squares and products matrix. It can be decomposed into the within-class sums of squares and products matrix W, given here by $(n - 2)S$, and the between-class matrix of sums of squares and products. The latter is given by

$$\sum_{i=1}^{2} n_i(\bar{y}_i - \bar{y})(\bar{y}_i - \bar{y})^T,$$

which can be expressed as

$$(n_1 n_2/n)(\bar{y}_1 - \bar{y}_2)(\bar{y}_1 - \bar{y}_2)^T.$$

Concerning the right-hand side of (6.36), it equals

$$\begin{aligned}
\sum_{j=1}^{n}(z_{1j} - \bar{z}_1)(y_j - \bar{y}) &= \sum_{j=1}^{n} z_{1j}(x_j - \bar{y}) \\
&= n_1(\bar{y}_1 - \bar{y}) \\
&= (n_1 n_2/n)(\bar{y}_1 - \bar{y}_2)
\end{aligned}$$ (6.37)

On using (6.37) and substituting

$$V = (n - 2)S + (n_1 n_2/n)(\bar{y}_1 - \bar{y}_2)(\bar{y}_1 - \bar{y}_2)^T$$

into the left-hand side of (6.36), it follows that

$$(n - 2)Sa = (\bar{y}_1 - \bar{y}_2)[(n_1 n_2/n)\{1 - (\bar{y}_1 - \bar{y}_2)^T a\}],$$ (6.38)

which shows that the least-squares estimate of a satisfies

$$a \propto S^{-1}(\bar{y}_1 - \bar{y}_2).$$

It can be confirmed by using (6.37) and (6.38), that

$$t_v^2/(n - 2) = \rho_v^2/(1 - \rho_v^2),$$ (6.39)

where ρ_v is the correlation between the vth feature variable and the class label z_1 and where

$$t_v = \frac{\bar{y}_{1v} - \bar{y}_{2v}}{s_v\sqrt{1/n_1 + 1/n_2}},$$ (6.40)

$\bar{y}_{iv}$ denotes the sample mean of the n_i expression values for the jth gene in class C_i $(i = 1, 2)$ and $s_v^2 = (\boldsymbol{S})_{vv}$ $(v = 1, \ldots, p)$.

Thus ranking the feature variables in terms of the correlation of the feature variables with their class labels is equivalent to ranking them in terms of the values of the usual (pooled) two-sample t-statistic. We note this as there is a tendency in the biological and medical literature (for example, van 't Veer et al., 2002) to express the selection of the genes in terms of their correlation with the class labels of the tissue samples rather than in terms of the t-statistic, which is commonly used in the statistics literature.

6.11 LOGISTIC DISCRIMINATION

In Section 6.9 we have considered a fully parametric approach to discriminant analysis, where the class-conditional densities $f_i(\boldsymbol{y})$ are assumed to have specified functional forms except for a finite number of parameters to be estimated. Logistic discrimination can be viewed as a partially parametric approach, as it is only the ratios of the densities $\{f_i(\boldsymbol{y})/f_j(\boldsymbol{y}), \ i \neq j\}$ that are being modeled. For simplicity of discussion we consider first the case of $g = 2$ classes.

The fundamental assumption of the logistic approach to discrimination is that the log of the class-conditional densities is linear; that is,

$$\log\{f_1(\boldsymbol{y})/f_2(\boldsymbol{y})\} = \beta_0 + \boldsymbol{\beta}^T \boldsymbol{y}, \tag{6.41}$$

where β_0 and $\boldsymbol{\beta} = (\beta_1, \ldots, \beta_p)^T$ constitute $p + 1$ parameters to be estimated. Let

$$\beta_0^* = \beta_0 + \log(\pi_1/\pi_2).$$

The assumption (6.41) is equivalent to taking the log (posterior) odds to be linear, as under (6.41), we have that the posterior probability of an entity with $Y = y$ belonging to class C_1 is

$$\tau_1(\boldsymbol{y}) = \exp(\beta_0^* + \boldsymbol{\beta}^T \boldsymbol{y})/\{1 + \exp(\beta_0^* + \boldsymbol{\beta}^T \boldsymbol{y})\}, \tag{6.42}$$

and so

$$\begin{aligned}
\text{logit}\{(\tau_1(\boldsymbol{y})\} &= \log\{\tau_1(\boldsymbol{y})/\tau_2(\boldsymbol{y})\} \\
&= \beta_0^* + \boldsymbol{\beta}^T \boldsymbol{y}.
\end{aligned} \tag{6.43}$$

Conversely, (6.41) is implied by the linearity (6.42) of the log odds. The linearity here is not necessarily in the basic variables; transforms of these may be taken.

To define the logistic model in the case of $g > 2$ classes, let

$$\boldsymbol{\beta}_i = (\beta_{1i}, \ldots, \beta_{pi})^T \qquad (i = 1, \ldots, g-1),$$

be a vector of p parameters. Corresponding to the conventional but arbitrary choice of C_g as the base class, the logistic model assumes that

$$\log\{f_i(\boldsymbol{y})/f_g(\boldsymbol{y})\} = \beta_{0i} + \boldsymbol{\beta}_i^T \boldsymbol{y} \qquad (i = 1, \ldots, g-1), \tag{6.44}$$

or equivalently, that

$$\tau_i(\boldsymbol{y}) = \exp(\beta_{0i}^* + \boldsymbol{\beta}_i^T \boldsymbol{y}) / \sum_{h=1}^{g-1} \{1 + \exp(\beta_{0h}^* + \boldsymbol{\beta}_h^T \boldsymbol{y})\} \quad (i = 1, \dots, g-1), \quad (6.45)$$

where

$$\beta_{0i}^* = \beta_{0i} + \log(\pi_i / \pi_g) \qquad\qquad (i = 1, \dots, g-1).$$

6.12 NEAREST-CENTROID RULE

It can be seen from (6.22) and (6.29) that to be able to apply the sample versions of the NQDR or the NLDR, we need to be able to invert the sample class-covariance matrices $\boldsymbol{S}_i$ or the pooled within-class sample covariance matrix $\boldsymbol{S}$. In order for the $\boldsymbol{S}_i$ to be nonsingular, we need

$$n_i \geq p+1 \quad (i = 1 \dots, g), \qquad\qquad (6.46)$$

while for $\boldsymbol{S}$ to be nonsingular, we require

$$n \geq p + g. \qquad\qquad (6.47)$$

These inequalities will not hold for high-dimensional feature data. One way to proceed with such data is to ignore the correlations between the feature variables; that is, to take the sample class-covariance matrices $\boldsymbol{S}_i$ to be diagonal,

$$\boldsymbol{S}_i = \text{diag}\,(s_{i1}, \dots, s_{ip}), \qquad\qquad (6.48)$$

where $s_{iv} = (\boldsymbol{S}_i)_{vv}$ is the vth diagonal element of $\boldsymbol{S}_i$ $(v = 1, \dots, p; i = 1, \dots, g)$.

Under (6.48), this sample version of the NQDR rule can be expressed as

$$r(\boldsymbol{y}; \boldsymbol{t}) = \arg \min_i \sum_{v=1}^{p} (y_v - \bar{y}_{iv})^2 / s_{iv}^2 - \log(\hat{\pi}_i), \qquad\qquad (6.49)$$

where y_v is the vth element of the feature vector $\boldsymbol{y}$ and $\bar{y}_{iv} = (\bar{\boldsymbol{y}}_i)_v$, and $\bar{\boldsymbol{y}}_i$ is the sample mean of the training data from the ith class.

This simplified rule (6.49) is referred to as the nearest-centroid rule in Tibshirani et al. (2002a, 2003). In the case where the estimated prior probabilities π_i are equal and the sample class variances s_{iv} are taken to have the common value s^2 ($s_{iv} = s^2$ for $i = 1, \dots, g; v = 1 \dots, p$), the rule (6.48) assigns the feature vector $\boldsymbol{y}$ to the class whose mean (centroid) is closest in Euclidean distance.

In the case of high-dimensional data, Tibshirani et al. (2002a, 2003) proposed a modified version of (6.49), which they called the nearest-shrunken centroid rule. We shall consider it in the next chapter in the context of class prediction for microarray data.

It is noted that for a microarray data matrix $\boldsymbol{A}$ in which the columns (but not the rows) have been standardized, the Euclidean squared distance between a feature vector $\boldsymbol{y}_o$ and the centroid of the ith class $\overline{\boldsymbol{y}}_i$ is proportional to the negative of the (sample) correlation between $\boldsymbol{y}_o$ and $\overline{\boldsymbol{y}}_i$. Thus in those situations in the biological literature where the allocation of a signature expression vector of a tissue is expressed in terms of maximizing the correlation between the signature vector and the class centroid (van 't Veer et al., 2002), a nearest-centroid rule is effectively being used.

6.13 SUPPORT VECTOR MACHINES

The use of a support vector machine is an example of the algorithmic approach to statistical modeling. It contrasts with the mainstream statistical approach where the data are assumed to have been generated from a stochastic model. A nice account of these two approaches may be found in Breiman (2001) and its discussion.

6.13.1 Two Classes

Vapnik (1998) considered support vector machines (SVMs) in the case of $g = 2$ classes. In this section, we replace the vector $\boldsymbol{z}_j$ of zero-one class labels by the scalar z_j, where in the usual notation with SVMs, $z_j = 1$ or -1, according as $\boldsymbol{y}_j$ belongs to the class C_1 or C_2. Support vector machines are a direct implementation of the Structural Risk Minimization Principle. The SVM learning algorithm (with linear kernel) aims to find the separating hyperplane

$$\boldsymbol{\beta}^T \boldsymbol{y} - \beta_0 = 0$$

that is maximally distant from the training data of the two classes.

When the classes are linearly separable, the hyperplane is located so that it has maximal margin (that is, so that there is maximal distance between the hyperplane and the nearest point in any of the classes), which should lead to better performance on test data. When the data are not separable, there is no separating hyperplane; in this case, we still try to maximize the margin but allow some classification errors subject to the constraint that the total error (distance from the hyperplane on the wrong side) is less than a constant.

The standard SVM achieves this by solving the following optimization problem:

$$\min_{(\beta,\beta_0,\xi)\in\mathbb{R}^{(p+1+n)}} \boldsymbol{\beta}^T\boldsymbol{\beta}, \text{ subject to } \begin{cases} y_i(\boldsymbol{\beta}^T\boldsymbol{y} - \beta_0) \geq 1 - \xi_i, & \forall i, \\ \\ \xi_i \geq 0, & \sum \xi_i \leq \text{constant}, \end{cases}$$

where $\boldsymbol{\xi} = (\xi_1, \ldots, \xi_n)^T$ is the vector of so-called slack variables. It is easy to prove that

$$\boldsymbol{\beta} = \sum_{j=1}^{n} \alpha_j z_j \boldsymbol{y}_j,$$

$$\beta_0 = \frac{\sum_{j=1}^{n}(\alpha_j z_j, \boldsymbol{\beta}^T \boldsymbol{y}_j) - (1 - \epsilon_j)}{\sum_{j=1}^{n} \alpha_j z_j}, \tag{6.50}$$

where the α_i $(j = 1, \ldots, n)$ are the Lagrange multipliers corresponding to the $z_j(\boldsymbol{\beta}^T\boldsymbol{y} - \beta_0) \geq 1 - \xi_j$ constraints.

Given the solutions of the optimization problem above, the corresponding discriminant rule is

$$r(\boldsymbol{y};\,\boldsymbol{t}) = \text{sign}(\sum_{j=1}^{n}\alpha_j z_j \boldsymbol{y}_j^T \boldsymbol{y} - \beta_0).$$

The standard SVM can be extended to nonlinear decision functions by using the kernel technique, which consists of mapping the data to a higher-dimensional space $\mathcal{H}$ by means of the function

$$\phi : \mathbb{R}^d \to \mathcal{H}.$$

For a particular choice of ϕ, the inner product $K(\boldsymbol{y}_j, \boldsymbol{y}_k) = (\phi(\boldsymbol{y}_j) \cdot \phi(\boldsymbol{y}_k))$ can be easily computed and the discriminant rule becomes

$$r(\boldsymbol{y};\,\boldsymbol{t}) = \text{sign}(\sum_{j=1}^{n}\alpha_j z_j K(\boldsymbol{y}_j, \boldsymbol{y}) - \beta_0).$$

In Figure 6.1, we display the separating surfaces for two types of kernel functions:

- the linear kernel $K(\boldsymbol{y}_j, \boldsymbol{y}_k) = (\boldsymbol{y}_j, \boldsymbol{y}_k)$.

- the Gaussian kernel $K(\boldsymbol{y}_j, \boldsymbol{y}_k) = e^{-\frac{|y_j - y_k|^2}{\sigma^2}}$.

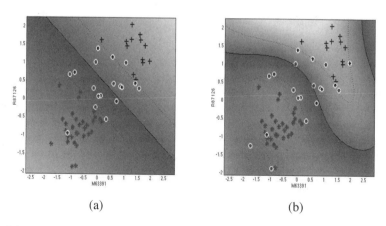

(a)   (b)

Fig. 6.1 Separating surface spawned by SVM with (a) linear kernel (b) Gaussian kernel. See the insert for a color representation.

6.13.2 Selection of Feature Variables

For the SVM with linear kernel, a ranking of the feature variables can be obtained on the basis of the magnitude of the coefficients of the variables. For high-dimensional problems, Guyon et al. (2002) have proposed that the feature variables be selected according to a backward elimination process, which they call recursive feature elimination (RFE). It is a recursive process. Starting by considering all available features, each step consists in ranking the features according to the order of magnitude of their associated coefficients (weights) β_i, and then discarding the bottom-ranked variables. The process continues by considering the remaining features. As noted by Guyon et al. (2002), removing one variable at a time is more accurate than removing chunks of variables at a time. However, as they observed that there are only significant differences for smaller subsets of variables (less than 100), they suggested that, without trading accuracy for speed, one can use RFE by removing chunks of features in the first few iterations and then remove one feature at a time once the feature set size reaches a few hundred. In the application of the SVM with RFE to microarray data, we first discarded enough bottom-ranked genes so that the number retained was the greatest power of 2 (less than the original number of genes). We then sequentially proceeded to discard half the current number of genes on each subsequent step. This practice was adopted in the examples in Guyon et al. (2002). More recently, Furlanello et al. (2003) have proposed an accelerated version of RFE.

6.13.3   Multiple Classes

Several solutions have been proposed for generalizing two-class SVMs to multiclass SVMs. They can be divided in two types of approaches:

- combination of binary SVM classifiers;

- consideration of all classes at once in a general optimization problem.

The binary approaches were first implemented. Among them, the one-against-all approach may be the earliest. It constructs g SVM models, where g is the number of classes. Each classifier is trained using the examples of a given class against all other classes. A new observation is assigned to the class with the largest decision function output among the g estimated functions. Another binary approach is the one-against-one, where $g(g-1)/2$ binary classifiers are trained, from all possible two-class problems. The appropriate class is found by a voting scheme: each binary classifier votes for one class. Other more sophisticated binary approaches have been devised, such as Direct Acyclic Graph SVM (DAGSVM), where the estimation phase is identical to the one-against-one classifier, but the assignment phase uses a directed acyclic graph whose nodes are the $g(g-1)/2$ binary SVM classifier and leaves the g classes (Platt et al., 2000).

The g-class SVM has been proposed independently under various formulations in Vapnik (1998), Bredensteiner and Bennett (1999), Weston and Watkins (1999), and Guermeur et al. (2000). As with the one-against-all approach, g decision functions are estimated, but they are all obtained by solving one problem.

A comparison of these methods on large-scale problems can be found in Hsu and Lin (2002). It appears that all-together methods need less support vectors but more training time than one-against-one and DAG approaches. Considering their relative performances, there is not a significant difference between the methods, and Hsu and Lin (2002) thus conclude that one-against-one and DAG may be more suitable for practical use. Greater differences may show up in problems with a large number of classes with few data points.

6.13.4 Computer Software

We have implemented the Recursive Feature Elimination method for two- and multiclass problems in a R language package that we call **rfe**. The source code is available from the Comprehensive R Archive Network (CRAN) under the section *Download* from

`http://www.r-project.org/`

or alternatively from

`http://www.hds.utc.fr/~ambroise/RFE/`.

As noted above, the Recursive Feature Elimination Algorithm was first proposed by Guyon et al. (2002) for selecting relevant features in the two class problem using linear SVM.

For multiclass problems, the RFE procedure needs some adaptation. First, one has to choose the approach for dealing with multiple classes using SVM. Second, it is necessary to adapt the ranking criterion. In the **rfe** package, the following solutions are considered:

- Multiclass classification is achieved using the one-against-one approach, in which $g(g - 1)/2$ binary classifiers are trained, from all possible two-class problems. The appropriate class is found by a voting scheme: each binary classifier votes for one class. The observation to be classified is assigned to the class having the maximum number of votes. This procedure was first used by Friedman (1996) and Kreßel (1999). An interesting comparison with other multiclassification SVM algorithms can be found in Hsu and Lin, 2002.

- Each two-class SVM classifier is described by a weight vector. The g-class classifier based on the "one-against-one" approach is characterized by $g(g - 1)/2$ weight vectors. The sum of the absolute value of the weight vector coordinate is used to characterize the discriminatory power of the associated feature.

The **rfe** package depends on the e1071 package for the implementation of linear Support Vector Machines (SVM). The underlying C++ code for linear SVM is provided by Chang, Chih-Chung and Lin, Chih-Jen:

LIBSVM: a library for Support Vector Machines

`http://www.csie.ntu.edu.tw/~cjlin/libsvm`

6.14 VARIANTS OF SUPPORT VECTOR MACHINES

Recently, Marron and Todd (2002) and Benito et al. (2004) have considered a variant of the support machine for the discrimination of data that are of very high dimension relative to the sample size, as in a typical microarray analysis. Their method of "distance weighted discrimination" aims to reduce the "data piling" at the margin with applications of the SVM to high-dimensional data.

In other work for high-dimensional data, Bhattacharyya et al. (2003) have formulated the minimax probability machine (MPM) as a viable alternative to the SVM. The MPM forms a discriminant rule by the minimization of an upper bound on the overall error rate. They describe LIKNON, a specific implementation of a statistical approach for creating a rule and identifying a small number of relevant features simultaneously. Given two-class data, LIKNON estimates a sparse linear discriminant rule by exploiting the simple and well-known property that minimizing the L_1 norm (via linear programming) yields a sparse hyperplane.

6.15 NEURAL NETWORKS

We consider the case of $g = 2$ classes, but the definition extends in a straightforward manner to $g > 2$ classes. In this section, we replace the class-indicator vector z_j of the feature vector y_j by the scalar z_j, where z_j is one or zero, according as y_j belongs to class C_1 or C_2.

For a feedforward neural network with one hidden layer of m units (see Figure 6.2), we can analytically represent the output $\tilde{z}$ corresponding to the input y as

$$\tilde{z} = a_2(\alpha_{02} + \boldsymbol{\alpha}_2^T \boldsymbol{w}), \tag{6.51}$$

where $\boldsymbol{w} = (w_1, \ldots, w_m)^T$ and where $w_h = a_1(\alpha_{h01} + \boldsymbol{\alpha}_{h1}^T \boldsymbol{y})$ is the hth hidden variable $(h = 1, \ldots, m)$. Here α_{h01} is the bias term, $\boldsymbol{\alpha}_{h1}$ is the vector of weights used to produce the hth hidden variable, α_{02} is the bias term, and $\boldsymbol{\alpha}_2$ is the m-dimensional vector of weights used to produce the (scalar) output $\tilde{z}$; see, for example, Bishop (1995).

The activation function a_1 transforming the linear sum $\boldsymbol{\alpha}_{h1}^T \boldsymbol{y}$ to give the hth hidden neuron, is usually taken to be the sigmoidal function $s(u)$, where

$$s(u) = \exp(u)/\{1 + \exp(u)\}.$$

The activation function a_2 for the output units is also usually taken to be the sigmoidal (logistic) function $s(u)$. In the present context of classification problems where $a_2(\cdot) = s(\cdot)$, the consequent output $\tilde{z}$ represents the conditional probability that the entity belongs to class C_1 given the input y (that is, it is the posterior probability of class membership of C_1). In those instances where a_2 is taken to be a threshold function, the output $\tilde{z}$ is a direct estimate of the membership label z of y.

For a feedforward network, the conventional learning rule uses the back-propagation algorithm to minimize a target function such as the sum of the squared errors,

$$\sum_{j=1}^{n}(\tilde{z}_j - z_j)^2,$$

where $\tilde{z}_j = a_2(\alpha_{02} + \boldsymbol{\alpha}_2^T \boldsymbol{w}_j)$ and where now $\boldsymbol{w}_j = (w_{1j}, \ldots, w_{mj})^T$, and

$$w_{hj} = a_1(w_{h01} + w_{h1}^T \boldsymbol{y}_j) \tag{6.52}$$

is the hth hidden variable corresponding to the input $\boldsymbol{y}_j$ ($h = 1, \ldots, m; j = 1, \ldots, n$). Recently, Ng and McLachlan (2004) have considered the use of the EM algorithm in the training of neural networks.

An example of an application of neural networks to the classification of microarray gene-expression data has been given by Khan et al. (2001)

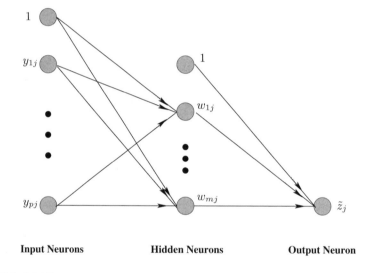

Fig. 6.2 Multilayer perceptron neural networks.

6.16 NEAREST-NEIGHBOR RULES

6.16.1 Introduction

As noted by Fix and Hodges (1951), nearest-neighbor allocation is based on a variant of nearest-neighbor density estimation of the class-conditional densities; see McLachlan (1992, Chapter 9). Suppose that $f_1(\boldsymbol{y})$ is estimated by the naive kernel density estimator $f_1^{(K)}(\boldsymbol{y})$, using as its kernel the uniform density over the neighborhood

$\mathcal{N}_1$ taken large enough so as to contain a given number k of points in the combined sample $\boldsymbol{y}_j$ $(j = 1, \ldots, n)$. The same neighborhood $\mathcal{N}_1$ is then used to construct the naive kernel density $\hat{f}_2^{(K)}(\boldsymbol{y})$ of $f_2(\boldsymbol{y})$. Then it can be seen that the allocation rule based on the relative size of $\hat{f}_1^{(K)}(\boldsymbol{y})$ and $\hat{f}_2^{(K)}(\boldsymbol{y})$ is equivalent to that based on the relative size of k_1/n_1 and k_2/n_2, where k_i is the number out of the k points from the combined sample in the neighborhood $\mathcal{N}_1$ of $\boldsymbol{y}$, that come from C_i $(i = 1, 2)$.

For $k = 1$, the rule based on the relative size of k_1/n_1 and k_2/n_2 is precisely the 1-NN discriminant rule (nearest-neighbor rule of order 1), which assigns an entity with feature vector $\boldsymbol{y}$ to the class of its nearest neighbor in the training set. For $k > 1$, their approach incorporates a variety of so-called k-NN discriminant rules, including simple majority vote among the classes of a point's k nearest neighbors and modifications that take differing class-sample sizes into account.

6.16.2 Definition of a k-NN Rule

We now give a formal definition of a kth nearest neighbor (k-NN) rule within the regression framework, as adopted by Stone (1977) in his important paper on consistent nonparametric regression. For an unclassified entity with feature vector $\boldsymbol{y}$, consider estimates of its posterior probabilities of class membership having the form

$$\hat{\tau}_i(\boldsymbol{y};\, \boldsymbol{t}) = \sum_{j=1}^{n} w_{nj}(\boldsymbol{Y};\boldsymbol{t}_y)z_{ij} \quad (i = 1, \ldots, g), \tag{6.53}$$

where $w_{nj}(\boldsymbol{y};\boldsymbol{t}_y)$ are nonnegative weights which sum to unity and which may depend on $\boldsymbol{y}$ and the training feature vectors $\boldsymbol{y}_1, \ldots, \boldsymbol{y}_n$, but not their associated class-indicator vectors $\boldsymbol{z}_1, \ldots, \boldsymbol{z}_n$. As previously, the classified training data $\boldsymbol{t}$ are defined by (6.12), while

$$\boldsymbol{t}_y = (\boldsymbol{y}_1^T, \ldots, \boldsymbol{y}_n^T)^T.$$

Rank the n training observations $\boldsymbol{y}_1, \ldots, \boldsymbol{y}_n$ according to increasing values of $||\boldsymbol{y}_j - \boldsymbol{y}||$ to obtain the n indices $R_1, \ldots, R_n$. The training entities j with $R_j \le k$ define the k nearest neighbors of the entity with feature vector $\boldsymbol{y}$. Ties among the $\boldsymbol{y}_j$ can be broken by comparing indices; that is, if $||\boldsymbol{y}_{j_1} - \boldsymbol{y}|| = ||\boldsymbol{y}_{j_2} - \boldsymbol{y}||$, then $R_{j_1} < R_{j_2}$ if $j_1 < j_2$; otherwise, $R_{j_1} > R_{j_2}$. In Stone's (1977) work, any weight attached to kth nearest neighbors is divided equally when there are ties.

If w_{nj} is a weight function such that $w_{nj}(\boldsymbol{y};\, \boldsymbol{t}_y) = 0$ for all $R_j > k$, it is called a k-NN weight function. The sample rule obtained by using the estimate $\hat{\tau}_i(\boldsymbol{y};\, \boldsymbol{t})$ with these weights in the definition (1.4.3) of the Bayes rule is a k-NN rule. For example, if $k = 1$, then

$$
\begin{aligned}
w_{nj}(\boldsymbol{y};\, \boldsymbol{t}_y) \quad &= \quad 1 \qquad && \text{if } R_j = 1, \\
&= \quad 0 \qquad && \text{if } R_j > 1,
\end{aligned}
$$

so that

$$\hat{\tau}_i(\boldsymbol{y};\, \boldsymbol{t}) = z_{iR_j} \quad (i = 1, \ldots, g).$$

These estimates of the posterior probabilities of class membership imply that in an outright allocation of the unclassified entity with feature vector y, it is assigned to the class of its nearest neighbor in the training set.

In the case of $k > 1$ with uniform weights, where

$$
\begin{aligned}
w_{nj}(y; t) &= 1/k & \text{if } R_j \leq k, \\
&= 0 & \text{if } R_j > k,
\end{aligned}
$$

we have that

$$
\hat{\tau}_i(y; t) = \sum_{j=1}^{k} z_{iR_j}/k,
$$

implying that the entity with feature vector y is allocated on the basis of simple majority voting among its k nearest neighbors.

The obvious metric on R^p to use is the Euclidean metric. This metric, however, is inappropriate if the feature variables are measured in dissimilar units; in which case, the feature variables should be scaled before applying the Euclidean metric. The influence of data transformations and metrics on the k-NN rule has been considered by Todeschini (1989). Also, Myles and Hand (1990) have considered the choice of metric in NN rules for multiple classes.

6.17 CLASSIFICATION TREES

Tree-structured rules are the subject of the book by Breiman et al. (1984). Their contributions are known by the acronym CART (classification and regression trees). Another popular methodology is the program ID3 and its later versions, C4.5 and C5.0, as developed by Quinlan (1986, 1993). There is also the method known as FACT proposed by Loh and Vanichsetakul (1988), among others.

An attractive feature of tree-based classifiers is their interpretability. A classification tree partitions the space into a set of hypercubes, and each decision can be interpreted using AND-OR rules in a straightforward manner. In this chapter we describe and use the CART method, which has had a seminal influence in the field.

It is easy to grow a tree having a zero apparent error rate (see Figure 6.3) with only a few genes. Thus, some pruning of the tree is required. With the CART method, a large tree is grown in the first instance, which ensures against stopping too early. The bottom nodes are then recombined or "pruned" upward to give the final tree. The degree of pruning is determined by cross-validation using a cost-complexity function that balances the apparent error rate with the tree size. An example of a pruned tree produced by CART is given in Figure 6.4; see also Figure 6.5. The data being used here are the colon cancer data of Alon et al. (1999), as described in Section 2.6.

Because CART recursively partitions the training data into subsets of ever decreasing size, it requires n to be very large in order to maintain reasonable sample sizes at each successive node. Breiman et al. (1984) proved that under mild regularity conditions, rules based on recursive partitioning are Bayes risk consistent.

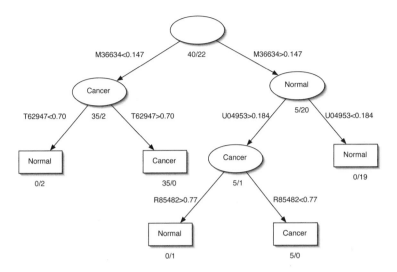

Fig. 6.3 A tree with zero apparent error rate in its application to the Alon data.

Trees are different from previously considered classifiers because they are learning and selecting features simultaneously. Consequently, an optimal tree is defined for a given set of features. A characteristic of binary tree classifiers is the embedding of the feature selection within the learning procedure. Usually, the optimal tree complexity and the estimation of the error rate are considered at the same time using the same sample.

6.18 ERROR-RATE ESTIMATION

We consider now the estimation of the error rates associated with a discriminant rule $r(y; t)$ formed from some realized training data t, as defined by (6.12). There is a very extensive subfield of discriminant analysis and the reader is referred to McLachlan (1992, Chapter 10) and Schiavo and Hand (2000) for an in-depth account. As discussed in Section 6.6, it is the conditional or actual error rates of $r(y; t)$ that are of central interest once the training data t have been obtained. We let $ec(t)$ denote the overall conditional error rate of $r(y; t)$, as defined in Section 6.6. This error rate, which is conditional on the training data t, also depends on the class-conditional distributions. But this dependence is suppressed here for simplicity of notation.

6.18.1 Apparent Error Rate

An obvious and easily computed nonparametric estimator of the conditional error rate $ec(t)$ of $r(y; t)$ is the apparent error rate A of $r(y; t)$ in its application to the observations in t. That is, A is the proportion of the observations in t misallocated

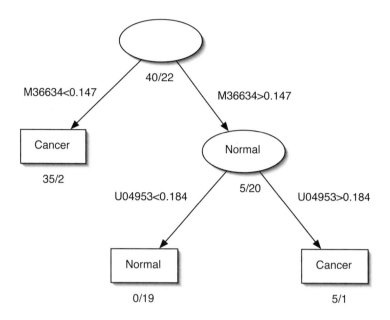

Fig. 6.4 Optimal pruned tree for Alon data obtained by CART method. The ten-fold cross-validated error rate is 15%.

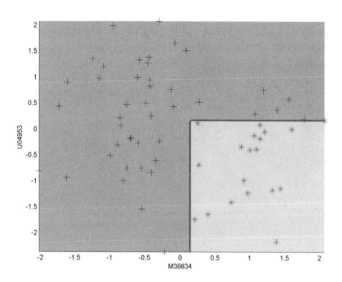

Fig. 6.5 Partition of the Alon data using the optimal pruned tree.

by $r(\boldsymbol{y}; \boldsymbol{t})$. Thus we can write

$$A = \frac{1}{n} \sum_{i=1}^{g} \sum_{j=1}^{n} z_{ij} Q[i, r(\boldsymbol{y}_j; \boldsymbol{t})], \tag{6.54}$$

where for any u and v, $Q[u, v] = 0$ for $u = v$ and 1 for $u \neq v$.

The apparent error rate, or resubstitution estimator as it is often called, was first suggested by Smith (1947) in connection with the sample NQDR. As the apparent rate is obtained by applying the rule to the same data from which it has been formed, it provides an optimistic assessment of the true conditional error rates. In particular, for complicated discriminant rules, overfitting is a real danger, resulting in a grossly optimistic apparent error. Although the optimism of the apparent error rate declines as n increases, it usually is of practical concern.

6.18.2   Bias Correction of the Apparent Error Rate

We now proceed to consider some nonparametric methods for correcting the apparent error rate for bias. One way of avoiding the bias in the apparent error rate as a consequence of the rule being tested on the same data from which it has been formed (trained), is to use a holdout method as considered by Highleyman (1962), among others. The available data are split into disjoint training and test subsets. The discriminant rule is formed from the training subset and then assessed on the test subset. Clearly, this method is inefficient in its use of the data. There are, however, methods of estimation, such as cross-validation, the Quenouille–Tukey jackknife, and the bootstrap of Efron (1979), that obviate the need for a separate test sample. An excellent account of these three methods has been given by Efron (1982), who has exhibited the close theoretical relationship between them.

6.19   CROSS-VALIDATION

6.19.1   Leave-One-Out(LOO) Estimator

One way of almost eliminating the bias in the apparent error rate is through the leave-one-out (LOO) technique as described by Lachenbruch and Mickey (1968) or cross-validation (CV) as discussed in a wider context by Stone (1974) and Geisser (1975). For the estimation of $ec(\boldsymbol{t})$ by the apparent error rate A, the leave-one-out cross-validated estimate is given by

$$A^{(CV)} = \frac{1}{n} \sum_{i=1}^{g} \sum_{j=1}^{n} z_{ij} Q[i, r(\boldsymbol{y}_j; \boldsymbol{t}_{(j)}], \tag{6.55}$$

where $\boldsymbol{t}_{(j)}$ denotes $\boldsymbol{t}$ with the point $\boldsymbol{y}_j$ deleted ($j = 1, \ldots, n$). Hence before the sample rule is applied at $\boldsymbol{y}_j$, it is deleted from the training set and the rule recalculated on the basis of $\boldsymbol{t}_{(j)}$. This procedure at each stage can be viewed as the extreme version of the holdout method where the size of the test set is reduced to a single entity.

According to Stone (1974), the refinement of this type of assessment appears to have been developed by Lachenbruch (1965) following a suggestion in Mosteller and Wallace (1963). Toussaint (1974) in his bibliography on error-rate estimation traces the idea back to at least 1964 in the Russian literature on pattern recognition.

It can be seen that, in principle at least, cross-validation requires a considerable amount of computing, as the sample rule has to be formed n times in the computation of $A^{(CV)}$. For some rules, however, it is possible to calculate $A^{(CV)}$ with little additional effort.

6.19.2 q-Fold Cross-Validation

As remarked by Efron (1983), cross-validation is often carried out, removing large blocks of observations at a time. Suppose, for example, that the training set is divided into, say q blocks, each consisting of m data points where, thus, $n = qm$ $(m \geq 1)$. Let now

$$t_{(k)} = (y_1^T, \ldots, y_{(k-1)m}^T, y_{km+1}^T, \ldots, y_n^T)^T;,$$

that is, the training set after the deletion of the kth block of m observations. Then the q-fold cross-validated error rate is given by

$$A^{(CVq)} = \sum_{i=1}^{g} \sum_{j=1}^{m} \sum_{k=1}^{q} z_{ij} Q[i, r(y_{(k-1)m+j}; t_{(k)}]/n, \qquad (6.56)$$

which requires only q recomputations of the rule.

The choice of $q = n$ (leave-one-out) does not perturb the data enough and results in higher variance. With $q = 2$, the training sets are too small relative to the full training sets. The values $q = 5$ or 10 are a good compromise.

6.20 ERROR-RATE ESTIMATION VIA THE BOOTSTRAP

6.20.1 The 0.632 Estimator

As shown by Efron (1979, 1983), suitably defined bootstrap procedures can reduce the variability of the leave-one-out error in addition to providing a direct assessment of variability for estimated parameters in the discriminant rule. Also, if we take the number of bootstrap replications K to be less than n, it will result in some saving in computation time relative to leave-one-out cross-validation.

As discussed by Efron and Tibshirani (1997), a bootstrap smoothing of leave-one-out cross-validation is given by the leave-one-out bootstrap error $B^{(1)}$, which predicts the error at a point y_j only from bootstrap samples that do not contain the point y_j. To define $B^{(1)}$ more precisely, suppose that K bootstrap samples of size n are obtained by resampling with replacement from the original set of n classified tissue samples. We let $r(y; t_k^*)$ be the bootstrap version of the rule $r(y; t)$ formed from the kth bootstrap sample in exactly the same manner that $r(y; t)$ was formed from the original training set t. Then the (Monte Carlo) estimate of $B^{(1)}$ on the basis of the K bootstrap samples is given by

$$B^{(1)} = \sum_{j=1}^{n} E_j/n, \qquad (6.57)$$

where

$$E_j = \sum_{k=1}^{K} I_{jk} \sum_{i=1}^{g} z_{ij} Q[i, r(\boldsymbol{y}_j; \boldsymbol{t}_k^*)] / \sum_{k=1}^{K} I_{jk}$$

and where I_{jk} is one if $\boldsymbol{y}_j$ is not contained in the kth bootstrap sample and is zero otherwise.

Typically, $B^{(1)}$ is based on about $0.632n$ of the original data points and Efron (1983) confirmed that it closely agrees with the *half-sample cross-validation* (that is, two-fold) error rate $A^{(CV2)}$. Thus $B^{(1)}$ is upwardly biased and Efron (1983) proposed the 0.632 estimator,

$$B^{(0.632)} = 0.368\,A + 0.632\,B^{(1)} \qquad (6.58)$$

for correcting the upward bias in $B^{(1)}$ with the downwardly biased apparent error A.

As $B^{(1)}$ is almost the same as $A^{(CV2)}$, $B^{(.632)}$ is almost the same as

$$0.368\,A + 0.632\,A^{(CV2)}. \qquad (6.59)$$

Estimators of this type were considered previously by Toussaint and Sharpe (1975) and McLachlan (1977) in the context of choosing the weight w so that

$$A^{(w)} = (1 - w)A + \omega A^{(CVq)} \qquad (6.60)$$

where, as above, $A^{(CVq)}$ denotes the estimated rate after cross-validation, removing $m = n/q$ observations at a time.

For the rule based on the sample NLDF with zero cutoff point, McLachlan (1977) calculated the value w_o of w for which $A^{(w)}$ has zero first-order bias. For $q = 2$, McLachlan (1977) showed that w_o ranged from 0.6 to 0.7 for the combinations of the parameters considered. Hence, at least under the homoscedastic normal model, the estimator $A^{(w_o)}$ is about the same as the 0.632 estimator of Efron (1983). The latter therefore should have almost zero first-order bias under (6.24), at least for the sample NLDR.

6.20.2   Mean Squared Error of the Estimated Error Rate

We have seen in the previous sections that the actual error rate of a discriminant rule can be estimated nonparametrically by the use of cross-validation and that the bootstrap provides a viable alternative with less variation. In practice, it is also of interest to provide a guide to the variability in the error-rate estimator. Here we outline how the root mean squared error (RMSE) of the cross-validated estimator can be estimated by the bootstrap.

Let $r(y; t)$ be some rule formed from the training data t and let $A^{(CV)}$ be its cross-validated error rate as defined by (6.55). Its RMSE can be estimated by the bootstrap as follows.

Let $r(y; t_k^*)$ be the rule formed from the kth bootstrap sample t_k^* obtained by sampling with replacement from the original training data t $(k = 1, \ldots, K)$. The conditional or actual error rate ec_k of $r(y; t_k^*)$ when it is applied to a randomly chosen observation from the bootstrap distribution F_n is given by

$$ec_k = \frac{1}{n} \sum_{i=1}^{g} \sum_{j=1}^{n} z_{ij} Q[i, r(y_j; t_k^*)]. \tag{6.61}$$

Here F_n denotes the empirical distribution function that places mass $1/n$ at each point $(y_j^T, z_j^T)^T$ in the training data t as given by (6.12). We let $A_k^{(CV)}$ be the cross-validated error rate of $r(y; t_k^*)$ formed from t_k^*.

Then the RMSE of $A^{(CV)}$ can be approximated by

$$\text{RMSE}(A^{(CV)}) = [\sum_{k=1}^{K} (A_k^{(CV)} - ec_k)^2 / K]^{\frac{1}{2}}. \tag{6.62}$$

For q-fold cross-validation, there is an easily computed estimate of the standard errors of the estimated error rates $A^{(CVq)}$. It is given by the standard error of the q apparent error rates that are obtained when the discriminant rule is applied to the q validation subsamples during the cross-validation. This standard error of $A^{(CVq)}$ is

$$\text{SE}(A^{(CVq)}) = [\frac{1}{q(q-1)} \sum_{h=1}^{q} (A_h^{(CVq)} - A^{(CVq)})^2]^{\frac{1}{2}}, \tag{6.63}$$

where $A_h^{(CVq)}$ is the apparent error rate when the rule formed from the hth training subsample is applied to the hth validation subsample. Similarly, the standard errors of the individual q-fold cross-validated error rates can be formed.

6.21 SELECTION OF FEATURE VARIABLES

With high-dimensional feature data, consideration has to be given to the choice of feature variables to be used in the formation of the sample discriminant rule. For some rules, like the nearest-neighbor centroid or a SVM, it is possible to construct the rule using all the feature variables. But with, say, Fisher's linear discriminant rule, it cannot be formed unless the number of feature variables is sufficiently small for the inequality (6.47) to hold. Also, even if it is possible to form the rule using all the feature variables, the use of all the variables may harm the performance of the sample rule. This is because for training samples of finite size, the performance of a given discriminant rule in a frequentist framework does not keep on improving as the number p of feature variables is increased. Rather, its overall unconditional error rate will stop decreasing and start to increase as p is increased beyond a certain threshold, depending on the particular situation. Consider, for example, the case of $g = 2$ classes under the homoscedastic normal model (6.24). Although the deletion of some variables from the feature vector y can never increase the Mahalanobis distance Δ between the classes C_1 and C_2, the overall unconditional error rate of the sample NLDR may be reduced if the subsequent reduction in the distance is sufficiently small relative to the number of variables deleted (McLachlan, 1992, Chapter 12). This peaking phenomenon will be demonstrated further in the next chapter with the application of Fisher's rule and SVMs to some microarray data.

Of course there may be other reasons for not wishing to use the entire set of variables. For example, the primary aim of the discriminant analysis might be to determine the set of variables most relevant for identification of the underlying class structure of the problem. This is a common aim in the analyses of microarray data, where the aim is to find so-called marker genes, which are genes that are most useful in assigning an unclassified tissue to its correct class of origin.

Numerous heuristic search algorithms have been proposed in the literature. Excellent reviews have been given recently by Kudo and Sklansky (2000) and Molina, Belanche, and Nebot et al. (2002). All these approaches involve searching for an optimal or near optimal subset of features that optimize a given criterion. If the selection process takes place before the construction of the discriminant rule, the method is said to follow a filter approach. If the computation of the criterion uses the rule as a subroutine, the method is said to follow a wrapper approach (Kohavi and John, 1997). Also, feature-selection schemes are classified as to whether they are embedded or not. An embedded scheme is one for which a rule has its own feature-selection procedure, such as with tree classifiers and nearest-shrunken centroids. The latter is to be defined in the next chapter.

Feature subset selection can also be classified into three categories according to the strategy used for searching the feature subset space: they are (1) exhaustive search, (2) heuristic search, and (3) random search.

In many situations attention is focused not on just one subset of the available feature variables. Rather the intention is to find the best subset in some sense. Thus consideration of a number of subsets has to be undertaken. Ideally, the performance of the discriminant rule should be assessed on the basis of the specified criterion for each possible subset. But unless the total number p of variates is small, an exhaustive search is computationally prohibitive unless the number of features is small enough, or the evaluation criterion is known to be monotonic.

As a consequence, stepwise selection procedures, either forward or backward, are commonly employed. The two basic algorithms are Sequential Forward Selection (SFS) and Sequential Backward Selection (SBS). SFS starts with an empty set of features and adds the best possible feature at each stage. SBS starts with the complete set of features and removes the worst feature at each stage. Both algorithms fail to take into account the complexity in the dependence relationships of the features and are only optimal at each stage. Usually backwards algorithms perform better than their counterpart, but they are computationally more expensive. Among all the alternative search strategies available, the sequential floating strategies of Pudil et al. (1994) rank among the best.

Random search algorithms offer an alternative to heuristic search to avoid being stuck on local optima of the chosen criterion. Genetic algorithms have proved to be effective tools in variable selection.

6.22   ERROR-RATE ESTIMATION WITH SELECTION BIAS

6.22.1   Selection Bias

Caution has to be exercised in selecting a small number of variables from a large set, as there will be a selection bias associated with choosing the optimal of a large number of possible subsets, regardless of the criterion used. This problem in discriminant analysis has been considered by Murray (1977), Gong (1986), Ganeshanandam and Krzanowski (1989), and Snapinn and Knoke (1989), among others. Miller (1984, 1990) makes a number of allied points regarding the selection of variables in multiple regression.

Consider the selection of the subvector $y^{(S)}$ of the full feature vector y. Suppose that s_o defines the subset of feature variables of some specified size p_{s_o} that minimizes the adopted estimate $\hat{e}(t^{(S)})$ over all possible $\begin{pmatrix} p \\ p_{s_o} \end{pmatrix}$ distinct subsets s of size p_{s_o}. Although $\hat{e}(t^{(S)})$ may be an unbiased estimator of the overall conditional error rate, $\hat{e}(t^{(S)})$ is obviously not as it is obtained by taking the smallest of the estimated error rates after they have been ordered. As noted by Murray (1977), the situation is exacerbated if an optimistic estimator of the error rate, such as the apparent error rate, is used. There is then a double layer of overoptimistic bias inherent in the assessment.

6.22.2   External Cross-Validation

Suppose that $y^{(s_o)}$ contains the subset of feature variables selected as being the best of size p_{s_o}, according to some criterion. Let $r(y^{(s_o)}; t^{(s_o)})$ denote some arbitrary sample discriminant rule formed from the classified training data $t^{(s_o)}$ on the feature vector $y^{(s_o)}$.

The overall apparent error rate of this rule may be expressed as

$$A(t^{(s_o)}) = \frac{1}{n} \sum_{i=1}^{g} \sum_{j=1}^{n} z_{ij} Q[i, r(y_j^{(s_o)}; t^{(s_o)})], \qquad (6.64)$$

where the Q-function is defined by (6.7). The optimism arising from the use of the apparent error rate may be almost eliminated using cross-validation. The (leave-one-out) cross-validated estimate is

$$A^{(CV)}(t^{(S_o)}) = \frac{1}{n} \sum_{i=1}^{g} \sum_{j=1}^{n} z_{ij} Q[i, r(y_j^{(S_o)}; t_{(j)}^{(S_o)})], \qquad (6.65)$$

where $t_{(j)}^{(S_o)}$ denotes the training data $t^{(S_o)}$ with $(y_j^{(S_o)^T}, z_j^T)^T$ deleted.

In order to reduce the selection bias which is still present in the estimate (6.65), an external cross-validation should be performed whereby the selection process is undertaken for each deletion of a feature vector from the training set. This external cross-validated estimate of the overall error rate of $r(y^{(S_o)}; t^{(S_o)})$ is given by

$$A^{(CVE)}(t) = \frac{1}{n} \sum_{i=1}^{g} \sum_{j=1}^{n} z_{ij} Q[i, r(y_j^{(S_{oj})}; t_{(j)}^{(S_{oj})})], \qquad (6.66)$$

where s_{oj} denotes the optimal subset, according to the adopted selection criterion applied to the training data $t_{(j)}$ without $(y_j^T, z_j^T)^T$. As the notation implies, the selected subset s_{oj} for the allocation of the jth entity may be different for each j $(j = 1, \ldots, n)$.

This way of overcoming the selection bias involves a high computing penalty. But there are many situations in practice where its implementation is computationally feasible to implement. An illustration of its use is to be given in Chapter 7 on the supervised classification of microarray data.

6.22.3 The 0.632+ Estimator

An alternative method for error-rate estimation in the presence of selection bias is the use of the so-called 0.632+ estimator, $B^{(0.632+)}$. Efron and Tibshirani (1997) proposed this estimator for highly overfit rules like nearest neighbors, where the apparent error rate is zero. It puts relatively more weight on the leave-one-out bootstrap error $B^{(1)}$. Ambroise and McLachlan (2002) subsequently applied it in the context of microarray data, where the prediction rule is an overfit as a consequence of its being formed from a very large number of genes relative to the number of tissues.

The .632+ estimate $B^{(.632+)}$ is defined as

$$B^{(.632+)} = (1 - w)A + wB^{(1)}, \qquad (6.67)$$

where the weight w is given by

$$w = \frac{.632}{1 - .368u},$$

and where

$$u = \frac{B^{(1)} - A}{\gamma - A} \qquad (6.68)$$

is the relative overfitting rate and γ is the no-information error rate that would apply if the distribution of the class-membership label of the jth feature vector did not depend on its feature vector $\boldsymbol{y}_j$. It is estimated by

$$\gamma = \sum_{i=1}^{g} p_i(1 - q_i), \qquad (6.69)$$

where p_i is the proportion of the (original) training data from the ith class and q_i is the proportion of them assigned to the ith class by $r(\boldsymbol{y}; \boldsymbol{t})$. The rate u may have to be truncated to ensure that it does not fall outside the unit interval $[0,1]$. The weight w ranges from 0.632 when $u = 0$ (yielding $B^{(.632)}$) to 1 when $u = 1$ (yielding $B^{(1)}$).

The $B^{(.632+)}$ estimate hence puts more weight on the bootstrap leave-one-out error $B^{(1)}$ in situations where the amount of overfitting as measured by $B^{(1)} - A$ is relatively large. It is thus also applicable in situations where the discriminant rule $r(\boldsymbol{y}; \boldsymbol{t})$ is overfitted due to feature selection. This is to be illustrated in the next chapter.

7

Supervised Classification of Tissue Samples

7.1 INTRODUCTION

In Chapter 4, we have considered the unsupervised classification (cluster analysis) of microarray data. In this chapter, we apply some of the methods of discriminant analysis described in the previous chapter to carry out supervised classification of tissue samples.

As explained by Xiong et al. (2001), there is increasing interest in changing the emphasis of tumor classification from morphologic to molecular. In this context, the problem is to construct a discriminant (prediction) rule $r(y)$ that can accurately predict the class of origin of a tumor tissue with feature vector y, which is unclassified with respect to a known number $g\,(\geq\,2)$ of distinct tissue types, denoted here by $C_1, \ldots, C_g$. Here the feature vector y contains the expression levels on a very large number N of genes (features). In applications concerned with the diagnosis of cancer, one class C_1 may correspond to cancer and the other (C_2) to benign tumors. In applications concerned with patient survival following treatment for cancer, one class (C_1) may correspond to the good-prognosis group and the other C_2 to the poor-prognosis group. Also, there is interest in the identification of "marker" genes that characterize the different tissue classes. This is the feature selection problem as discussed in Section 6.21.

A recent study on breast cancer suggested that a set of 70 genes could better predict the clinical outcome in patients than could standard clinical criteria (van de Vijver et al., 2002). Following these results, the first microarray-based prognostic screening of cancer patients has been set up in the Netherlands and will be used to guide treatment (Schubert, 2003).

In order to train the discriminant (prediction) rule, there are available training data t consisting of $n = M$ tissue samples of known classification. These data are obtained from M microarrays, where the jth microarray experiment gives the expression levels of the $p = N$ genes in the jth tissue sample $\boldsymbol{y}_j$ of the training set. In the notation introduced in Section 4.2, $\boldsymbol{y}_j$ is the gene expression signature vector for the jth tissue. The class of origin of the jth tissue sample $\boldsymbol{y}_j$ is denoted by the g-dimensional vector of zero-one class labels $\boldsymbol{z}_j$ $(j = 1, \ldots, n)$. As in Chapter 3, we write the sample rule formed from the training data t as $r(\boldsymbol{y}; t)$ to show its dependence on t, which can be represented by (6.12). In the sequel, we shall refer to the tissue sample $\boldsymbol{y}_j$ simply as a tissue, since in statistics the collection of the n tissue samples from $\boldsymbol{y}_1, \ldots, \boldsymbol{y}_n$ that belong to a given class would be referred to as a sample.

There are many techniques available for supervised learning, some of which have been described in Chapter 6, in a general context. However, most of these techniques are not likely to work "off the shelf", as expression data present special challenges. The difficulty is that the number of feature variables (genes) is large compared with the number of observations (tissue samples), and they tend to be highly correlated.

The supervised learning problem for the class prediction of a tissue sample is easier than in an unsupervised context, as considered in Chapter 4. But it is still a challenging problem, due to the difficulty of prediction in molecular biology. For example, in the classification of liver cancer, both normal and liver tumors are complex tissues composed of diverse specialized cells. Indeed, whenever we assay gene expression in a tissue, we are observing a mixture of cell types (Churchill, 2003), Further, not all interesting class distinctions are determined by gene expression levels alone. Cellular differences may be regulated by alternative splice variants or by methylation, neither of which would necessarily be evident from expression chip data (Slonim, 2002).

7.2 REDUCING THE DIMENSION OF THE FEATURE SPACE OF GENES

In a standard discriminant analysis, the number of training observations n is usually much larger than the number of feature variables p. But in the present context of microarray data, the number of tissue samples $(n = M)$ is typically between 10 and 100, and the number of genes $(p = N)$ is in the thousands. This presents a number of problems. Firstly, the prediction rule $r(\boldsymbol{y}; t)$ may not be able to be formed using all p available genes. For example, the pooled within-class sample covariance matrix S required to form Fisher's linear discriminant function (6.34) is singular if $n < g + p$. Secondly, even if all the genes can be used as, say, with the nearest-centroid rule or a support vector machine (SVM), the use of all the genes may allow the noise associated with genes of little or no discriminatory power, to inhibit and degrade the performance of the rule $r(\boldsymbol{y}; t)$ in its application to unclassified data. That is, although the apparent error rate A (the proportion of the training tissues misallocated by $r(\boldsymbol{y}; t)$) will decrease as it is formed from more and more genes, its error rate in classifying tissues outside of the training set will eventually increase. That is, the generalization error of $r(\boldsymbol{y}; t)$ will be increased if it is formed from a sufficiently large number of genes. The conditional and unconditional forms of the generalization error have been defined in Section 6.6. Hence, in practice, consideration has to be given to implementing some procedure for reducing the dimension of the feature vector of genes to be used in constructing the rule $r(\boldsymbol{y};\ \boldsymbol{t})$.

7.2.1 Principal Components

A common approach is to carry out a principal component analysis (PCA) and work with the leading components. The PCA can be implemented via a singular value decomposition as outlined in Section 3.19; see, for example, West et al. (2001) and Liu et al. (2003). The disadvantages of this approach are that the PCA does not take into account the class structure of the genes, and genes that show a large variation across the tissues may not be differentially expressed. Also, as the principal components are linear combinations of the original number of genes, biological interpretation of the components is not straightforward.

7.2.2 Partial Least Squares

One method that does take into account the class structure of the tissue samples in reducing the dimension of the feature space is partial least squares, as explained in Section 3.21. However, it still suffers from the same interpretation difficulties as with principal components, as the components are linear combinations of all the genes.

Partial least squares for the supervised classification of tissue samples has been considered by Nguyen and Rocke (2001, 2002a, 2002b), using logistic discrimination and the normal-based quadratic rule. The response vector is taken to consist of the class-indicator variables. Nguyen and Rocke (2002a) demonstrated in their study that if the top genes for discrimination purposes were selected before performing the principal component analysis, then it would give similar results to partial least squares.

There are other methods of dimension reduction; for example, (Antoniadis, et al., 2003) consider the MAVE method of dimension reduction.

7.2.3 Ranking of Genes

One common way of approaching the gene selection problem is to perform a preliminary ranking of genes on the basis of a fast computable criterion and then arbitrarily select a number of the best-ranked genes. Then either a discriminant rule is formed on the basis of these selected genes or further selection is undertaken before constructing the rule.

A commonly used criterion for ranking the individual genes $y_v = (\boldsymbol{y})_v$ $(v = 1, \ldots, p)$ is the ratio of the between-class sum of squares to the within-class sum of squares,

$$F_v = (\boldsymbol{B})_{vv}/(\boldsymbol{S})_{vv}, \tag{7.1}$$

where the sums of squares and products matrices $\boldsymbol{B}$ and $\boldsymbol{S}$ are defined by (3.51) and (3.55), respectively. Under the null hypothesis that the vth gene has the same variance in each class, the statistic F_v has an F-distribution with $g - 1$ and $n - g$ degrees of freedom. The use of (7.1) is equivalent to the likelihood ratio statistic $-2 \log \lambda$ for the test of no differences between the means of the classes under the assumption of the homoscedastic model (6.24) for the class-conditional distributions of the genes. Also, in the case of $g = 2$ classes, it is equivalent to the usual two-sample (pooled) Studentized t-statistic.

Another criterion is to use the apparent error rate A_v of the rule $r(\boldsymbol{y}_v; \boldsymbol{t}_v)$, where the latter is formed using just the training data on the vth gene (Braga-Neto et al., 2004). Alternatively, we may use the (leave-one-out) cross-validated error rate $A_v^{(CV)}$. A further criterion is to rank the genes on the basis of the absolute values of their coefficients in the linear form of $r(\boldsymbol{y}; \boldsymbol{t})$ for an SVM formed with linear kernel. This is to be discussed further in the next section.

There are also rules where the ranking is being done implicitly in their construction; for example, nearest-shrunken centroids to be considered in Section 7.9.

7.2.4 Grouping of Genes

Another way to handle the problem of having to form a discriminant rule from a very large number of feature variables (genes) is to put the genes into groups either by some clustering method or by some supervised selection procedure that makes use of their known class labels. There is now a variety of ways proposed in the literature for the grouping of the genes, which shall be outlined in the final section of this chapter. Having so grouped the genes, a discriminant rule can be formed from the genes (metagenes) selected to represent each group. Recent papers that make use of this approach include Dettling and Bühlmann (2002), Liu et al. (2002). Díaz-Uriarte (2003), Hastie et al. (2001a), and Goh, Song, and Kasabov (2004).

7.3 SVM WITH RECURSIVE FEATURE ELIMINATION (RFE)

In Section 6.13.1, we briefly described the support vector machine (SVM) as a discriminant rule. Support vector machines are becoming increasingly popular classifiers in many areas, including microarrays (Brown et al., 2000; Furey et al. 2000; Guyon et al., 2002; Ramaswamy et al., 2001).

Advantages of an SVM in the present context, where the number of feature variables (genes) p is so large relative to the sample size n, are that it is able to be fitted to all the genes and that its performance appears not to be too affected by using the full set of genes. However, in practice, some form of gene selection would generally be contemplated. Another advantage of the SVM (with a linear kernel) is that gene selection can be undertaken fairly simply using the vector of weights as the criterion.

For an SVM with linear kernel, the rule $r(\boldsymbol{y}; \boldsymbol{t})$ can be written as

$$r(\boldsymbol{y}; \boldsymbol{t}) = \text{sign} \, (\hat{\beta}_0 + \hat{\boldsymbol{\beta}}^T \boldsymbol{y}), \tag{7.2}$$

where $\hat{\beta}_v = (\hat{\boldsymbol{\beta}})_v$ denotes the coefficient of the expression level y_v for gene v.

As shown by Guyon et al. (2002), a good guide to the relative importance of the genes in this SVM is given by the relative size of the absolute values of their fitted coefficients $\hat{\beta}_v$ (that is, the weights). Hence a ranking of the discriminatory power of the genes can be given by ranking the genes from top to bottom on the basis of the absolute values of the weights $\hat{\beta}_v$.

We consider here the selection procedure of Guyon et al. (2002), who used a backward selection procedure, which they termed recursive feature elimination (RFE). This recursive selection procedure has been described in Section 6.13.2. It considers initially all the available genes, which are ranked according to their weights and the bottom-ranked genes discarded. The SVM is then refitted to the remaining genes, which are then reranked according to their new weights. Again, the bottom-ranked genes are discarded, and so on.

In the applications to follow on microarray data, we first discarded enough bottom-ranked genes so that the number retained was the greatest power of 2 (less than the original number of genes). We then proceeded sequentially to discard half the current number of genes on each subsequent step. Initially, the error rate usually falls as genes are deleted, but generally, it will start to rise once a sufficiently large number of genes have been deleted.

The error rate at any stage can be assessed by undertaking an external cross-validation as described in Section 6.22.2. An alternative is to use the 0.632+ estimator of Efron and Tibshirani (1997); see Section 6.22.3. As explained in Section 6.22.3, there may be a considerable selection bias present in the apparent error rate when a reduced number of genes is selected from a very large number. Thus it is not sufficient to use an ordinary (internal) cross-validation, as employed by several authors in the past, including Guyon et al. (2002). With an external cross-validation of the error rate at a given stage of the selection process, at each split of the original training data into training and validation subsets, the selection process has to be implemented from the beginning on the basis of the training subset. That is, with the present selection process of RFE, the SVM has to be fitted to all the genes and then the process continued until the present stage of the selection process has been reached. Then the rule for this newly selected subset of genes is applied to the validation subset.

We note that results in the bioinformatics literature on the relative performances of error-rate estimators in the case of a small number of genes do not apply to the present case, where there may be thousands of genes. For example, Braga-Neto and Dougherty (2004) considered results for only $p=2$ and 5 genes in their recent study.

Before discussing other discriminant rules for the prediction of class membership of tissues, we shall report the results of Ambroise and McLachlan (2002) to illustrate this selection bias. These results will also serve to demonstrate the performance of the SVM with RFE. The size of the selection bias has also been investigated by Nguyen and Rocke (2002b) for multiclass discrimination via the logistic and the normal quadratic discriminant rules, using partial least squares.

7.4 SELECTION BIAS: SVM WITH RFE

Ambroise and McLachlan (2002) investigated the magnitude of the selection bias and its correction for an SVM (with linear kernel) and Fisher's linear discriminant function in their application to two cancer data sets. They were the colon data of Alon et al. (1999) and the leukemia data of Golub et al. (1999), as considered in an unsupervised context in Chapter 4. The former set consists of 40 tumor tissues and 22 normal tissues, while the latter set consists of 72 tissues from 47 patients with acute lymphoblastic leukemia (ALL) and 25 patients with acute myeloid leukemia (AML).

To illustrate the size of the selection bias for the colon data set, Ambroise and McLachlan (2002) split it into a training set and a test set, each of size 31, by sampling without replacement from the 40 tumor and 22 normal tissues separately, so that each set contained 20 tumor and 11 normal tissues. The training set is used to carry out gene selection and to form the apparent error rate A, the (leave-one-out) cross-validated error rate $A^{(CV)}$ using just internal validation, and the external ten-fold cross-validated rate $A^{(CV10E)}$ for a selected subset of genes. An unbiased error-rate estimate is given by the test error (T), equal to the proportion of tissues in the test set misallocated by the rule. They calculated these quantities for 50 such splits of the colon data into training and test sets. For the leukemia data set, they divided the set of 72 tissues into a training set of 38 tissues (25 ALL, 13 AML) and a test set of 34 tissues (22 ALL, 12 AML) for each of the 50 splits. The average values of the error-rate estimates are plotted in Figure 7.1, while the corresponding averages for the leukemia data are plotted in Figure 7.2. The error bars on T refer to the 95% confidence limits. The 0.632+ bootstrap error estimate, $B^{(0.632+)}$, was formed using $K = 30$ bootstrap replications for each of the 50 splits of a full training set. In Figures 7.1 and 7.2 and subsequent figures, the apparent error A, the (leave-one-out) cross-validated error $A^{(CV)}$, the external ten-fold cross-validated error $A^{(CV10E)}$, the 0.632+ bootstrap error estimate $B^{(0.632+)}$, and the test error are denoted by A, CV, CV10E, B.632+, and T, respectively.

It can be seen from Figure 7.1 that the true prediction error rate as estimated by T is not negligible, being above 15% for all selected subsets. The lowest value of 17.5% occurs for a subset of 2^6 genes. Similarly, for the leukemia data set, it can be seen from Figure 7.2 that the selection bias cannot be ignored when estimating the true prediction error, although it is smaller (about 5%) for this second set as the two leukemia classes ALL and AML are more easily separated.

Concerning the estimation of the prediction error by external ten-fold cross-validation and the bootstrap, it can be seen that $A^{(CV10E)}$ has little bias for both data sets. For the colon data set, the bootstrap estimate $B^{(0.632+)}$ is more biased. However, $B^{(0.632+)}$ was found to have a slightly smaller root mean squared error than $A^{(CV10E)}$ for the selected subsets of both data sets. Efron and Tibshirani (1997) comment that future research might succeed in producing a better compromise between the unbiasedness of cross-validation and the reduced variance of the leave-one-out bootstrap. It would be of interest to consider this for the present problem where there is also feature-selection bias to be corrected for.

Ambroise and McLachlan (2002) observed comparable behavior of the SVM rule formed using all the available tissue samples (62 and 72 tissue samples for the colon and leukemia data sets, respectively). As to be expected for training samples of twice or nearly twice the size, the (estimated) bias was smaller: between 10 and 15% for the colon data set and between 2 and 3% for the leukemia data set.

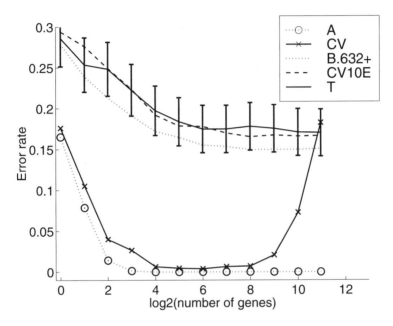

Fig. 7.1 Error rates of the SVM rule with RFE procedure averaged over 50 random splits of the 62 colon tissue samples into training and test subsets of 31 samples each.

It can be seen from Figures 7.1 and 7.2 that the estimated prediction rate according to $B^{(0.632+)}$ and $A^{(CV10E)}$ remains essentially constant as genes are deleted in the SVM, until around about 64 or so genes when these estimates start to rise sharply. The internal cross-validated error $A^{(CV)}$, which is uncorrected for selection bias, also starts to rise then. Hence feature selection provides essentially little improvement in the performance of the SVM rule for the two considered data sets. But it does show that the number of genes can be greatly reduced without increasing the prediction error.

It is of interest to note that the plot in Guyon et al. (2002) for their leave-one-out cross-validated error is very similar to the plot of $A^{(CV10E)}$ in Figure 7.1. However, for this data set, Guyon et al. transformed the normalized (logged) data further by a squashing function to "diminish the importance of outliers." This could have an effect on the selection bias. For example, for this data set there is some doubt (Moler, 2000) as to the validity of the labels of some of the tissues, in particular, tumor tissues on patients 30, 33, and 36, and normal tissues on patients 8, 34, and 36, as labeled in Alon et al. (1999). When Ambroise and McLachlan (2002) deleted these latter

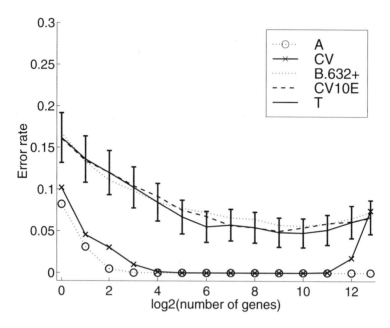

Fig. 7.2   Error rates of the SVM rule with RFE procedure averaged over 50 random splits of the 72 leukemia tissue samples into training and test subsets of 38 and 34 samples, respectively.

tissue samples, the selection bias of the rule was estimated to be almost zero. But in a sense, this is to be expected, for if all tissue samples that are difficult to classify are deleted, then the rule should have a prediction error that is close to zero regardless of the selected subset of genes.

7.5   SELECTION BIAS: FISHER'S RULE WITH FORWARD SELECTION

Ambroise and McLachlan (2002) also considered the selection bias incurred with a sequential forward selection procedure for the rule based on Fisher's linear discriminant function. For the selection of genes for Fisher's rule, they first reduced the set of available genes to 400 for each data set in order to reduce the computation time. This was undertaken by selecting the top 400 genes as ranked in terms of increasing order of the average of the maximum (estimated) posterior probabilities of class membership. Ambroise and McLachlan (2002) noted that this initial selection incurs some (small) bias, which they ignored in their illustrative examples. The forward selection procedure was applied with the decision to add a feature (gene) based on the leave-one-out cross-validated error. For this, Ambroise and McLachlan (2002) used all the available tissue samples (62 for the colon data set and 72 for the leukemia) to train the Fisher rule, and so there was no test set available in each case. The cross-validation and bootstrap estimates of the prediction error of Fisher's rule formed via forward selection of the genes are plotted in Figures 7.3 and 7.4 for the colon and leukemia data sets, respectively.

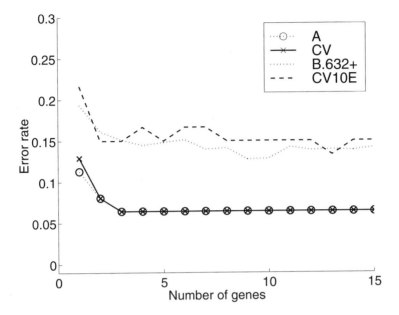

Fig. 7.3 Error rates of Fisher's rule with stepwise forward selection procedure using all of the colon data (62 samples).

Considering the colon data set, it can be seen in Figure 7.3 that the leave-one-out cross-validated error $A^{(CV)}$ (with the selection bias still present) is optimized (6.5%) for only three genes. However, the external cross-validated error $A^{(CV10E)}$, which has the selection bias removed, is approximately equal to 15% for more than four genes. The fact that the estimates $B^{(0.632+)}$ and $A^{(CV10E)}$ are around 15% for 10 or so genes for the colon data set would appear reasonable given that this set has six tissues whose class of origin is in some doubt.

Similarly, it can be seen in Figure 7.4 that the selection bias incurred with forward selection of genes for Fisher's rule on the leukemia data set is not negligible. As a consequence, although the leave-one-out error $A^{(CV)}$ error is zero for only three selected genes, the external ten-fold cross-validated error $A^{(CV10E)}$ and the bootstrap $B^{(0.632+)}$ error estimate are about 5%.

The results above for $A^{(CV)}$ for Fisher's rule formed via forward selection of the genes are in agreement with the results of Xiong et al. (2001). Their error rate corresponding to our $A^{(CV)}$, was reported to be equal to 6.5 and 0% for the colon and leukemia data sets, respectively; see Table 2 in Xiong et al. (2001).

Comparing the performance of forward selection with Fisher's rule to backward elimination with SVM, Ambroise and McLachlan (2002) found that the latter procedure leads to slightly better results, with a 2% or 3% improved error rate for both data sets. When trying forward selection with SVM, they found results very similar

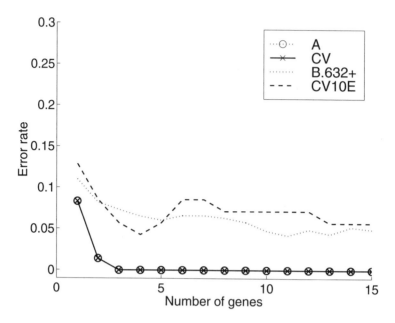

Fig. 7.4 Error rates of Fisher's rule with stepwise forward selection procedure using all of the leukemia data (72 samples).

to those obtained using Fisher's rule with a comparable number of genes. Hence it seems that the selection method and the number of selected genes are more important than the classification method for constructing a reliable prediction rule.

Consideration has been given recently in the bioinformatics literature (Braga-Neto et al., 2004) to whether the apparent error rate is competitive with cross-validation in ranking the genes or sets of genes. It can be seen from Figures 7.1 to 7.4 that the use of the apparent error rate would not be sufficiently reliable for deciding when the gene selection process should be terminated with the SVM and Fisher's linear discriminant rule.

7.6 SELECTION BIAS: NONINFORMATIVE DATA

To further illustrate this selection bias, Ambroise and McLachlan (2002) generated a no-information training set by randomly permuting the class labels of the colon tissue samples. For each of 20 no-information sets so obtained, an SVM rule was formed by selecting genes by the RFE method and the apparent error A and the leave-one-out cross-validated error $A^{(CV)}$ were calculated. The average values of these two error rates and the no-information error γ (6.69) over the 20 sets are plotted in Figure 7.5, with the average value of the $A^{(CV10E)}$ and $B^{(.632+)}$ error estimates that correct for the selection bias. It can be seen that, although the feature vectors have been generated independently of the class labels, we can form an SVM rule that has not only an average zero apparent error A but also an average $A^{(CV)}$ error close to zero for a subset of 128 genes and around 20% for only eight genes in the selected subset. It is reassuring to see that the error estimates $A^{(CV10E)}$ and $B^{(0.632+)}$, which correct for the selection bias, are between 0.40 and 0.45, consistent with the fact that we are forming a prediction rule on the basis of a no-information training set.

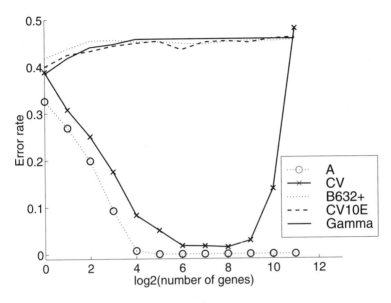

Fig. 7.5 Error rates of the SVM rule averaged over 20 noninformative samples generated by random permutations of the class labels of the colon tumor tissues.

As noted in Section 6.21, some rules such as tree classifiers have an embedded selection procedure whereby feature selection is implicitly undertaken in the formation of the rule. Thus an ordinary cross-validation of such a rule should correct adequately for the selection bias provided that the original rule is formed by a method that allows for the selection bias. This will be true with the tree classifier CART (Breiman et al., 1984). To illustrate this, we generated another no-information training set by randomly permuting the class labels of the colon tissue samples. For each of these 20 no-information sets, a tree (CART) was grown and the average values of the apparent error A and the ordinary ten-fold cross-validated error rate $A^{(CV10)}$ were calculated for the subtrees on the validation trials. They are plotted versus the average size of the corresponding trees in Figure 7.6, along with the no-information error rate γ. It can be seen that the curve for $A^{(CV10)}$ is close to the curve representing γ.

Thus concerning the CART rule formed from thousands of genes, selection bias is not an issue as with ordinary cross-validation of the SVM or Fisher's linear discriminant function. An explanation of this is that the tree classifier CART is a rule formed via an embedded feature selection scheme that almost eliminates the selection bias in its growing and pruning of the tree.

Nearest-shrunken centroids is an example of another rule with an embedded feature selection scheme which, for the soft thresholding version, automatically corrects for the selection bias during the cross-validation trials conducted to choose the threshold.

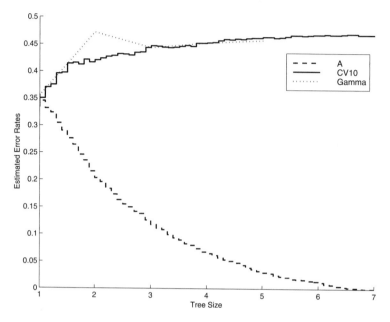

Fig. 7.6 Error rates of the tree rule (CART) averaged over 20 noninformative samples generated by random permutations of the class labels of the colon tumor tissues.

7.7 DISCUSSION OF SELECTION BIAS

For two data sets commonly analyzed in the microarray literature, Ambroise and McLachlan (2002) have demonstrated that it is important to correct for the selection bias in estimating the prediction error for Fisher's rule or SVM formed using a subset of genes selected from a very large set of available genes.

From these examples presented, it can be seen that it is important to recognize that a correction for the selection bias be made in estimating the prediction error of a rule formed using genes selected from a very large set of available genes. It is also important to note that if a test set is used to estimate the prediction error, then there will be a selection bias if this test set was used also in the gene-selection process. Thus the test set must play no role in the feature selection process for an unbiased estimate to be obtained.

Given that there are usually only a limited number of tissue samples available for the training of the prediction rule, it is not practical for a subset of tissue samples to be so put aside for testing purposes. However, we can correct for the selection bias either by cross-validation or by the bootstrap, as implemented above in the examples. Concerning the former approach, an internal cross-validation does not suffice. That is, an external cross-validation must be performed whereby at each stage of the validation process with the deletion of a subset of the observations for testing, the rule must be trained on the retained subset of observations by performing the same feature selection procedure used to train the rule in the first instance on the full training set.

7.8 SELECTION OF MARKER GENES WITH SVM

7.8.1 Description of van de Vijver Breast Cancer Data

In Section 7.4, we have seen that the subset of variables selected during the external cross-validation of the error rate of an SVM formed from a reduced set of genes can vary considerably for each split of the original (full) training set into training and validation subsets. Concerning the importance of a gene with the use of the support vector machines (SVMs), we can note the number of times a gene is chosen in the selected subset on each split of the training data during the external cross-validation.

To demonstrate this approach, we consider the application of a SVM with RFE to the breast cancer data set in van de Vijver et al. (2002). In Section 4.11, we considered in an unsupervised context some breast cancer data from van 't Veer et al. (2002). Following on from this, van de Vijver et al. (2002) studied a larger series of breast cancer patients. The original study of van 't Veer et al. (2002) included 78 sporadic (non-BRCA carrier) breast tumors, which had been specifically selected on the basis of outcome; 34 patients developed distant metastases within 5 years, and 44 patients remained disease-free after 5 years. Using supervised classification methods, van 't Veer et al. (2002) were able to identify a set of 70 genes with expression profiles associated with the risk of early metastasis. This selection was carried out on the basis of the correlation between the gene expression profile and the class label, which is equivalent to using the (pooled) two-sample t-statistic; see Section 6.10.2 on this point. They called these 70 genes the prognostic marker genes. In the larger study of van de Vijver et al. (2002), 61 of the 78 original patients were included, as well as an additional 234 patients not chosen on the basis of outcome. van de Vijver et al. (2002) ran oligonucleotide arrays, using $N = 25,000$ genes (as in van 't Veer et al., 2002) for the total of $M = 295$ tissues. They used the expressions of the 70 marker genes to classify the larger set of tumors into either good- (no metastases within 5 years) or poor- (early metastasis) prognosis categories defined in the study of van 't Veer et al. (2002).

For each of the tumors, they calculated the correlation coefficient of its expression signature vector comprised of the 70 marker genes with the class label (defining the two prognosis categories); that is, they effectively used a nearest-centroid rule (see Section 6.12 on this point). For the 234 new patients, those with a correlation coefficient above a threshold of 0.4 (this value gave a 10% rate of false negatives in the original study of van 't Veer et al., 2002) were assigned into the good-prognosis category and all other patients were assigned to the poor-prognosis category. For the 61 patients from the original study, a cutoff of 0.55 was used, which gave a 10% rate of false negatives in the cross-validation performed in van 't Veer et al. (2002). This resulted in $n_1 = 115$ tumors classified into the good-prognosis class (C_1) and $n_2 = 180$ tumors classified into the poor-prognosis class (C_2).

7.8.2 Application of SVM with RFE

The SVM with RFE was applied to this van de Vijver data, with half of the genes being eliminated at each stage of the RFE procedure. The external ten-fold cross-validated error rate $A^{(CV10E)}$ is plotted in Figure 7.7 as the number of genes is reduced from 70 to 64, and then successively halved; see also Table 7.1. It can be seen that $A^{(CV10E)}$ starts to increase as the number of genes is reduced, but only slightly down to 32 genes. So we decided to adopt as our prediction rule the SVM based on 32 genes.

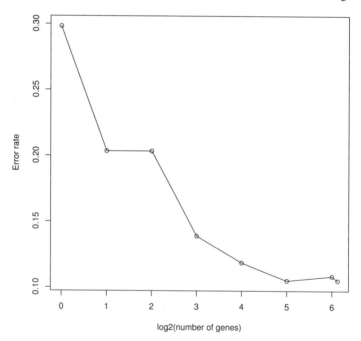

Fig. 7.7 External ten-fold cross-validated error rate of the 295 breast cancer tissue samples on 70 genes in the van de Vijver data.

To provide some guide to the relative importance of the genes in the formation of this rule, we noted how many times a gene was selected in the ten subset of size 32 selected for each of the ten splits of the training data. The frequencies are reported in Table 7.2 for the 44 genes that were selected for at least one of the ten splits.

Table 7.1 Results of SVM with RFE Applied to 295 Breast Cancer Tissue Samples on 70 Genes in the van de Vijver Data

Number of Genes	Overall Error Rate
1	0.298
2	0.203
4	0.203
8	0.138
16	0.118
32	0.105
64	0.108
70	0.105

Table 7.2 Selection Frequencies of Genes in External Ten-Fold Cross-Validation of SVM with RFE Applied to 295 Breast Cancer Tissue Samples on 70 Genes in the van de Vijver Data

Gene No.	Frequency of Selection on Ten-Fold Validation
66 65 64 60 57 55 52 51 49 46 42 40 39 38 33 22 21 15 9 8 7 4 3 2	10
63 59 23	9
68 13	8
1	6
61 25 20 10	4
43	3
54 27 14	2
53 50 48 44 34 12	1

In Table 7.2, the gene number refers to its numbering in the data set as supplied at the supplementary website (http://www.rii.com/publications/2002/nejm.htm) in van de Vijver et al. (2002). This gene number is the rank of the gene when the 70 marker genes are ordered on the basis of the magnitude of their correlation with the class label for the good-prognosis class (that is, the F-ratio), using the 78 tissue samples in van 't Veer et al. (2002).

To contrast these frequencies with the rankings of the genes considered individually on the basis of the criterion (7.1), equivalent to the F-ratio (or the pooled two-sample t-statistic since $g = 2$), we list in Table 7.3 the genes in descending order of rank according to this criterion. Only those genes (59 in total) for which the criterion exceeded the 95th percentile of the F-distribution with 1 and 293 degrees of freedom have been listed. For each gene, the criterion (7.1) was calculated using the $n = 295$ tissue samples. It can be seen that several of the top genes in terms of the F-statistic in Table 7.3 had high frequencies of selection in the ten-fold cross-validation process, as summarized in Table 7.2.

For the highly selected genes, we searched for functional annotation in order to link possible biological mechanisms with their apparent role in early tumor metastasis. Out of the 24 genes selected 10 times in Table 7.3, we found 12 with gene products of known function. These included cell nucleus proteins: CENPA, PRC1, RAMP, NUSAP1, and MELK (many of which seem to be involved in chromatin structure), and also the protein, ORC6L, essential for the initiation of replication, suggesting a role for these genes in tumor cell division. In addition, we identified cell signaling proteins; including IGFBP5 (which appears twice), a known potent inhibitor of growth of breast cancer cells in vitro and in vivo, as well as others involved in cell cycle control and tumorigenesis (FGF18 and TGFB3), and also CEGP1, a secreted protein expressed in vascular endothelium, suggesting a possible role for this gene in angiogenesis as part of tumor metastasis. Finally, we identify the protein AP2B1, involved in the cellular processes of endocytosis and Golgi assembly as it forms part of the clathrin coat assembly.

7.9 NEAREST-SHRUNKEN CENTROIDS

7.9.1 Definition

For high-dimensional data, Tibshirani et al. (2002a, 2003) have considered a modification to the nearest-centroid rule, as defined in Section 6.12. It is directly applicable in the present context of the supervised classification of tissue samples. With this approach, termed nearest-shrunken centroids by Tibshirani et al. (2002a, 2003), the usual estimates of the class means $\overline{\boldsymbol{y}}_i$ are shrunk toward the overall mean $\overline{\boldsymbol{y}}$ of the data, where

$$\overline{\boldsymbol{y}}_i = \sum_{j=1}^{n} z_{ij} \boldsymbol{y}_j / n_i$$

and

$$\overline{y} = \sum_{j=1}^{n} y_j / n.$$

Table 7.3 Rank on Basis of F-Ratio of Frequently Selected Genes in External Ten-Fold Cross-Validation of SVM with RFE Applied to 295 Breast Cancer Tissue Samples on 70 Genes in the van de Vijver Data

Rank	Gene No.	F-Ratio	Rank	Gene No.	F-Ratio
1	64	108.49	31	12	20.87
2	15	93.37	32	50	20.65
3	49	92.10	33	48	19.45
4	39	75.49	34	43	18.89
5	9	74.10	35	57	18.53
6	40	69.61	36	2	18.26
7	51	63.58	37	22	18.24
8	21	62.19	38	70	17.98
9	4	60.67	39	46	16.70
10	33	60.18	40	53	16.60
11	30	58.69	41	65	16.04
12	8	57.81	42	26	15.64
13	7	57.33	43	35	12.72
14	3	52.25	44	36	11.18
15	66	52.00	45	20	10.75
16	54	45.52	46	37	10.11
17	68	45.19	47	59	7.32
18	34	44.35	48	16	7.14
19	14	42.28	49	44	6.09
20	13	35.57	50	24	6.07
21	52	35.45	51	6	6.05
22	23	32.00	52	45	5.73
23	25	31.02	53	56	5.47
24	38	30.06	54	5	5.22
25	42	29.74	55	11	5.20
26	55	25.89	56	47	5.05
27	1	25.84	57	61	4.79
28	60	22.95	58	31	4.37
29	10	22.37	59	28	4.06
30	32	22.27			

From Section 6.12, the nearest-centroid rule is given by

$$r(\boldsymbol{y};\, \boldsymbol{t}) = \arg\min_i \sum_{v=1}^{p} (y_v - \bar{y}_{iv})^2 / s_{iv}^2 - \log(\hat{\pi}_i), \tag{7.3}$$

where y_v is the vth element of the feature vector $\boldsymbol{y}$ and $\bar{y}_{iv} = (\bar{\boldsymbol{y}})_v$. In the definition (7.3) of the nearest-shrunken centroid rule, we replace the sample mean $\bar{y}_{iv}$ of the vth gene by its shrunken estimate

$$\bar{y}_{iv}^* = \bar{y}_{iv} + m_i s_i d_{iv}^* \quad (i = 1, \ldots, g;\ v = 1, \ldots, p), \tag{7.4}$$

where

$$d_{iv}^* = \text{sign}(d_{iv})(|d_{iv}| - k)_+ \tag{7.5}$$

and

$$d_{iv} = \frac{\bar{y}_{iv} - \bar{y}_v}{m_i s_v}, \tag{7.6}$$

and where $m_i = (n_i^{-1} - n^{-1})^{\frac{1}{2}}$. In (7.6), s_v is the pooled within-class sample standard deviation of gene v; that is, s_v^2 is the ith diagonal of the pooled (bias-corrected) sample covariance matrix $\boldsymbol{S}$ defined by (3.55). Also, in (7.5), the subscript plus means *positive part*; that is, $a_+ = a$, if $a > 0$, and zero otherwise. The denominator $m_i s_v$ in (7.6) is the standard error of the numerator $\bar{y}_{iv} - \bar{y}_v$.

This shrinkage proposed by Tibshirani et al. (2002a, 2003) is called *soft thresholding*. The absolute value of each d_{iv} is reduced by an amount k and is set to zero if the result is less than zero. As explained by Tibshirani et al. (2003), since many of the $\bar{y}_{iv}$ will be noisy and close to the overall mean $\bar{y}_i$, soft thresholding usually produces "better" (more reliable) estimates of the true class means μ_{iv}.

An attractive property of this shrunken approach is that many of the genes are eliminated as far as their contribution to the sample rule if the threshold k is chosen sufficiently large. For if k causes d_{iv} to shrink to zero, then $\bar{y}_{iv}^*$ is the same as the overall mean $\bar{y}_v$, and so gene v does not contribute to the class decision based on (7.4).

Tibshirani et al. (2003) use ten-fold cross-validation to choose the value of the threshold k. It is external in the sense that for a given value of k the selection of genes is carried out separately on each of the ten cross-validation trials. They also consider adaptive choice of thresholds, soft versus hard thresholding, and ways to capture heterogeneity in the class training data.

Tibshirani et al. (2003) propose forming estimates of the posterior probabilities $\tau_i(\boldsymbol{y})$ of class membership of a new observation $\boldsymbol{y}$ by taking the estimate of the ith class-conditional density of $\boldsymbol{y}$ to be given by

$$\tau_i(\boldsymbol{y}) = \frac{\hat{\pi}_i \exp\{-\sum_{v=1}^{p} \frac{1}{2}(y_v - \bar{y}_{iv}^*)^2 / s_{iv}^2\}}{\sum_{h=1}^{g} \hat{\pi}_h \exp\{-\sum_{v=1}^{p} \frac{1}{2}(y_v - \bar{y}_{hv}^*)^2 / s_{hv}^2\}}.$$

7.10 COMPARISON OF NEAREST-SHRUNKEN CENTROIDS WITH SVM

To illustrate the application of nearest-shrunken centroids to a real data set, we now apply it to the colon data of Alon et al. (1999), using the program as supplied in Tibshirani et al. (2003). This data set was analyzed in an unsupervised context in Section 4.10. Also, we apply the (linear) SVM with RFE to this data set, as it will be of interest to contrast the performance of nearest-shrunken centroids with this method. Note that the results given earlier in Section 7.4 for the SVM applied to this data set did not use the full set for training, as test sets were put aside in order to estimate the true error rates. So for comparative purposes, we train the SVM with RFE on the full data set. We also apply nearest-shrunken centroids and SVM to the van de Vijver cancer data considered in Section 7.8.

7.10.1 Alon Data

The results for nearest-shrunken centroids applied to the colon data are given in in Table 7.4, where the overall (estimated) error rate versus the number of retained genes is reported. In this table, the "No. of Genes" refers to the number of active genes in the formation of the shrunken rule; that is, the number of genes with at least one d_{iv}^* nonzero for each gene.

The overall and class-specific error rates (with their standard errors) are plotted in Figure 7.8. A "cross" in the plot of the overall error rate denotes its minimum value over the number of active genes. The corresponding results for the SVM with RFE with its overall and class-specific error rates estimated by external ten-fold cross-validation are displayed in Figure 7.9 and Table 7.5.

On comparing the results in Tables 7.4 and 7.5, it can be seen that the SVM with the genes selected by RFE has a similar error rate over the genes to nearest shrunken-centroids.

As noted in Guyon et al. (2002) on the use of RFE with SVM, removing one variable at a time is more accurate than removing chunks of variables at a time. So we reapplied the SVM with the RFE implemented as before down to 128 genes, but from then on eliminating only one gene at a time. It led to a slightly better error rate of 10% for 10 to 18 genes on the Alon data.

7.10.2 van de Vijver Data

To compare further SVM with nearest-shrunken centroids, we consider now the breast cancer data set from van de Vijver et al. (2002). The compilation of this set has been described in Section 7.8. It consists of $n = 195$ breast cancer tumor samples of $p = 70$ genes with $n_1 = 115$ tumors in the good-prognosis class (C_1) and $n_2 = 180$ tumors in the poor-prognosis class (C_2).

We trained the SVM with RFE on these $n = 295$ tumors using the $p = 70$ available genes. The external ten-fold cross-validated error rate $A^{(CV10E)}$ is plotted in Figure 7.10 as the number of genes is reduced from 70 to 64, and then successively

Table 7.4 Overall Error Rate for Nearest-Shrunken Centroids Applied to the Alon Data

Iteration	Threshold	No. of Genes	Error
1	0.00	2,000	0.15
2	0.12	1,784	0.15
3	0.23	1,601	0.15
4	0.34	1,432	0.15
5	0.46	1,257	0.15
6	0.57	1,095	0.15
7	0.69	928	0.15
8	0.80	786	0.15
9	0.92	671	0.15
10	1.03	556	0.15
11	1.15	475	0.15
12	1.26	397	0.15
13	1.38	312	0.13
14	1.49	254	0.13
15	1.61	201	0.13
16	1.72	160	0.13
17	1.84	123	0.13
18	1.95	88	0.13
19	2.07	68	0.13
20	2.18	49	0.15
21	2.29	35	0.13
22	2.41	23	0.16
23	2.52	16	0.27
24	2.64	8	0.32
25	2.75	6	0.34
26	2.87	4	0.35
27	2.98	4	0.35
28	3.10	3	0.35
29	3.21	2	0.35
30	3.33	0	0.35

halved. It can be seen that $A^{(CV10E)}$ starts to increase as the number of genes is reduced, but only slightly, down to 32 genes, after which it starts to increase markedly.

The results of the nearest-shrunken centroid rule applied to this data set are displayed in Figure 7.11. It can be seen that SVM and nearest-shrunken centroids have comparable (estimated) error rates. For economy of space, we have not listed the class-specific cross-validated errors here, but they are similar for both rules too.

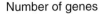

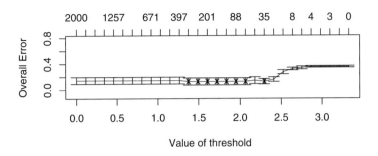

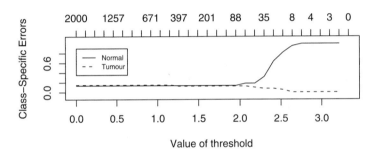

Fig. 7.8 Plot of overall and class-specific error rates for nearest-shrunken centroids applied to Alon data.

Table 7.5 Ten-Fold External Cross-Validated Error Rates of SVM with RFE Applied to the Alon Data

No. of Genes	CV10E	SE	AE	Normal	Tumor
1	0.55	0.06	0.52	0.82	0.40
2	0.18	0.06	0.11	0.27	0.12
4	0.17	0.05	0.08	0.27	0.10
8	0.20	0.06	0.05	0.32	0.12
16	0.22	0.07	0.05	0.27	0.17
32	0.17	0.04	0.02	0.23	0.12
64	0.12	0.04	0.02	0.14	0.10
128	0.13	0.05	0.02	0.18	0.10
256	0.13	0.05	0.02	0.18	0.10
512	0.13	0.05	0.03	0.18	0.10
1,024	0.13	0.05	0.03	0.18	0.10
2,000	0.16	0.05	0.03	0.32	0.07

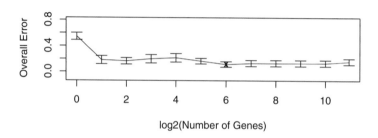

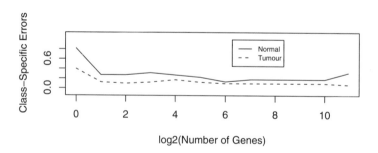

Fig. 7.9 Overall and class-specific error rates for SVM with RFE applied to Alon data.

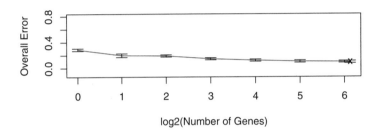

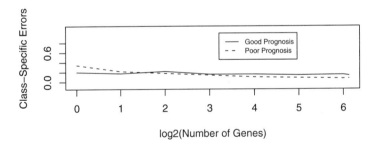

Fig. 7.10 Plot of overall and class-specific error rates for SVM with RFE applied to the 295 tissue samples of 70 genes in the van de Vijver data.

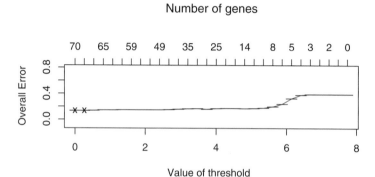

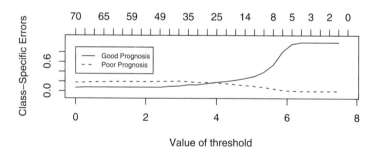

Fig. 7.11 Plot of overall and class-specific error rates for nearest-shrunken centroids applied to the 295 tissue samples of 70 genes in the van de Vijver data.

Finally, we trained the model-based rule, using a mixture of g factor analyzers with $q = 4$ factors to model the class-conditional densities for the two classes. We took $g = 1$ for the good-prognosis class and $g = 2$ for the poor-prognosis class. In the latter case, we took the component-factor models to have the same diagonal matrices but unrestricted loading matrices. The (leave-one-out) cross-validated error rate was equal to 12.8%. Thus these three rules (SVM with RFE, nearest-shrunken centroids, and mixtures of factor analyzers) have comparable error rates on this set.

As explained in Section 7.8, given the way the training data have been compiled, they are not really the observed outcomes of random samples drawn from the two classes. Indeed, since 234 of the tumors have been assigned to the two classes on the basis of a rule formed from 61 of the tumors in the training set, the error rates could be expected to be biased downward. Moreover, the rule based on the 61 tumors was actually formed using the top 70 genes, which was based on 98 tissues that contained these 61 tumors. This also would serve to introduce a further bias. Ideally, we should try to estimate the error rate by performing an external cross-validation using all 5,422 genes, as to be discussed in the next section. This could not be done here, as we had access only to the top 70 genes.

We now proceed to look at the bias that can result in using the error rate of a discriminant rule formed from a optimal subset of genes.

7.11 SELECTION BIAS WORKING WITH THE TOP 70 GENES

7.11.1 Bias in Error Rates

We return now to the breast cancer data from van 't Veer et al. (2002) that were analyzed in an unsupervised context in Section 4.11. As explained in Section 7.8, they selected a set of 70 top genes (marker genes) on the basis of the 78 tissues in their study.

To demonstrate the bias in estimating the error rate of a rule formed from a subset set of genes without using the full set during the cross-validation, Zhu, Ambroise, and McLachlan (2004) applied an SVM with RFE to the 78 tissue samples on the top 70 genes in the van 't Veer et al. (2002) data. In Section 4.11, we obtained 4,869 genes from the prefiltering process applied to the 24,881 genes in the 98 tissue samples, consisting of 78 sporadic and and 22 breast cancer tumors. Zhu et al. (2004) applied the same filtering process to the 24,881 genes in the tissue samples for just the 78 sporadic breast cancer tumors, which resulted in 5,422 genes being retained.

At each stage of the feature elimination process with the SVM, Zhu et al. (2004) estimated the overall error rate using ten-fold cross-validation. They performed the latter, using both internal and external cross-validation. For internal cross-validation, the top 70 genes were fixed during the validation process, and so it ignores the selection bias in working with the top 70 genes from the set of 5,422 genes.

In the external cross-validation, this bias is corrected for by going back to the full set of 5,422 genes and selecting the top 70 genes on the training subset at each stage

Table 7.6 Number of Genes and Error Rates with and without Corrections for Selection Bias

Number of Genes	Error Rate for Top 70 Genes (without Correction for Selection Bias as Top 70)	Error Rate for Top 70 Genes (with Correction for Selection Bias as Top 70)	Error Rate for 5,422 Genes (with Correction for Selection Bias)
1	0.50	0.53	0.56
2	0.32	0.41	0.44
4	0.26	0.40	0.41
8	0.27	0.32	0.43
16	0.28	0.31	0.35
32	0.22	0.35	0.34
64	0.20	0.34	0.35
70	0.19	0.33	—
128	—	—	0.39
256	—	—	0.33
512	—	—	0.34
1,024	—	—	0.33
2,048	—	—	0.37
4,096	—	—	0.40
5,422	—	—	0.44

of cross-validation. Then the SVM with RFE is applied to this selected set of top 70 genes, which may have little in common with the original set of top 70 genes.

The results for the internal ten-fold cross-validated overall error rate $A^{(CV10)}$ and the corresponding external rate $A^{(CV10E)}$ are listed in Table 7.6 and plotted in Figure 7.12.

It can be seen from Table 7.6 and Figure 7.12 that the selection bias in ignoring the fact that the SVM is being applied to the top 70 genes from a total of 5,422 is approximately 12%.

We have also listed in Table 7.6 the external cross-validated error for the SVM with RFE, starting with the full set of 5,422 genes. It can be seen that it is similar to that of the external cross-validated error rate of the rule starting with the top 70 genes.

7.11.2 Bias in Comparative Studies of Error Rates

This example also serves to make the point that care must be exercised in comparing the error rates of two discriminant rules formed from the same tissue samples of different sets of genes. For example, one rule r_1 may be formed from a training set of $n = M$ tissue samples of $p = N$ genes, while another rule r_2 might be formed using a subset of these N genes, say, the top 100 genes. If a fair comparison is to be made between the error rates of these two rules, then the error rate of the second

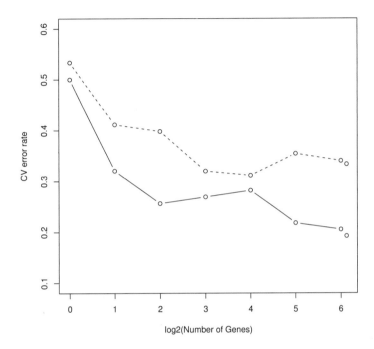

Fig. 7.12 The solid line is the ten-fold cross-validated error rate of the SVM with RFE applied to the top 70 genes in the 295 tissue samples in the van de Vijver data, calculated without correction for selection bias due to using top 70 genes. The dashed line is the corresponding rate with correction for this bias.

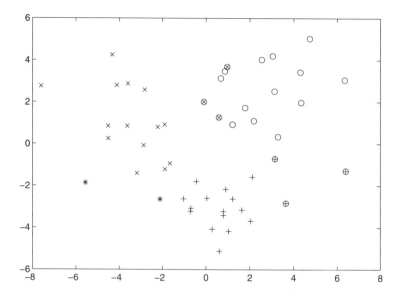

Fig. 7.13 Plot of the first two components for 60 observations, with 20 from each of three classes (class 1 ×, class 2 +, class 3 ∘) calculated on the top 100 genes according to the F-ratio.

rule r_2 should not be estimated by just working with the top 100 genes during the cross-validation. Rather, one should start initially with the full set of N genes and select the top 100 genes on each stage of the training of r_2 in the cross-validation trials.

7.11.3 Bias in Plots

This selection bias also is present in plots of the expression signatures formed from the top genes taken from a much larger set of genes. To illustrate this bias, we generated three random samples, each consisting of $n_i = 20$ observations of dimension $p = 2,000$ drawn from a normal distribution with mean zero and covariance matrix equal to the identity matrix. We then ranked the feature variables (corresponding to the genes) in terms of the F-ratio for a one-way analysis of variance for $g = 3$ classes. We then performed a principal component analysis of the combined sample of 60 observations, using the top 100 features. In Figure 7.13, we have plotted the first two principal components so obtained for the 60 observations. This plot by itself would suggest that the three classes are widely separated, whereas in fact they represent observations with identical distributions. Maindonald (2004) has suggested that to circumvent this selection bias a cross-validation be performed, using Procrustes analysis (see Sibson, 1978) to allow for the varying reference frames obtained at each stage of the process. Note that if we fitted a g-normal mixture model to the PCs in Figure 7.13 and then used a resampling approach to test the null hypothesis that $g = 1$, it would be retained, thus providing no evidence for any group structure.

7.12 DISCRIMINANT RULES VIA INITIAL GROUPING OF GENES

As mentioned in Section 7.2, one way to handle the formation of a discriminant rule for the supervised classification on the basis of a very large number of genes is to first put the genes into groups. This can be effected by either clustering the genes or selecting groups of genes in a supervised manner based on the known classification of the tissues. Then a reduced feature vector can be formed consisting of representatives from each of the groups (metagenes).

We consider next some methods of supervised classification that first cluster the genes.

7.12.1 Supervised Version of EMMIX-GENE

We can implement a model-based approach to supervised classification that makes use of the first steps in the EMMIX-GENE procedure of McLachlan et al. (2002) in a supervised manner. The select-genes step is run to rank the genes with respect to the F-ratio based on the specified classification of the tissues.

Then the cluster-genes step is run to cluster all genes ranked above a specified threshold; for example, a specified percentile of the F-distribution with $g - 1$ and $n - g$ degrees of freedom, where g is the number of classes. The genes are clustered by fitting in equal proportions a mixture of normal components with common spherical component-covariance matrices, which is equivalent to a "soft" version of k-means. The intent is to have highly correlated genes within the same group. A discriminant rule can then be formed by modeling the class-conditional densities of the metagenes by normal mixtures or mixtures of factor analyzers, as discussed in Section 6.9.6. In the current version of EMMIX-GENE, the metagenes are taken to be the sample means of the gene clusters.

7.12.2 Bayesian Tree Classification

A similar approach to supervised classification has been adopted by Huang et al. (2003). They apply k-means clustering after an initial screen to remove genes that show little variation. Each cluster of genes is represented by a metagene taken to be the first principal component of the cluster of genes. This approach varies from the one above in that it clusters a much larger number of genes; that is, it does not undertake a preliminary step (beyond the initial screen) corresponding to the select-genes step in EMMIX-GENE. It thus produces a much larger number of metagenes (for example 495 in the analysis in Huang et al., 2003). The discriminant rule is formed from the metagenes using a Bayesian classification tree analysis.

7.12.3 Tree Harvesting

Another approach that first clusters the genes is the procedure proposed by Hastie et al. (2001a), which they call "tree harvesting". They carry out the clustering of the genes using hierarchical clustering (average linkage). They note that this strategy has two advantages. Firstly, hierarchical clustering has become a standard descriptive tool for expression data. Secondly, by using clusters as input, they bias the inputs toward correlated sets of genes. As they explain, this reduces the rate of overfitting of the model.

They consider all $2p - 1$ clusters in the hierarchy produced by their hierarchical clustering, and they build a model with the aim of finding additive and interaction structure among the clusters of genes in their relation to an outcome measure. The latter can take many forms besides class labels, including censored survival times. Their model is fitted using a forward stepwise strategy, involving sums and products of the average gene expression of chosen clusters (metagenes). They prefer this model as it provides interpretable, biologically plausible models. However, their procedure, and probably any procedure with similar aims, requires a large number of samples to uncover such structure successfully.

Their initial clustering of the genes is biased toward large clusters, because they are more likely to be biologically meaningful. As they explain, large clusters can result from a pathway of genes involved in a biological process, or a heterogeneous experimental sample containing different cell types. In addition, the finding of a large cluster correlated with the outcome is less likely to be spurious than that of a small cluster, because there are many more smaller clusters than larger clusters.

7.12.4 Block PCA

Liu et al. (2002) propose dividing the genes into blocks according to correlation using cluster analysis so as to obtain blocks of high correlation among the genes in a block. These clusters of genes are formed without using information from a dependent variable (such as the class labels). Within each block they perform a principal component analysis (PCA) to select "important" genes. Then they use a second PCA with the selected genes from the previous step. Finally, they use these components for classification. Thus, Liu et al. (2002) explicitly try to obtain blocks with high internal correlation.

7.12.5 Grouping of Genes via Supervised Procedures

We now discuss procedures that have been proposed recently for the formation of discriminant rules from groups of genes that have been formed using the known classification of the tissues. They involve forming groups of genes to ensure that highly correlated genes are not duplicated and that the clusters are not very large. These procedures contrast with the approaches described earlier.

With the method developed by Dettling and Bühlmann (2002), the aim is to identify sparse linear combinations of genes whose average expression level is uniformly low for one response class and uniformly high for the other class(es). A cluster of genes is grown incrementally by adding one gene after another. Subsequent cleaning steps are carried out to remove spurious genes that were incorrectly added to the cluster at earlier stages. As with the CAST procedure of Ben-Dor et al. (1999), the growth and removal steps are repeated until the cluster stabilizes. Then a new cluster is started. Each cluster consists of genes for which the metagene is a linear combination of the genes (with weights of 1 or minus 1). The genes are chosen to provide a good discriminant rule as assessed in terms of a specified objective function that makes use of the known classification of the tissues in the training data. For this, they use the Wilcoxon statistic (1945). The use of possibly a minus one in forming the metagene from a cluster of genes is to allow for negatively correlated genes. The number of clusters has to be specified by the user. In summary, the aim is to identify sparse linear combinations of genes whose average expression level is uniformly low for one response class and uniformly high for the other class(es). In their approach, Dettling and Bühlmann (2002) used a nearest-neighbor rule and a normal-based linear rule with diagonal component-covariance matrices. They have since investigated performing feature selection with the nonparametric scoring method of Park et al. (2001) in conjunction with the use of boosting in forming discriminant rules. They adopted the LogitBoost procedure of Friedman et al. (1998).

With the method proposed by Díaz-Uriarte (2003), the intent is to form tight groups of genes so far as the coexpression of the genes. Initially, a seed gene with good discriminatory capacity is selected. Then a search is conducted for a group of genes that are highly correlated with the seed gene, shows tight coexpression, and has good discriminatory power. The latter is assessed by using internal cross–validation. This set of genes is reduced to a vector of metagenes (signature component in the terminology of Díaz-Uriarte, 2003), using principal components analysis. The process is repeated until the addition of a metagene vector does not reduce the assessed error rate by a specified threshold. It is claimed that the suggested method can recover signatures present in the data and has predictive performance comparable to state-of-the-art methods.

More recently, Goh, Song, and Kasabov (2004) have proposed using an evolving connectionist system (ECOS) with the genes put into groups using a supervised measure in conjunction with the degree of their correlation with other genes.

8

Linking Microarray Data with Survival Analysis

8.1 INTRODUCTION

In clinical medicine, determining the stage of disease is crucial in the management of cancer patients. Stage is defined using a combination of clinical parameters (tumor size, lymph node involvement and the presence of metastases). Together with other criteria (such as age, the histologic type and pathological grade of cancer, or hormone-receptor status) stage provides a measure for future recurrence and overall survival, and is used to guide treatment. However, patients with the same stage of a particular cancer can have very different treatment responses and also clinical outcomes. This clinical heterogeneity is probably due to the genetic complexity of individual tumors, with changes in the expression of many genes that drive tumor growth, invasion and metastasis. Microarrays allow the simultaneous measurement of complex multigene expression patterns in cancer. Currently there is much interest in directly linking gene expression data with patient outcomes, and to determine whether microarrays can be used to provide better prognostic tools for cancer.

A common approach in this type of survival analysis is to first cluster the tissue samples, usually by some hierarchical method, and then to compare the survival curves for each cluster so obtained using a nonparametric Kaplan–Meier analysis (Alizadeh et al., 2000). On approaches that attempt to include gene expressions in models in survival analysis, there is the tree harvesting approach of Hastie et al. (2001a), as discussed in Section 7.12.3, used as a method of supervised classification when the response vector defines the class memberships of the tissues. It also applies when the response variable is the survival time. But as noted there, their procedure requires a very large number of samples. West et al. (2001) used principal components and then used probit regression with Bayesian regularization. Nguyen and Rocke (2002c) and Park et al. (2002) used partial least squares (PLS) with the proportional hazards model of Cox (1972). In their study, the continuous response variable was taken to represent the survival times, some of which may be censored. Hence the PLS components constructed may contain some bias, depending on the amount of censoring. Also, as noted by Nguyen and Rocke (2002c), interpretation of the fitted parameters of the PLS gene-component profiles in terms of the original gene expression profiles does not appear to be feasible directly. This is due to the fact that the PLS gene components are linear combinations of all the predictor genes. Park et al. (2002) showed that the issues can be circumvented by reformulating the problem as a standard Poisson regression problem. More recently, Li and Gui (2004) have developed a partial Cox regression method for constructing mutually uncorrelated components based on gene expression data for predicting the survival of future patients.

In their recent study of breast cancer patients, van de Vijver et al. (2002) showed that the gene expression profiles were a better predictor of clinical outcome than any of the currently used criteria. Following on from these promising results, the CAMDA'03 challenge (`http://www.camda.duke.edu/camda03`) set the task of linking microarray gene expressions from four lung cancer data sets with patient outcomes. In this chapter, we consider the approach of Ben-Tovim Jones et al. (2004a,b) to this problem. They initially used model-based clustering to classify the tumor tissues, and then examined the association between the tissue clusters and patient survival (or recurrence) times. They went on to show that the clustering provides significant prognostic information beyond that based on stage of disease at presentation.

8.2 FOUR LUNG CANCER DATA SETS

The CAMDA'03 challenge included a total of four lung cancer data sets. Each data set was made up of different numbers and types of lung tumors; see Table 8.1. These four data sets had been analyzed previously by Wigle et al. (2002), Garber et al. (2001), Bhattacharjee et al. (2001), and Beer et al. (2002). They were referred to there as the Ontario, Stanford, Harvard, and Michigan data sets, respectively, and we shall continue with this labeling here for convenience. The first two used cDNA arrays, while the latter two used different versions of Affymetrix oligonucleotide arrays. These data may be downloaded from the CAMDA'03 contest website as above.

Ben-Tovim Jones et al. (2004a) used the available genes for the two cDNA arrays (2,880 for the Ontario set and 918 for the Stanford set). For the Affymetrix arrays, they started with the 3,312 genes in the Harvard set and the 4,965 genes in the Michigan set, but these had outlier values. Therefore, in the Harvard set, they imposed a floor of 1 and a ceiling of 3,000, leaving 3,190 genes. For the Michigan data set, they imposed a floor of minus 1 and a ceiling of 26,000, leaving 4,728. In all four data sets, missing values were replaced by values obtained via the method in Dudoit et al. (2002a). In each data set, the data were logged except for the Michigan data set, in which the generalized log transformation (2.2) was used with $c = 1$. Each

Table 8.1 Tumor Types in the Four Data Sets

Tumor Type	Number of Samples			
	Ontario	Stanford	Harvard	Michigan
Adenocarcinoma	19	41	127	86
Squamous Cell	14	16	21	0
Large Cell	4	5	0	0
Carcinoid	1	0	20	0
Small Cell	0	5	6	0
Normal	0	5	17	10
Other	1	1	12	0
TOTAL	39	73	203	96

data set was subsequently column and then row standardized, although no column standardization was carried out in the Harvard data set, as it had been column-adjusted in its originally available form.

By comparing Unigene identifiers, Ben-Tovim Jones et al. (2004a) found that the cDNA arrays had at least 105 genes in common, while the Affymetrix arrays had at least 1,257 genes in common in the input data sets. The Harvard data set has by far the most tumors. However, in the Harvard, Stanford, and Michigan data sets, clinical data were available only for the adenocarcinoma (AC) patients. In the Ontario study the outcome was provided for both AC and non-AC patients, in the form of tumor recurrence versus nonrecurrence.

8.3 STATISTICAL ANALYSIS OF TWO DATA SETS

In the analyses of Ben-Tovim Jones et al. (2004a), the EMMIX-GENE procedure was initially applied to cluster all the tumor types in an unsupervised context. They were able to retrieve the histological classification for at least the non-AC tumors in all data sets, except for the Ontario set. For the other three data sets, they focused on just the AC tumors (for which clinical data were available).

We now consider an analysis of the two cDNA lung cancer sets from the CAMDA'03 contest (the Ontario and the Stanford sets) to demonstrate how a model-based cluster analysis in conjunction with a survival analysis can be used to assess the prognostic information in the microarray data. We follow the same approach as in Ben-Tovim Jones et al. (2004a,b). We shall first perform a model-based clustering using the EMMIX-GENE algorithm (McLachlan et al., 2002). Then we shall perform a survival analysis to determine the association between clusters formed and patient survival (or recurrence) times. In the survival analysis, only tissues from patients with known survival times (or recurrence times) are considered.

8.4 ONTARIO DATA SET

8.4.1 Cluster Analysis

We firstly consider the Ontario data set, which consists of $n = 39$ tumor samples on $p = 2,880$ genes. With the Ontario data set, the outcome is defined as the time between surgery and the recurrence. The tumors labeled 1 to 24 here are those for which there has been a recurrence of the cancer, while those labeled 25-39 have had no recurrence before the end of the study; that is, their times to recurrence are censored. After applying the screening step of EMMIX-GENE to this data set, 766 genes remained. The top genes included immunoglobulin lambda light chain IGL, hypothetical protein FLJ10404, HLA-B associated transcript 2 D6S51E, Friend leukemia virus integration 1 FLI1, and ATP-binding cassette ABCD3.

The latter were then clustered into $N_o = 20$ groups on the second step of this algorithm. The means of these 20 groups (the metagenes) can be used to represent the initial set of 2,880 genes.

The final step of EMMIX-GENE can be used to cluster tissues in terms of the metagenes. Given the very small number of tumors ($n = 39$) available here relative to the number of genes or, indeed, metagenes, some constraints need to be imposed on the component-covariance matrices in fitting a normal mixture model to cluster these tumors. We considered fitting to all 20 metagenes (a) mixtures of normals with equal component-covariance matrices; (b) mixtures of normals with (unrestricted) diagonal component-covariance matrices; and (c) mixtures of factor analyzers with equal component-covariance matrices for $q = 6$ factors.

All three models led to the same clustering, namely

$$C = C_1 \cup C_2, \tag{8.1}$$

where

$$C_1 = \{15, 30 - 32, 34, 35, 37, 39\}.$$

and

$$C_2 = \{1 - 14, 16 - 29, 33, 36, 38\}$$

Thus cluster C_1 corresponds to the good-prognosis group with seven patients who are recurrence-free plus one patient who had experienced relapse of the tumor. This patient, however, was still alive at the end of the follow-up period. The second cluster C_2 corresponds to the poor-prognosis group, as it contains 23 of the 24 patients with recurrence of their tumor plus eight patients having censored recurrence times.

That is, in one cluster C_1, it put one recurrence tumor (tumor 15) with seven of the tumors with censored recurrence times (30–32, 34, 36, 38). In the other cluster (C_2), it put 23 of the 24 tumors with recurrence along with eight tumors (25–29, 33, 36, and 38) with censored recurrence times. Thus cluster C_1 (apart from tumor 15) would appear to correspond to the recurrence-free group, whereas cluster C_2 corresponds to the recurrence group.

To provide further support that the first cluster C_1 can be viewed as corresponding to the recurrence-free group, we follow Ben-Tovim Jones et al. (2004b) and consider

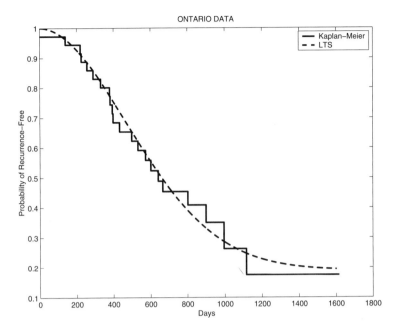

Fig. 8.1 Fitted LTS model versus the Kaplan–Meier estimate. From Ben-Tovim Jones et al. (2004b).

the long-term survival model

$$S(t) = \pi_1 + \pi_2 S_2(t), \tag{8.2}$$

where t is the time to recurrence, $S_2(t)$ is the conditional survival function for time to recurrence given that recurrence will occur, and $\pi_2 = 1 - \pi_1$ is the probability of a recurrence. Under (8.2), a proportion π_1 of the patients will not have a recurrence; that is, their recurrence time is at infinity. The survival function $S_2(t)$ is taken to have the Weibull form,

$$S_2(t) = \alpha \lambda t^{\alpha-1} \exp(-\lambda t^{\alpha}). \tag{8.3}$$

The exact recurrence times of two of the patients in C_2 were unknown and so they were excluded from this and the subsequent survival analyses. It meant that there were 37 patients with 15 censored (recurrence-free and still alive at the end of the follow-up period).

In Figure 8.1, we have plotted the fitted Weibull-based long-term survival model $\hat{S}(t)$ along with the Kaplan–Meier estimate. It can be seen that there is excellent agreement between the nonparametric estimate as given by the Kaplan–Meier estimate and the parametric estimate $\hat{S}(t)$. In particular, it can be seen from the asymptote of these curves in Figure 8.1 that the probability π_1 of a patient being recurrence-free is approximately 0.2. Thus on average, one would expect to have approximately eight recurrence-free patients in a set of 39. Here the cluster C_1, which is conjectured as

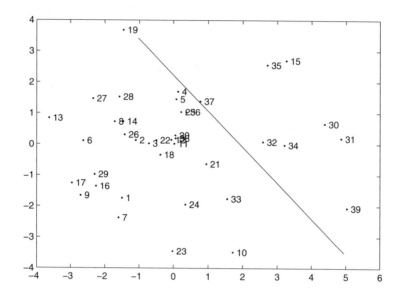

Fig. 8.2 PCA of tissues based on 20 metagenes. From Ben-Tovim Jones et al. (2004b).

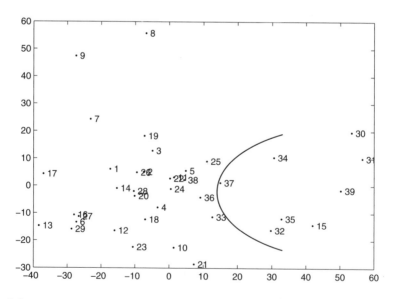

Fig. 8.3 PCA of tissues based on all genes (via SVD). From Ben-Tovim Jones et al. (2004b).

Table 8.2 Nonparametric Survival Analysis for the Ontario Data Set

Cluster	No. of Patients (Censored)	Mean Time to Recurrence ($\pm$ SE)
C_1	8 (7)	$1{,}388 \pm 155.7$
C_2	29 (8)	665 ± 85.9

Source: From Ben-Tovim Jones et al. (2004b).

corresponding to the recurrence-free group, has eight members. Interestingly, five of the censored patients from the poor-prognosis cluster C_2, were also put together in a cluster corresponding to early recurrence in the hierarchical clustering of Wigle (2002).

This long-term survival model (8.2) can be used also to estimate the posterior probability that a patient with a censored recurrence time will be recurrence-free. Unfortunately, unless the censored time is very long, these estimated posterior probabilities are equal, being around 0.5. Patient 25 (P81 AC), who has a censored time of 1,161 days, has a high posterior probability of being recurrence-free, so her membership of cluster C_2 would appear to be atypical.

To further investigate the validity of their clusters, Ben-Tovim Jones et al. (2004b) also considered clustering the tumors by just looking at the first two principal components (PCs) of the tumors obtained by a singular-value decomposition, as explained in Section 3.19. The scatter plots for the first two PCs based on (a) the 20 metagenes and (b) all the genes are shown in Figures 8.2 and 8.3, respectively. In each of the plots, the allocation boundary corresponding to the clusters obtained previously in (8.1) is imposed. In each case, it can be seen that this boundary represents a reasonable partition of the data into two clusters in the space of the first two PCs.

8.4.2 Survival Analysis

Among the 37 patients with survival data available, the clustering described in the preceding section showed that 29 were classified as poor prognosis and eight were classified as good prognosis. The Kaplan–Meier estimate is used to provide an estimate of the overall probability of being recurrence-free following surgery. Given that there is only one recurrence in the good-prognosis cluster C_1, it should have a significantly better Kaplan–Meier estimate than that of the poor-prognosis cluster C_2, and this is confirmed in Table 8.2. These two Kaplan–Meier estimates are plotted in Figure 8.4. The Kaplan–Meier curves were compared using the log-rank test.

We also fitted the proportional hazards model of Cox (1972), using covariates to represent the clinical data and a zero-one indicator variable to represent membership or lack of membership in cluster C_2. The fit for the final form of this model is given in Table 8.3. The significance of estimated hazard ratios were tested using the Wald test. All calculations in the survival analysis were performed with the S-Plus statistical package. It can be seen that membership of cluster C_2 (the poor-prognosis

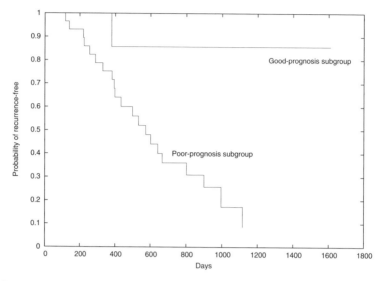

Fig. 8.4 Kaplan–Meier curves of recurrence-free probability for the two clusters (Ontario data set). From Ben-Tovim Jones et al. (2004b).

Table 8.3 Multivariate Cox Hazards Analysis of the Risk of Recurrence (Ontario Data Set)

Variable	Hazard Ratio (95% CI)	P-Value
Poor-prognosis cluster (vs. good-prognosis)	6.8 (0.9–51.8)	0.06
Stage 2 or 3 (vs. Stage 1)	1.1 (0.4–2.7)	0.88

Source: From Ben-Tovim Jones et al. (2004b).

cluster) was the only factor approaching significance ($P = 0.06$) in its effect on the event of being recurrence-free.

8.4.3 Discriminant Analysis

We have seen in the last section, that at the 6% level, there is additional prognostic information in the microarray gene expressions beyond that provided by the clinical criteria. To further investigate the predictive power of the microarrays, we fitted a support vector machine (SVM) with recursive feature elimination (RFE) to the two clusters. We estimated the error (generalization) rate of this discriminant rule, using external ten-fold cross-validation as defined in Section 6.22.2. That is, we treated the two clusters C_1 and C_2 as if they represented random training samples from good- and poor-prognosis groups. This approach is therefore somewhat self-serving in that we are treating the two clusters as if they are representative samples of the two underlying groups of interest. The error rate of this discriminant rule is thus more optimistic than that which would be achieved for properly constituted training

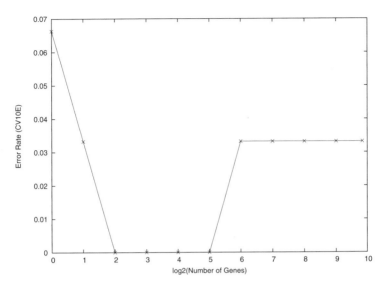

Fig. 8.5 External ten-fold cross-validated error rate of SVM with RFE applied to $g = 2$ clusters of tumors in Ontario data set. From Ben-Tovim Jones et al. (2004b).

samples. But it may be useful in that if the error rate of this clustering-based rule is not small, then the error rate of a rule based on random trainings samples will be higher. Also, it should be noted that the discriminant rule formed here from these clusters is a nonparametric one and starts with all the available genes, which lessens the incestuous nature of this exercise. Another use for this prediction rule is that it provides a method for revealing potential marker genes, as it can be noted how many times a gene was selected in the final form of the SVM on each of the ten subsamples during the ten-fold cross-validation. In this use, it complements the "cluster-genes" step of EMMIX-GENE, which can be used to find marker genes.

In Figure 8.5, we have plotted the external ten-fold cross-validated error rate $A^{(CV10E)}$ of the SVM with RFE applied to the Ontario data set. It can be seen that $A^{(CV10E)}$ is zero up to eight genes after which the error rate starts to rise as further genes are eliminated from the SVM.

8.5 STANFORD DATA SET

We now consider the Stanford lung cancer data set, which is the second of the two cDNA array sets. It includes measurements for 67 lung tumor samples, of which the majority were classified histologically as adenocarcinoma (AC) tumors (see Table 8.1). Eleven of the tumors were sampled twice (or in one case three times) from the same patient (such as a primary tumor/lymph node pair), and the 41 AC tumor samples came from a total of 34 patients. Using the EMMIX-GENE procedure, Ben-

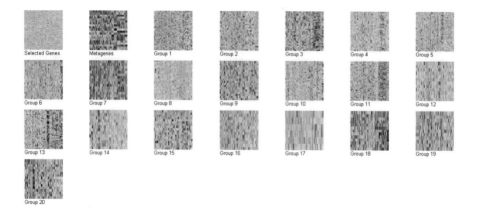

Fig. 8.6 Heat maps for all of the genes, the 20 metagenes, and the 20 gene groups on the 73 tissues in the Stanford data. Tissues are ordered by their histological classification: Adenocarcinoma (1-41), Fetal Lung (42), Large cell (43-47), Normal (48-52), Squamous cell (53-68), Small cell (69-73). From Ben-Tovim Jones et al. (2004b). See the insert for a color representation.

Tovim Jones et al. (2004a) reduced the number of genes from 918 to 453, which were then clustered into 20 groups; see the heat maps in Figure 8.6.

On the basis of the 20 metagenes so obtained, they clustered the 73 tumors into five clusters, corresponding to adenocarcinoma (AC), large cell lung cancer (LCLC), normals, squamous cell (SCC) and small cell lung cancer (SCLC). Six of the adeno-carcinoma samples were put into non-AC clusters (tumors 5, 6 and 26 as LCLC; 7 and 29 as SCLC; 40 as SCC). These samples were subsequently excluded, leaving 35 AC tumors to be analyzed. In the study of Garber et al. (2001), two non-AC tumors (43 LCLC and 68 SCC) were clustered as AC tumors and so were included with the latter for analysis. These were were not included in the study of Ben-Tovim Jones et al. (2004a), as they were put in the cluster corresponding to LCLC.

8.5.1 Cluster Analysis of AC Tumors

As in Ben-Tovim Jones et al. (2004b), we reapplied the select-genes and cluster-genes options of the EMMIX-GENE procedure to the 35 AC tumors, reducing the number of genes from 918 to 219, which were then clustered into 15 groups. The top genes among the retained 219 genes included CD36 antigen, signal transducer and activator of transcription 4, aldo-keto reductase family 1 member C1, and kynureninase. The heat maps for these 15 groups are displayed in Figure 8.7. A plot of the expression profiles of the metagenes from the first four groups of genes is given in Figure 8.8. Here we order the 35 AC tumors according to groups obtained by Garber et al. (2001), using hierarchical clustering. When EMMIX-GENE was used to cluster the AC tumors into $g = 3$ clusters based on the metagenes, we find our clusters essentially agree with those of Garber et al. (2001). Our cluster C_1 corresponds to

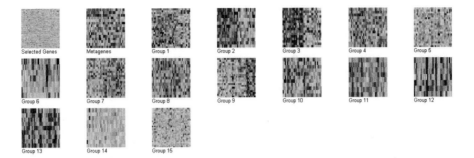

Fig. 8.7 Heat maps for all of the genes, the 15 metagenes, and the 15 gene groups on the 35 tissues in the Stanford data. Tissues are ordered according to the AC groups of Garber et al. (2001): AC group 1 (1-19), AC group 2 (20-26), AC group 3 (27-35). From Ben-Tovim Jones et al. (2004b). See the insert for a color representation.

their good-prognosis group (AC group 1: tumors 1 to 19), cluster C_2 to the long-term survivors (AC group 2: tumors 20 to 26), and C_3 to the poor-prognosis group (AC group 3: tumors 27 to 35). Only tissue 15 is put into a different cluster (C_2) by us. Indeed, cluster C_2 had only one patient that died. Therefore for the survival analysis, the tissues were reclustered into $g = 2$ clusters. This gave a single good-prognosis cluster C_1 which corresponded to AC groups 1 and 2, and a poor-prognosis cluster C_2 which corresponded to AC group 3. There was one exception; now patient 29 appeared in cluster C_1. The expression profiles for patients 15 and 29 are shown in pink in Figure 8.8.

8.5.2 Survival Analysis

With the Stanford data set, there were clinical data available for 26 AC tumor samples, with four tumor pairs derived from the same patients. In the analysis, each tumor pair was treated as one observation. This gave a total of 22 observations, with 10 censored.

The Kaplan–Meier survival curves (Figure 8.9) show a significant difference in the probability of overall survival between the good-prognosis and poor-prognosis clusters ($P < 0.001$). The mean ($\pm$SE) survival times are 37.5 ± 5.0 and 5.2 ± 2.3 months, respectively. Table 8.4 shows the results of the multivariate Cox regression analysis. It is evident that the two prognosis clusters are different after the adjustment for the clinical factors ($P = 0.002$). The estimated hazard ratio for overall survival in the poor-prognosis cluster as compared with the good-prognosis cluster is 15.5 (95% confidence interval: 2.7 to 90.2). In Table 8.5, we present the univariate Cox proportional hazards analysis for each of the top 15 metagenes.

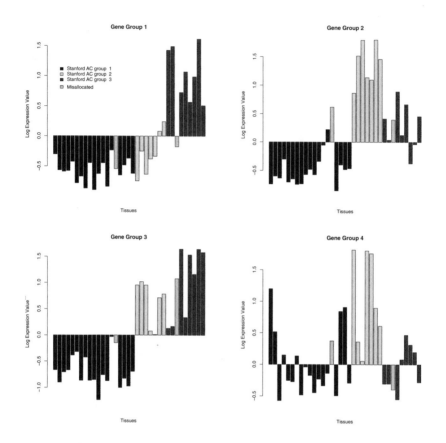

Fig. 8.8 Expression profiles for top metagenes (Stanford 35 AC tissues). From Ben-Tovim Jones et al. (2004b). See the insert for a color representation.

Table 8.4 Multivariate Cox Proportional Hazards Analysis of the Risk of Death (Stanford Data Set)

Variable	Hazard Ratio (95%CI)	*P*-Value
Poor-prognosis cluster (vs. good-prognosis)	15.5 (2.7-90.2)	0.002
Tumor grade 3 (vs. grade 1 or 2)	1.8 (0.4 - 9.2)	0.47
Tumor size 1 (vs. sizes 2 to 4)	0.5 (0.03-7.4)	0.59
Presence of tumor in lymph nodes	4.4 (0.4-48.6)	0.23
Presence of metastases	4.3 (0.8-24.6)	0.10

Source: From Ben-Tovim Jones et al. (2004b).

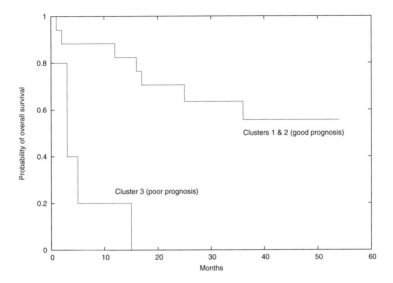

Fig. 8.9 Kaplan–Meier survival curves for the two prognosis clusters (Stanford data set). From Ben-Tovim Jones et al. (2004b).

Table 8.5 Univariate Cox Proportional Hazards Analysis of the Metagenes (Stanford Data Set)

Metagene	Coefficient (SE)	P-value
1	1.37 (0.44)	0.002
2	−0.24 (0.31)	0.44
3	0.14 (0.34)	0.68
4	−1.01 (0.56)	0.07
5	0.66 (0.65)	0.31
6	−0.63 (0.50)	0.20
7	−0.68 (0.57)	0.24
8	0.75 (0.46)	0.10
9	−1.13 (0.50)	0.02
10	0.73 (0.39)	0.06
11	0.35 (0.50)	0.48
12	−0.55 (0.41)	0.18
13	−0.61 (0.48)	0.20
14	0.22 (0.36)	0.53
15	1.70 (0.92)	0.06

Source: From Ben-Tovim Jones et al. (2004b).

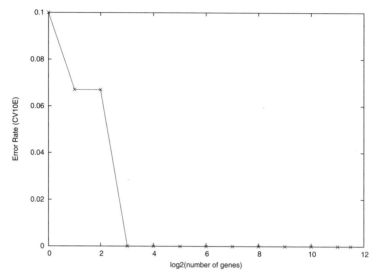

Fig. 8.10 External ten-fold cross-validated error rate of SVM with RFE applied to $g = 2$ clusters of AC tumors in Stanford data set. From Ben-Tovim Jones et al. (2004b).

8.5.3 Discriminant Analysis

As with the Ontario data set, we considered the Stanford data set in a discriminant analysis context by treating the three clusters as obtained in Section 8.5.1 as if they were randomly obtained training samples. Again, we formed an SVM with RFE to progressively reduce the number of genes. The external ten-fold cross-validated error rate $A^{(CV10E)}$ is plotted in Figure 8.10 versus the retained number of genes. It can be seen that $A^{(CV10E)}$ is zero provided that the number of genes is between 4 and 32.

8.6 DISCUSSION

In the work of Ben-Tovim Jones et al. (2004a,b) considered partially above, a model-based clustering approach has been applied to classify tumor tissues, using their gene signatures, into clusters corresponding to tumors of a given subtype. The clusters were then able to be identified with clinical outcomes such as recurrence versus nonrecurrence of tumor and death versus long-term survival. It was able to be established that the gene expression data provide prognostic information, beyond clinical indicators such as stage.

A limiting factor in the analyses was the small numbers of tumors available. In addition, the clinical data were available for only subsets of the tumors (for example, only for one tumor type, adenocarcinoma, in the Stanford data set). Further, the high proportion of censored observations limited the comparison of survival rates.

References

Accelrys Inc. (2004). InsightII Modeling Environment, Release 2000.1, San Diego: Accelrys Inc.

Adams, M.D., Kelley, J.M, Gocayne, J.D.., Dubnick, M., Polymeropoulos, M.H., Xiao, H., Merril, C.R., Wu, A., Olde, B., Moreno, R.F., Kerlavage, A.R., McCombie, W.R., and Venter, J.C. (1991). Complementary DNA sequencing: expressed sequence tags and the human genome project. *Science* **252**, 1651–1656.

Adams, R. and Bischof, L. (1994). Seeded region growing. *IEEE Transactions on Pattern Analysis and Machine Intelligence* **16**, 641–647.

Ahmed, A.A., Vias, M., Iyer, N.G., Caldas, C., and Brenton, J.D. (2004). Microarray segmentation methods significantly influence data precision. *Nucleic Acids Research* **32**, e50.

Aitkin, M., Anderson, D., and Hinde, J. (1981). Statistical modelling of data on teaching styles (with discussion). *Journal of the Royal Statistical Society A* **144**, 414–461.

Aitman, T.J. (2001). DNA microarrays in medical practice. *British Medical Journal* **323**, 611–615.

Alizadeh, A., Eisen, M., Davis, R.E., Ma, C., Lossos, I.S., Rosenwal, A., Boldrick, J.C., Sabet, H., Tran, T., Yu, X., Powell, J.I., Yang, L., Marti, G., Moore, T., Hudson, J., Lu, L., Lewis, D.B., Tibshirani, R., Sherlock, G., Chan, W.C., Greiner, T.C., Weisenburger, D.D., Armitage, J.O., Warnke, R., Levy, R., Wilson, W., Grever, M.R., Byrd, J.C., Botstein, D., Brown, P.O., and Staudt, L.M. (2000). Distinct types of diffuse large B-cell lymphoma identified by gene expression profiling. *Nature* **403**, 503–511.

Allison, D.B., Gadbury, G.L., Heo, M., Fernández, J.R., Lee, C.-K., Prolla, T.A., and Weindruch, R. (2002). A mixture model approach for the analysis of microarray gene expression data. *Computational Statistics and Data Analysis* **39**, 1–20.

Alon, U., Barkai, N., Notterman, D.A., Gish, K., Ybarra, S., Mack, D., and Levine, A.J. (1999). Broad patterns of gene expression revealed by clustering analysis of tumor and normal colon tissues probed by oligonucleotide arrays. *Proceedings of the National Academy of Sciences USA* **96**, 6745–6750.

Alter, O., Brown, P.O., and Botstein, D. (2000). Singular value decomposition for genome-wide expression data processing and modeling. *Proceedings of the National Academy of Sciences USA* **97**, 10101–10106.

Ambroise, C. and McLachlan, G.J. (2002). Selection bias in gene extraction on the basis of microarray gene-expression data. *Proceedings of the National Academy of Sciences USA* **99**, 6562–6566.

Anderson, T.W. (1951). Classification by multivariate analysis. *Psychometrika* **16**, 31–50.

Antoniadis, A., Lambert-Lacroix, S., and Leblanc, F. (2003). Effective dimension reduction methods for tumor classification using gene expression data. *Bioinformatics* **19**, 563–570.

Åstrand, M. (2001). Normalizing oligonucleotide arrays. `http://www.math.chalmers.se/~magnusaa/maffy.html`.

Audic, S. and Claverie, J.-M. (1997). The significance of digital gene expression profiles. *Genome Research* **7**, 986–995.

Axon Instruments, Inc. (1999). *GenePix 4000A User's Guide*.

Baggerly, K.A., Coombes, K.R., Hess, K.R., and Stivers, D.N. (2002). Studentizing microarray data. In *Computational and Statistical Approaches to Genomics*. Norwell, MA: Kluwer, pp. 53–64.

Baggerly, K.A., Coombes, K.R., Hess, K.R., Stivers, D.N., Abruzzo, L.V., and Zhang, W. (2001). Identifying differentially expressed genes in cDNA microarray experiments. *Journal of Computational Biology* **8**, 639–659.

Bailey, T.L. and Elkan, C. (1995). Unsupervised learning of multiple motifs in biopolymers using expectation maximization. *Machine Learning* **21**, 51–80.

Baldi, P., Brunak, S., Chauvin, Y., Andersen, C.A.F., and Nielsen, H. (2000). Assessing the accuracy of prediction algorithms for classification: an overview. *Bioinformatics* **16**, 412–424.

Baldi, P. and Long, A.D. (2001). A Bayesian framework for the analysis of microarray expression data: regularized t-test and statistical inferences of gene changes. *Bioinformatics* **17**, 509–519.

Ballman, K.V., Grill, D., Oberg, A., and Therneau, T. (2004). Faster cyclic loess: normalizing DNA arrays via linear models. *Technical Report No. 68*, Mayo Foundation. `http://www.mayo.edu/hsr/techrpt/68.pdf`.

Banfield, J.D. and Raftery, A.E. (1993). Model-based Gaussian and non-Gaussian clustering. *Biometrics* **49**, 803–821.

Barash, Y. and Friedman, N. (2001). Context-specific Bayesian clustering for gene expression data. In *Proceedings of the Fifth Annual International Conference on*

Computational Biology, T. Lengauer and D. Sankoff (Eds.). Montreal, Quebec, Canada: CRM, pp. 12–21.

Barnard, G.A. (1963). Contribution to the discussion of paper by M.S. Bartlett. *Journal of the Royal Statistical Society B* **25**, 294.

Basford, K.E. and McLachlan, G.J. (1985). Estimation of allocation rates in a cluster analysis context. *Journal of the American Statistical Association* **80**, 286–293.

Beer, D.J., Kardia, S.L.R., Huang, C.-C., Giordano, T.J., Levin, A.M., Misek, D.E., Lin, L., Chen, G., Gharib, T.G., Thomas, D.G., Lizyness, M.L., Kuick, R., Hayasaka, S., Taylor, J.M.G., Iannettoni, M.D., Orringer, M.B., Hanash, S. (2002). Gene-expression profiles predict survival of patients with lung adenocarcinoma. *Nature Medicine* **8**, 816–824.

Bellman, R.E. (1961). *Adaptive Control Processes*. Princeton, NJ: Princeton University Press.

Ben-Dor, A., Bruhn, L., Friedman, N., Nachman, I., Schummer, M., and Yakhini, Z. (2000). Tissue classification with gene expression profiles. *Journal of Computational Biology* **7**, 559–584.

Ben-Dor, A., Shamir, R., and Yakhini, Z. (1999). Clustering gene expression patterns. *Journal of Computational Biology* **6**, 281–297.

Benito, M., Parker, J., Du, Q., Wu, J., Xiang, D., Perou, C.M., and Marron, J.S. (2004). Adjustment of systematic microarray data biases. *Bioinformatics* **20**, 105–114.

Benjamini, Y. and Hochberg, Y. (1995). Controlling the False Discovery Rate: A Practical and Powerful Approach to Multiple Testing. *Journal of the Royal Statistical Society B* **57**, 289–300.

Benjamini, Y. and Yekutieli, D. (2001). The control of the false discovery rate under dependency. *Annals of Statistics* **29**, 1165–1188.

Benson, D.A., Karsch-Mizrachi, I., Lipman, D.J., Ostell, J., and Wheeler, D.L. (2004). GenBank: update. *Nucleic Acids Research* **32**, D23–D26.

Ben-Tovim Jones, L., Ng, S.K., Ambroise, C., Monico, K., Khan, N., and McLachlan, G.J. (2004a). Use of microarray data via model-based classification in the study and prediction of survival from lung cancer. In *Methods of Microarray Data Analysis IV*, K.F. Johnson and S.M. Lin (Eds.). Dordrecht, The Netherlands: Kluwer. To appear.

Ben-Tovim Jones, L., Ng, S.K., Monico, K., and McLachlan, G.J. (2004b). Linking gene-expression experiments with survival-time data. *Technical Report*. Brisbane, Queensland, Australia: Department of Mathematics, University of Queensland.

Beucher, S. and Meyer, F. (1993). The morphological approach to segmentation: the watershed transformation. Mathematical morphology in image processing. *Optical Engineering* **34**, 433–481.

Bhattacharjee, A., Richards, W.G., Staunton, J., Li, C., Monti, S., Vasa, P., Ladd, C., Beheshti, J., Bueno, R., Gillette, M., Loda, M., Weber, G., Mark, E.J., Lander, E.S., Wong, W., Johnson, B.E., Golub, T.R., Sugarbaker, D.J., and Meyerson, M. (2001). Classification of human lung carcinomas by mRNA expression profiling reveals distinct adenocarcinoma subclasses. *Proceedings of the National Academy of Sciences USA* **98**, 13790–13795.

Bhattacharyya, C., Grate, L.R., Rizki, A., Radisky, D., Molina, F.J., Jordan, M.I., Bissell, M.J., and Mian, I.S. (2003). Simultaneous classification and relevant feature identification in high-dimensional spaces: application to molecular profiling data. *Signal Processing* **83**, 729–743.

Bickel, D.R. (2003). Robust cluster analysis of microarray gene expression data with the number of clusters determined biologically. *Bioinformatics* **19**, 818–824.

Biernacki, C., Celeux, G., and Govaert, G. (1998). Assessing a mixture model for clustering with the integrated classification likelihood. *IEEE Transactions on Pattern Analysis and Machine Intelligence* **22**, 719–725.

BioDiscovery Inc. (1997). ImaGene.

Bishop, C.M. (1995). *Neural Networks for Pattern Recognition*. Oxford: Oxford University Press.

Bittner, M., Meltzer, P., Chen, Y., Jiang, Y., Seftor, E., Hendrix, M., Radmacher, M., Simon, R., Yakhini, Z., Ben-Dor, A., Sampas, N., Dougherty, E., Wang, E., Marincola, F., Gooden, C., Lueders, J., Glatfelter, A., Pollock, P., Carpten, J., Gillanders, E., Leja, D., Dietrich, K., Beaudry, C., Berens, M., Alberts, D., Sondak, V., Hayward, N., and Trent, J. (2000). Molecular classification of cutaneous malignant melanoma by gene expression profiling. *Nature* **406**, 536–540.

Blekas, K., Fotiadis, D.I., and Likas, A. (2003). Greedy mixture learning for multiple motif discovery in biological sequences. *Bioinformatics* **19**, 607–617.

Bock, H.H. (1992). A clustering technique for maximizing ϕ-divergence, noncentrality and discriminating power. In *Analyzing and Modeling Data and Knowledge*, M. Schader (Ed). New York: Springer-Verlag, pp. 19–36.

Bolstad, B. (2001). Probe level quantile normalization of high density oligonucleotide array data.
`http://www.stat.Berkeley.EDU/~bolstad/stuff/qnorm.pdf`

Bolstad, B., Irizarry, R.A., Åstrand, M., and Speed, T.P. (2003). A comparison of normalization methods for high density oligonucleotide array data based on variance and bias. *Bioinformatics* **19**, 185–193.

Bortoluzzi, S., d'Alessi, F., Romualdi, C., and Danieli, G.A. (2001). Differential expression of genes coding for ribosomal proteins in different human tissues. *Bioinformatics* **17**, 1152–1157.

Bozinov, D. and Rahnenführer, J. (2002). Unsupervised technique for robust target separation and analysis of DNA microarray spots through adaptive pixel clustering. *Bioinformatics* **18**, 747–756.

Braga-Neto, U.M. and Dougherty, E. (2004). Is cross-validation valid for small-sample microarray classification? *Bioinformatics* **20**, 374–480.

Braga-Neto, U.M., Hashimoto, R., Dougherty, E., Nguyen, D. and Carroll, R. (2004). Is cross-validation better than resubstitution for ranking genes? *Bioinformatics* **20**, 253–258.

Brazma, A., Robinson, A., Cameron, G., and Ashburner, M. (2000). One-stop shop for microarray data. *Nature* **403**, 699–700.

Brazma, A. and Vilo, J. (2000). Gene expression data analysis. *Federation of European Biochemical Societies Letters* **480**, 17–24.

Bredensteiner, E.J. and Bennett, K.P. (1999). Multicategory classification by support vector machines. *Computational Optimization and Applications* **12**, 53–79.

Breiman, L. (2001). Statistical modeling: the two cultures (with discussion). *Statistical Science* **16**, 199–231.

Breiman, L., Friedman, J.H., Olshen, R.A., and Stone, C.J. (1984). *Classification and Regression Trees*. Belmont, CA: Wadsworth.

Broberg, P. (2003). Statistical methods for ranking differentially expressed genes. *Genome Biology* **4**, R41.1–R41.9.

Brody, J.P., Williams, B.A., Wold, B.J., and Quake, S.R. (2002). Significance and statistical errors in the analysis of DNA microarray data. *Proceedings of the National Academy of Sciences USA* **99**, 12975–12978.

Broët, P., Richardson, S., and Radvanyi, F. (2002). Bayesian hierarchical model for identifying changes in gene expression from microarray experiments. *Journal of Computational Biology* **9**, 671–683.

Brown, C.S., Goodwin, P.C., and Sorger, P.K. (2001). Image metrics in statistical analysis of DNA microarray data. *Proceedings of the National Academy of Sciences USA* **98**, 8944–8949.

Brown, M.P.S., Grundy, W.N., Lin, D., Cristianini, N., Sugnet, C.W., Furey, T.S., Ares, M., and Haussler, D. (2000). Knowledge-based analysis of microarray gene expression data by using support vector machines. *Proceedings of the National Academy of Sciences USA* **97**, 262–267.

Bryan, J., Pollard, K.S., and van der Laan, M.J. (2002). Paired and unpaired comparison and clustering with gene expression data, *Statistica Sinica* **12**, 87–110.

Buckley, M.J. (2000) *Spot User's Guide*. CSIRO Mathematical and Information Sciences, Sydney, Australia. `http://www.cmis.csiro.au/iap/Spot/spotmanual.htm`.

Bullinger, L., Döhner, K., Bair, E., Fröhling, S., Schlenk, R.F., Tibshirani, R., Döhner, H,, Pollack, J.R. (2004). Use of gene-expression profiling to identify prognostic subclasses in adult acute myeloid leukemia. *New England Journal of Medicine* **350**, 1605–1616.

Caliński, T. and Harabasz, J. (1974). A dendrite method for cluster analysis, *Communications in Statistics—Theory and Methods* **3**, 1–27.

Callow, M.J., Dudoit, S., Gong, E.L., Speed, T.P., and Rubin E.M. (2000). Microarray expression profiling identifies genes with altered expression in HDL-deficient mice. *Genome Research* **10**, 2022–2029.

Celeux, G. and Govaert, G. (1995). Gaussian parsimonious clustering model. *Pattern Recognition* **28**, 781–793.

Celeux, G., Martin, O., and Lavergne, C. (2004). Mixture of linear mixed models for clustering gene expression profiles from repeated microarray experiments. *Statistical Modelling*. To appear.

Chadt, E.E., Li, C., Su, C., and Wong, W.H. (2000). Analyzing high-density oligonucleotide gene expression array data. *Journal of Computational Biology* **80**, 192–202.

Chang, C.-C. and Lin, C.-J. (2003). LIBSVM: a library for support vector machines. `http://www.csie.ntu.edu.tw/~cjlin/libsvm`.

Chang, W.C. (1983). On using principal components before separating a mixture of two multivariate normal distributions. *Applied Statistics* **32**, 267–275.

Chen, Y., Dougherty, E.R., and Bittner, M.L. (1997). Ratio-based decisions and the quantitative analysis of cDNA microarray images. *Journal of Biomedical Optics* **2**, 364–374.

Chen, Y., Kamat, V., Dougherty, E.R., Bittner, M.L., Meltzer, P.S., and Trent, J.M. (2002). Ratio statistics of gene expression levels and applications to microarray data analysis. *Bioinformatics* **18**, 1207–1215.

Cheung, V.G., Morley, M., Aguilar, F., Massimi, A., Kucherlapati, R., and Childs, G. (1999). Making and reading microarrays. *Nature Genetics* **21(S)**, 15–19.

Chilingaryan, A. Gevorgyan, N., Vardanyan, A., Jones, D., and Szaboi, A. (2002). Multivariate approach for selecting sets of differentially expressed genes. *Mathematical Biosciences* **176**, 59–69.

Chow, M.L., Moler, E.J., and Mian, I.S. (2001). Identifying marker genes in transcription profiling data using a mixture of feature relevance experts. *Physiological Genomics* **5**, 99–111.

Chudin, E., Walker, R., Kosaka, A., Wu, S., Rabert, D., Chang, T.K., and Kreder, D.E. (2001). Assessment of the relationship between signal intensities and transcript concentration for Affymetrix GeneChip® arrays. *Genome Biology* **3**, 5.1–5.10.

Churchill, G.A. (2003). Contribution to the discussion of the paper by P. Sebastiani, E. Gussoni, I.S. Kohane,, and M.F. Ramoni. *Statistical Science* **18**, 64–69.

Cimons, M. (2003). Microarray to be used as routine clinical screen. *Nature Medicine* **9**, 9.

Cleveland, W.S. and Devlin, S.J. (1988). Locally-weighted regression: an approach to regression analysis by local fitting. *Journal of the American Statistical Association* **83**, 596–610.

Colantuoni, C., Henry, G., Zeger, S., and Pevsner, J. (2002). SNOMAD (Standardization and NOrmalization of MicroArray Data). *Bioinformatics* **18**, 1540–1541.

Coleman, D., Dong, X., Hardin, J., Rocke, D.M., and Woodruff, D.L. (1999). Some computational issues in cluster analysis with no a priori metric, *Computational Statistics and Data Analysis* **31**, 1–11.

Coombes, K.R., Highsmith, W.E., Krogmann, T.A., Baggerly, K.A., Stivers, D.N., and Abruzzo, L.V. (2002). Identifying and quantifying sources of variation in microarray data using high-density cDNA membrane arrays. *Journal of Computational Biology* **9**, 655–669.

Cope, L.M., Irizarray, R.A., Jaffee, H.A., Wu, Z., and Speed, T.P. (2004). A benchmark for Affymetrix GeneChip expression measures. *Bioinformatics* **20**, 323–331.

Cox, D.R. (1972). Regression models and life tables (with discussion). *Journal of the Royal Statistical Society B* **34**, 187–220.

Crammer, K. and Singer, Y. (2000). On the learnability and design of output codes for multiclass problems. In *Proceedings of the Thirteenth Annual Conference on Computational Learning Theory*, N. Cesa-Bianchi and S. Goldman (Eds.). San Francisco: Morgan Kaufmann, pp. 35–46.

Cui, X. and Churchill, G.A. (2003). Statistical tests for differential expression in cDNA microarray experiments. *Genome Biology* **4**, 210.1–210.9.

Dalton, R. (2001). Patent ruling could cut PCR enzyme prices. *Nature* **411**, 622.

Datta, S. (2001). Exploring relationships in gene expressions: a partial least squares approach. *Gene Expression* **9**, 249–255.

Datta, S. (2003). Statistical techniques for microarray data: a partial overview. *Communications in Statistics—Theory and Methods* **32**, 263–280.

Datta, S. and Datta, S. (2003). Comparisons and validation of statistical clustering techniques for microarray gene expression data. *Bioinformatics* **19**, 459–466.

Dawid, A.P. (1976). Properties of diagnostic data distributions. *Biometrics* **32**, 647–658.

De Francesco, L. (1998). Taking the measure of the message. *The Scientist* **12**, 20.

Dempster, A.P., Laird, N.M., and Rubin, D.B. (1977). Maximum likelihood from incomplete data via the EM algorithm (with discussion). *Journal of the Royal Statistical Society B* **39**, 1–38.

DeRisi, J.L., Iyer, V.R., and Brown, P.O. (1997). Exploring the metabolic and genetic control of gene expression on a genomic scale. *Science* **278**, 680–686.

DeRisi, J.L., Penland, L., Brown, P.O., Bittner, M.L., Meltzer, P.S., Ray, M., Chen, Y., Su, Y.A., and Trent, J.M. (1996). Use of a cDNA microarray to analyze gene expression patterns in human cancer. *Nature Genetics* **14**, 457–460.

Dettling, M. and Bühlmann, P. (2002). Supervised clustering of genes. *Genome Biology* **3**, research0069.1–0069.15.

Diaconis, P. and Ylvisaker, D. (1985). Quantifying prior opinion. In *Bayesian Statistics 2*, J.M. Bernardo, M.H. DeGroot, D.V. Lindley, and A. F. M. Smith (Eds.). New York: Wiley, pp. 133–156.

Díaz-Uriarte, R. (2003). A simple method for finding molecular signatures from gene expression. *Bioinformatics Unit Technical Report 004*, Madrid, Spain: Spanish National Cancer Center (CNIO).

Do, K.-A., Broom, B.M., and Wen, S. (2003). GeneClust. In *The Analysis of Gene Expression Data: Methods and Software*, New York: Springer-Verlag, 342–361.

Do, K.-A., Müeller, P. and Tang, F. (2003) A Bayesian mixture model for differential gene expression. *Technical Report*. Houston: Department of Biostatistics, University of Texas/M.D. Anderson Cancer Center.

Dobbin, K., Shih, J.H., and Simon, R. (2003). Statistical design of reverse dye microarrays. *Bioinformatics* **19**, 803–810.

Dobbin, K. and Simon, R. (2002). Comparison of microarray designs for class comparison and class discovery. *Bioinformatics* **18**, 1438–1445.

Dubes, R.C. and Jain, A.K. (1989). Random field models in image analysis. *Journal of Applied Statistics* **16**, 131–164.

Dudoit, S. and Fridlyand, J. (2002). A prediction-based resampling method for estimating the number of clusters in a dataset. *Genome Biology* **3**, research0036.1–0036.21

Dudoit, S. and Fridlyand, J. (2003). Classification in microarray experiments. In *Statistical Analysis of Gene Expression Microarray Data*, T. Speed (Ed.). Boca Raton, FL: Chapman & Hall/CRC, pp. 93–158.

Dudoit, S., Fridlyand, J., and Speed, T.P. (2002a). Comparison of discrimination methods for the classification of tumors using gene expression data. *Journal of the American Statistical Association* **97**, 77–87.

Dudoit, S., Popper Shaffer, J., and Boldrick, J.C. (2003). Multiple hypothesis testing in microarray experiments. *Statistical Science* **18**, 71–103.

Dudoit, S., Yang, Y.H., Callow, M.J., and Speed, T.P. (2002b). Statistical methods for identifying differentially expressed genes in replicated cDNA microarray experiments. *Statistica Sinica* **12**, 111–139.

Duggan, D.J., Bittner, M., Chen, Y., Meltzer, P., and Trent, J.M. (1999). Expression profiling using cDNA microarrays. *Nature Genetics* **21(S)**, 10–14.

Durbin, B.P., Hardin J.S., Hawkins, D.M., and Rocke, D.M. (2002). A variance-stabilizing transformation for gene-expression microarray data. *Bioinformatics* **18**, S105–S110.

Efron, B. (1982). *The Jackknife, the Bootstrap and Other Resampling Plans.* Philadelphia: SIAM.

Efron, B. (1983). Estimating the error rate of a prediction rule: improvement on cross-validation. *Journal of the American Statistical Association* **78**, 316–331.

Efron, B. (2003). Robbins, empirical Bayes and microarrays. *Annals of Statistics* **31**, 366–378.

Efron, B. and Tibshirani, R. (1993). *An Introduction to the Bootstrap.* London: Chapman & Hall.

Efron, B. and Tibshirani, R. (1997). Improvements on cross-validation: The .632+ bootstrap method. *Journal of the American Statistical Association* **92**, 548–560.

Efron, B., Tibshirani, R., Storey, J.D., and Tusher, V. (2001). Empirical Bayes analysis of a microarray experiment. *Journal of the American Statistical Association* **96**, 1151–1160.

Eisen, M.B. (1999). ScanAlyze. `http://rana.lbl.gov/EisenSoftware.htm`.

Eisen, M.B., Spellman, P.T., Brown, P.O., and Botstein, D. (1998). Cluster analysis and display of genome-wide expression patterns. *Proceedings of the National Academy of Sciences USA* **95**, 14863–14868.

Ekins, R.P. (1998). Ligand assays: from electrophoresis to miniaturized microarrays. *Clinical Chemistry* **44**, 2015–2030.

Ekins, R.P. and Chu, F.W. (1991). Multianalyte microspot immunoassay- microanalytical "compact disk" of the future. *Clinical Chemistry* **37**, 1955–1967.

Ekins, R. and Chu, F.W. (1999). Microarrays: their origins and applications. *Trends in Biotechnology* **17**, 217–218.

Elder, J.K., Johnson, M., Milner, N., Mir, K.U., Sohail, M., and Southern, E.M. (1999). Antisense oligonucleotide scanning arrays. In *DNA Microarrays: A Practical Approach*, M. Schena (Ed.). New York: Oxford University Press, pp. 1–3.

Epstein, J.R., Leung, A.P., Lee, K.H., and Walt, D.R. (2003). High-density, microsphere-based fiber optic DNA microarrays. *Biosensors and Bioelectronics* **18**, 541–546.

Evertsz, E., Starink, P., Gupta, R., and Watson, D. (2000). Technology and applications of gene expression microarrays. In *Microarray Biochip Technology*, M. Schena (Ed.). Natick, MA: Eaton Publishing.

Fisher, R.A. (1936). The use of multiple measurements in taxonomic problems. *Annals of Eugenics* **7**, 179–188.

Fix, E. and Hodges, J.L. (1951). Discriminatory analysis–nonparametric discrimination: consistency properties. *Report No. 4*. Randolph Field, TX: U.S. Air Force School of Aviation Medicine. (Reprinted in Agrawala, A.K., (1977), *Machine recognition of patterns*. New York: IEEE Press, pp. 261–279.)

Fodor, S.P.A., Read, J.L., Pirrung, M.C., Stryer, L., Lu, A.T., and Solas, D. (1991). Light-directed, spatially addressable parallel chemical synthesis. *Science* **251**, 767–773.

Foster, W.R. and Huber, R.M. (2002). Current themes in microarray experimental design and analysis. *Drug Discovery Today* **7**, 290–292.

Fowlkes, E.B. and Mallows, C.L. (1983). A method for comparing two hierarchical clusterings. *Journal of the American Statistical Association* **78**, 553–584.

Fraley, C. and Raftery, A.E. (1998). How many clusters? Which clustering method? Answers via model-based cluster analysis. *Computer Journal* **41**, 578–588.

Fraley, C. and Raftery, A.E. (2004). Model-based clustering, discriminant analysis, and density estimation. *Journal of the American Statistical Association* **97**, 611–631.

Frank, I.E. and Friedman, J.H. (1993). A statistical view of some chemometrics regression tools (with discussion). *Technometrics* **35**, 109–148.

Friedman, H.P. and Rubin, J. (1967). On some invariant criteria for grouping data. *Journal of the American Statistical Association* **62**, 1159–1178.

Friedman, J. (1996). Another approach to polychotomous classification. *Technical Report*. Stanford, CA: Department of Statistics, Stanford University. http://www-stat.stanford.edu/~jhf/ftp/poly.ps.Z.

Friedman, J., Hastie, T., and Tibshirani, R. (1998). Additive logistic regression: a statistical view of boosting. *Annals of Statistics* **28**, 337–407.

Friedman, J.H. and Meulman, J.J. (2004). Clustering objects on subsets of attributes (with discussion). *Journal of the Royal Statistical Society B* **66**. To appear.

Furey, T.S., Cristianini, N., Duffy, N., Bednarski, D.W., Schummer, M., and Haussler, D. (2000). Support vector machine classification and validation of cancer tissue samples using microarray expression data. *Bioinformatics* **16**, 906–914.

Furlanello, C., Serafini, M., Merler, S., and Jurman, G. (2003). An accelerated procedure for recursive feature ranking on microarray data. *Neural Networks* **16**, 641–648.

Gabriel, K.R. (1971). The biplot graphic display of matrices with applications to principal component analysis. *Biometrika* **58**, 453–467.

Gadbury, G.L., Page, G.P., Heo, M., Mountz, J.D., and Allison, D.B. (2003). Randomization tests for small samples: an application for genetic expression data. *Applied Statistics* **52**, 365–376.

Ganeshanandam, S. and Krzanowski, W.J. (1989). On selecting variables and assessing their performance in linear discriminant analysis. *Australian Journal of Statistics* **32**, 443–447.

Garber, M.E., Troyanskaya, O.G., Schluens, K., Petersen, S., Thaesler, Z., Pacyna-Gengelbach, M., van de Rijn, M., Rosen, G.D., Perou, C.M., Whyte, R.I., Altman, R.B., Brown, P.O., Botstein, D., and Petersen, I. (2001). Diversity of gene expression in adenocarcinoma of the lung. *Proceedings of the National Academy of Sciences USA*, **98**, 13784–13789.

Garrett, E.S. and Parmigiani, G. (2003). POE: statistical methods for qualitative analysis of gene expression. In *The Analysis of Gene Expression Data: Methods and Software*, G. Parmigiani, E.S. Garrett, R.A. Irizarry, and S.L. Zeger (eds.), Springer: New York, pp. 362–387.

Garthwaite, P.H. (1994). An interpretation of partial least squares. *Journal of the American Statistical Association* **89**, 122–127.

Gautier, L., Cope, L., Bolstad, B.M., Irizarry, R.A.. (2004). affy - Analysis of Affymetrix GeneChip data at the probe level. *Bioinformatics* **20**, 307–315.

Geisser, S. (1975). The predictive sample reuse method with applications. *Journal of the American Statistical Association* **70**, 320–328.

Geller, S.C., Gregg, J.P., Hagerman, P., and Rocke, D.M. (2003). Transformation and normalization of oligonucleotide microarray data. *Bioinformatics* **19**, 1817–1823.

Genovese, C.R. and Wasserman, L. (2002). Operating Characteristics and Extensions of the False Discovery Rate Procedure. *Journal of the Royal Statistical Society B* **64**, 499–517.

Gerhold, D. and Caskey, C.T. (1996). It's the genes! EST access to human genome content. *BioEssays* **18**, 973–981.

Getz, G. (2001). Personal communication.

Getz, G., Levine, E., and Domany, E. (2000). Coupled two-way clustering analysis of gene microarray data. *Cell Biology* **97**, 12079–12084.

Ghahramani, Z. and Hinton, G.E. (1997). The EM algorithm for factor analyzers. *Technical Report No. CRG-TR-96-1.* Toronto, Ontario, Canada: University of Toronto.

Ghosh, D. and Chinnaiyan, A.M. (2002). Mixture modelling of gene expression data from microarray experiments. *Bioinformatics* **18**, 275–286.

Gibbons, F.D. and Roth, F.P. (2002). Judging the quality of gene expression–based clustering methods using gene annotation. *Genome Research* **12**, 1574–1581.

Gibbons, R.D., Cox, D.R., Grayson, D.R., Bhaumik, D.K., Davis, J.M., and Sharma, R.P. (2001). Sequential prediction bounds for identifying differentially expressed genes in replicated microarray experiments. Preprint.

Goh, L., Kasabov, N., and Song, Q. (2004). A novel feature selection method to improve classification of gene expression data. In *Proceedings of the Second Asia-Pacific Bioinformatics Conference (APBC2004)*, Dunedin, New Zealand, CRPIT, 29, Y.-P. P. Chen (Ed.). Australian Computer Society, pp. 161–166.

Goldstein, D.R., Ghosh, D., and Conlon, E.M. (2002). Statistical issues in the clustering of gene expression data. *Statistica Sinica* **12**, 219–240.

Golub, T.R., Slonim, D.K., Tamayo, P., Huard, C., Gassenbeck, M., Mesirov, J.P., Coller, H., Loh, M.L., Downing, J.R., Caligiuri, M.A., Bloomfield, C.D., and Lander, E.S. (1999). Molecular classification of cancer: class discovery and class prediction by gene expression monitoring. *Science* **286**, 531–537.

Gong, G. (1986). Cross-validation, the jackknife, and the bootstrap: excess error estimation in forward logistic regression. *Journal of the American Statistical Association* **81**, 108–113.

Green, S.C., Pritchard, C., and Southern, E.M. (1999). Use of oligonucleotide arrays in enzymatic assays: assay optimization. In *DNA Microarrays: A Practical Approach*, M. Schena (Ed.). New York: Oxford University Press.

GSI Lumonics, Inc. (1999). QuantArray.

Guermeur, Y., Elisseeff, A., and Paugam-Moisy, H. (2000). A new multi-class SVM based on a uniform convergence result. In *International Joint Conference on Neural Networks IJCNN 2000*. Los Alamitos, CA: IEEE Computer Society, pp. IV-183–IV-188.

Guo, X., Qi, H.. Verfaillie, C.M., and Pan, W. (2003). Statistical significance analysis of longitudinal gene expression data. *Bioinformatics* **19**, 1628–1635.

Guyon, I., Weston, J., Barnhill, S., and Vapnik, V. (2002). Gene selection for cancer classification using support vector machines. *Machine Learning* **46**, 389–422.

Habbema, J.D.F. and Hermans, J. (1977). Selection of variables in discriminant analysis by F-statistic and error rate. *Technometrics* **19**, 487–493.

Hand, D.J., Mannila, H., and Smyth, P. (2001). *Principles of Data Mining*. Cambridge, MA: MIT Press.

Hand, D.J. and Yu, K. (2001). Idiot's Bayes - not so stupid after all? *International Statistical Review* **69**, 385–398.

Hansen, K.M. and Tukey, J.W. (1992). Tuning a major part of a clustering algorithm. *International Statistical Review* **60**, 21–43.

Hardin, J.S., Rocke, D.M., and Woodruff, D.L. (2000). Robust model-based clustering of genes in microarray data: are there gene clusters? *Technical Report*. Davis, CA: Department of Applied Science and Division of Biostatistics, University of California.

Hartigan, J.A. (1975). Statistical theory in clustering, *Journal of Classification* **2**, 63–76.

Hastie, T. and Tibshirani, R. (1996). Discriminant analysis by Gaussian mixtures. *Journal of the Royal Statistical Society B* **58**, 155–176.

Hastie, T., Tibshirani, R., Botstein, D., and Brown, P. (2001a). Supervised harvesting of expression trees. *Genome Biology* **2**, research0003.1–0003.12.

Hastie, T., Tibshirani, R., Eisen, M.B., Alizadeh, A., Levy, R., Staudt, L., Chan, W.C., Botstein, D., and Brown, P. (2000). 'Gene shaving' as a method for identifying distinct sets of genes with similar expression patterns. *Genome Biology* **1**, research0003.1–0003.21.

Hastie, T., Tibshirani, R., and Friedman, J. (2001b). *The Elements of Statistical Learning*. Basel: Springer-Verlag.

Hattori, M. and Taylor, T.D. (2001). The human genome: part three in the book of genes. *Nature* **414**, 854–855.

Hatzigeorgiou, A., Fiziev, P., and Reczko, M. (2001). DIANA-EST: a statistical analysis. *Bioinformatics* **17**, 913–919.

Hawkins, D.M., Muller, M.W., and ten Krooden, J.A. (1982). Cluster analysis. In *Topics in Applied Multivariate Analysis*, D.M. Hawkins (Ed.). Cambridge: Cambridge University Press, pp. 303–356.

Hedenfalk, I., Ringnér, M., Ben-Dor, A., Yakhini, Z., Chen, Y., Chebil, G., Ach, R., Loman, N., Olsson, H., Meltzer, P., Borg, A., and Trent, J. (2001). Molecular classification of familial non-BRCA1/BRCA2 breast cancer. *Proceedings of the National Academy of Sciences USA* **100**, 2532–2537.

Heller, M.J., Tu, E., Holmsen, A., Sosnowski, R.D., and O'Connell, J. (1999). Active microelectronic arrays for DNA hybridization analysis. In *DNA Microarrays: A Practical Approach*, M. Schena. (Ed.). New York: Oxford University Press.

Hennig, C. (2004). Breakdown points of maximum likelihood-estimators of location-scale mixtures. *Annals of Statistics* **32**. To appear.

Herrero, J., Valencia, A., and Dopazo, J. (2001). A hierarchical unsupervised growing neural network for clustering gene expression patterns. *Bioinformatics* **17**, 126–136.

Hess, K.R., Zhang, W., Baggerly, K.A, Stivers, D.N., and Coombes, K.R. (2001). Microarrays: handling the deluge of data and extracting reliable information. *Trends in Biotechnology* **19**, 463–468.

Highleyman, W.H. (1962). The design and analysis of pattern recognition experiments. *Bell Systems Technical Journal* **41**, 723–744.

Hill, A.A., Brown, E.L., Whitley, M.Z., Tucker-Kellogg, G., Hunter, C.P. and Slonim, D.K. (2001). Evaluation of normalization procedures for oligonucleotide array data based on spiked cRNA controls. *Genome Biology* **2**, research0055.1–0055.13.

Hinton, G.E., Dayan, P., and Revow, M. (1997). Modeling the manifolds of images of handwritten digits. *IEEE Transactions on Neural Networks* **8**, 65–73.

Hoaglin, D.C. (1985). Using quantiles to study shape. In *Exploring Data Tables, Trends, and Shapes*, D.C. Hoaglin, F. Mosteller, and J.W. Tukey (Eds.). New York: Wiley, pp. 417–460.

Hochberg, Y. and Benjamini, Y. (1990). More powerful procedures for multiple significance testing. *Statistics in Medicine* **9**, 811–818.

Hoffmann, R., Seidl, T., and Dugas, M. (2002). Profound effect of normalization on detection of differentially expressed genes in oligonucleotide microarray data analysis. *Genome Biology* **3**, research0033.1-0033.11.

Hollon, T. (2001). Comparing microarray data: what technology is needed? *Journal of the National Cancer Institute* **93**, 1126–1127.

Holm, S. (1979). A simple sequentially rejective multiple test procedure. *Scandinavian Journal of Statistics: Theory and Applications* **6**, 65–70.

Holmes, I. and Bruno, W.J. (2000). Finding regulatory elements using joint likelihoods for sequence and expression profile data. In *Proceedings of the Eighth Annual International Conference on Intelligent Systems for Molecular Biology*, R. Altman, T.L. Bailey, P. Bourne, M. Gribskov, T. Lengauer, and I. N. Shindyalov (Eds.). La Jolla, CA: AAAI Press, pp. 202–210.

Hope, A.C.A. (1968). A simplified Monte Carlo significance test procedure. *Journal of the Royal Statistical Society B* **30**, 582–598.

Horimoto, K. and Toh, H. (2001). Statistical extimation of cluster boundaries in gene expression profile data. *Bioinformatics* **17**, 1143–1151.

Hsu, C.-W. and Lin, C.-J. (2002). A comparison of methods for multi-class support vector machines. *IEEE Transactions on Neural Networks* **13**, 415–425.

Huang, E., Cheng, S.H., Dressman, H., Pittman, J., Tsou, M.-H., Horng, Ch.-F., Bild, A., Iversen, E.S., Liao, M., Chen, C.-M., West, M., Nevins, J.R., and Huang, A.T. (2003). Gene expression predictors of breast cancer outcomes. *Lancet* **361**, 1576–1577.

Huang, X. and Pan, W. (2003). Linear regression and two-class classification with gene expression data *Bioinformatics* **19**, 2072–2078.

Huber, W., von Heydebreck, A., Sueltmann, H., Poustka, A., and Vingron, M. (2002). Variance stabilization applied to microarray data calibration and to the quantification of differential expression. *Bioinformatics* **18**, S96–S104.

Huber, W., von Heydebreck, A., Sueltmann, H., Poustka, A., and Vingron, M. (2003). Parameter estimation for the calibration and variance stabilization of microarray data. *Statistical Applications in Genetics and Molecular Biology* **2(1)**, Article 3.

Hughes, T.R., Marton, M.J., Jones, A.R., Roberts, C.J., Stoughton, R., Armour, C.D., Bennett, H.A., Coffey, E., Dai, H., He, Y.D., Kidd, M.J., King, A.M., Meyer, M.R., Slade, D., Lum, P.Y., Stepaniants, S.B., Shoemaker, D.D., Gachotte, D., Chakraburtty, K., Simon, J., Bard, M., and Friend, S.H. (2000). Functional discovery via a compendium of expression profiles. *Cell* **102**, 109–126.

Hunt, L.A. and Jorgensen, M.A. (1999). Mixture model clustering: a brief introduction to the MULTIMIX program. *Australian & New Zealand Journal of Statistics* **41**, 153–171.

Ibrahim, J.G., Chen, M.-H., and Gray, R.J. (2002). Bayesian models for gene expression with DNA microarray data. *Journal of the American Statistical Association* **97**, 88–99.

Ideker, T., Thorsson, V., Siegel, A.F., and Hood, L.E. (2000). Testing for differentially-expressed genes by maximum-likelihood analysis of microarray data. *Journal of Computational Biology* **7**, 805–817.

Irizarry, R.A., Hobbs, B., Collin, F., Beazer-Barclay, Y.D., Antonellis, K.J., Scherf, U., and Speed, T. (2003a). Exploration, normalization, and summaries of high density oligonucleotide array probe level data. *Biostatistics* **4**, 249–264.

Iyer, V.R., Eisen, M.B., Ross, D.T., Schuler, G., Moore, T., Lee, J.C.F., Trent, J.M., Staudt, L.M., Hudson, J., Boguski, M.S., Lashkari, D., Shalon, D., Botstein, D., and Brown, P.O. (1999). The Transcriptional Program in the Response of Human Fibroblasts to Serum, *Science* **283** 83–87.

Jaeger, J., Sengupta, R., and Ruzzo, W.L. (2003). Improved gene selection for classification of microarrays. *Pacific Symposium on Biocomputing*, Kauai, HI, pp. 53–64.

Jain, A.K. and Dubes, R.C. (1988). *Algorithms for Clustering Data.* Englewood Cliffs: Prentice Hall.

Jin, W., Riley, R.M., Wolfinger, R.D., White, K.P., Passador-Gurgel, G., and Gibson, G. (2001). The contribution of sex, geneotype and age to transcriptional variance in *Drosophila melanogaster*. *Nature Genetics* **29**, 389–395.

Johnson, N.L. and Kotz, S. (1970). *Continuous Univariate Distributions*. Vol. 2. Wiley: New York.

Jordan, B. (2002). Historical background and anticipated developments. *Annals of the New York Academy of Science* **975**, 24–32.

Kallioniemi, A. (2002). Molecular signatures of breast cancer – predicting the future. *New England Journal of Medicine* **347**, 2067–2068.

Kan, T., Shimada, Y., Sato, F., Maeda, M., Kawabe, A., Kaganoi, J., Itami, A., Yamasaki, S., and Imamura, M. (2001). Gene expression profiling in human esophageal cancers using cDNA microarray. *Biochemical and Biophysical Research Communications* **286**, 792–801.

Karchin, R., Karplus, K., and Haussler, D. (2002). Classifying G-protein coupled receptors with support vector machines. *Bioinformatics* **18**, 147–159.

Kaufman, L. and Rousseeuw, P.J. (1990). *Finding Groups in Data*. New York: Wiley.

Kellam, P. (2001). Post-genomic virology: the impact of bioinformatics, microarrays and proteomics on investigating host and pathogen interactions. *Reviews in Medical Virology* **11**, 313–329.

Kendziorski, C.M., Newton, M.A., Lan, H., and Gould, M.N. (2003). On parametric empirical Bayes methods for comparing multiple groups using replicated gene expression profiles. *Statistics in Medicine* **22**, 3899–3914.

Kepler, T.B., Crosby, L., and Morgan, K.T. (2002). Normalization and analysis of DNA microarray data by self-consistency and local regression. *Genome Biology* **3**, research0037.1–0037.12.

Kerr, M.K. (2003). Design considerations for efficient and effective microarray studies. *Biometrics* **59**, 822–828.

Kerr, M.K., Afshari, C.A., Bennett, L., Bushel, P., Martinez, N.W., and Churchill, G.A. (2002). Statistical analysis of a gene expression microarray experiment with replication. *Statistica Sinica* **12**, 203–217.

Kerr, M.K. and Churchill, G.A. (2001). Bootstrapping cluster analysis: assessing the reliability of conclusions from microarray experiments. *Proceedings of the National Academy of Sciences USA* **98**, 8961–8965.

Kerr, M.K., Martin, M., and Churchill, G.A. (2000). Analysis of variance for gene expression microarray data. *Journal of Computational Biology* **7**, 819–837 .

Khan, J., Wei, J.S., Ringnér, M., Saal, L.H., Ladanyi, M., Westermann, F., Berthold, F., Schwab, M., Antonescu, C.R., Petersen, C., and Meltzer, P.S. (2001). Classification and diagnostic prediction of cancers using gene expression profiling and artificial neural networks. *Nature Medicine* **7**, 673–679.

Kimura, N., Oda, R., Inaki, Y., and Suzuki, O. (2004). Attachment of oligonucleotide probes to poly carbodiimide-coated glass for microarray applications. *Nucleic Acids Research* **32**, e68.

King, H.C. and Sinha, A.A. (2001). Gene expression profile analysis by DNA microarrays. *Journal of the American Medical Association* **286**, 2280–2288.

Knight, J. (2003). Promega changes tack in battle over patent. *Nature* **426**, 373.

Kohavi, R. and John, G.H. (1997). The wrapper approach. In *Feature Extraction, Construction and Selection: A Data Mining Perspective*, H. Liu and H. Motoda (Eds.). Norwell, MA: Kluwer.

Kooperberg, C., Fazzio, T.G., Delrow, J.J., and Tsukiyama, T. (2002). Improved background correction for spotted cDNA microarrays. *Journal of Computational Biology* **9**, 55–66.

Kreßel, U. (1999). Pairwise classification and support vector machines. In *Advances in Kernel Methods and Support Vector Learning*, B. Scholkopf, C.J.C. Burgess, and A.J. Smola (Eds.). Cambridge, MA: MIT Press, pp. 255–268.

Kricka, L.J. and Fortina, P. (2001). Microarray technology and applications. *Clinical Chemistry* **47**, 1479–1482.

Kroll, T.C. and Wölfl, S. (2002). Ranking: a closer look at globalisation methods for normalisation of gene expression arrays. *Nucleic Acids Research* **30**, e50.

Krzanowski, W.J. and Lai, Y. (1985). A criterion for determining the number of groups in a dataset using sum of squares clustering, *Biometrics* **44**, 23–34.

Kudo, M. and Sklansky, J. (2000). Comparison of algorithms that select features for pattern classifiers. *Pattern Recognition* **33**, 25–41.

Kuruvilla, F.A., Park, P.J., and Schreiber, S.L. (2002). Vector algebra in the analysis of genome-wide expression data. *Genome Biology* **3**, research0011.1–0011.11.

Lachenbruch, P.A. (1965). *Estimation of Error Rates in Discriminant Analysis*. Unpublished Ph.D. thesis, University of California, Los Angeles.

Lachenbruch, P.A. and Mickey, M.R. (1968). Estimation of error rates in discriminant analysis. *Technometrics* **10**, 1–11.

Lance, G.N. and Williams, W.T (1967). A generalized theory of classificatory sorting strategies: I. Hierarchical systems. *Computer Journal* **9**, 373–380.

Lapointe, J., Li, C. (2004). Gene expression profiling identifies clinically relevant subtypes of prostate cancer. *Proceedings of the National Academy of Sciences USA* **101**, 811–816.

Lazzeroni, L. and Owen, A.B. (2002). Plaid models for gene expression data. *Statistica Sinica* **12**, 61–86.

Lee, M.-L.T., Bulyk, M.L., Whitmore, G.A., and Church, G.M. (2002). A statistical model for investigating binding probabilities of DNA nucleotide sequences using microarrays. *Biometrics* **58**, 981–988.

Lee, M.-L.T., Kuo, F.C., Whitmore, G.A., and Sklar, J. (2000). Importance of replication in microarray gene expression studies: statistical methods and evidence from repetitive cDNA hybridizations. *Proceedings of the National Academy of Sciences USA* **97**, 9834–9838.

Lee, M.-L.T., Whitmore, G.A., and Yukhananov, R.Y. (2003). Analysis of unbalanced microarray data. *Journal of Data Science* **1**, 103–121.

Lee, Y. and Lee, C.-K. (2003). Classification of multiple cancer types by multicategory support vector machines using gene expression data. *Bioinformatics* **19**, 1132–1139.

Lehmann, E.L. (1959). *Testing Statistical Hypotheses*. Wiley: New York.

Li, C. and Wong, W.H. (2001a). Model-based analysis of oligonucleotide arrays: expression index computation and outlier detection. *Proceedings of the National Academy of Sciences USA* **98**, 31–36.

Li, C. and Wong, W.H. (2001b). Model-based analysis of oligonucleotide arrays: model validation, design issues and standard error applications. *Genome Biology* **2**, research0032.1–0032.11.

Li, F. and Stormo, D.G. (2001). Selection of optimal DNA oligos for gene expression arrays. *Bioinformatics* **17**, 1067–1076.

Li, H. and Gui, J. (2004). Partial Cox regression analysis for high-dimensional microarray gene expression data. Center for Bioinformatics and Molecular Biostatistics, Paper partialcox. `http://repositories.cdlib.org/cbmb/partialcox`.

Li, J.Q. and Barron, A.R. (2000). Mixture density estimation. *Technical Report*. New Haven, CT: Department of Statistics, Yale University.

Li, L., Weinberg, C.R., Darden, T.A., and Pedersen, L.G. (2001). Gene selection for sample classification based on gene expression data: study of sensitivity to choice of parameters of the GA/KNN method. *Bioinformatics* **17**, 1131–1142.

Li, Y., Campbell, C., and Tipping, M. (2002). Bayesian automatic relevance determination algorithms for classifying gene expression data. *Bioinformatics* **18**, 1332–1339.

Lipshutz, R.J., Fodor, S.P.A., Gingeras, T.R., and Lockhart, D.J. (1999). High density synthetic oligonucleotide arrays. *Nature Genetics* **21(S)**, 20–24.

Liu, A., Zhang, Y., Gehan, E., and Clarke, R. (2002). Block principal component analysis with application to gene microarray data application. *Statistics in Medicine* **21**, 3465–3474.

Liu, D.K., Yao, B., Fayz, B., Womble, D.D., and Krawetz, S.A. (2004). Comparative evaluation of microarray analysis software. *Molecular Biotechnology* **26**, 225–232.

Liu, J.S., Zhang, J.L., Palumbo, M.J., and Lawrence, C.E. (2003). Bayesian clustering with variable and transformation selections. In *Bayesian Statistics*, Vol. 7, J.M. Bernardo, M.J. Bayarri, J.O. Berger, A.P. Dawid, D. Heckerman, A.F.M. Smith, and M. West (Eds.). Oxford: Oxford University Press, pp. 249–275.

Liu, L., Hawkins, D.M., Ghosh, S., and Young, S.S. (2003). Robust singular value decomposition analysis of microarray data. *Proceedings of the National Academy of Sciences USA* **100**, 13167–13172.

Lockhart, D.J., Dong, H., Byrne, M.C., Follettie, M.T., Gallo, M.V., Chee, M.S., Mittmann, M., Wang, C., Kobayashi, M., Horton, H., and Brown, E.L. (1996). Expression monitoring by hybridization to high-density oligonucleotide arrays. *Nature Biotechnology* **14**, 1675–1680.

Lockhart, D.J. and Winzeler, E.A. (2000). Genomics, gene expression and DNA arrays. *Nature* **405**, 827–836.

Loh, W.-Y. and Vanichsetakul, N. (1988). Tree-structured classification via generalized discriminant analysis (with discussion). *Journal of the American Statistical Association* **83**, 715–728.

Lönnstedt, I. and Speed, T.P. (2002). Replicated microarray data. *Statistica Sinica* **12**, 31–46.

Lukashin, A.V. and Fuchs, R. (2001). Analysis of temporal gene expression profiles: clustering by simulated annealing and determining the optimal number of clusters. *Bioinformatics* **17**, 405–414.

MacQueen, J. (1967). Some methods for classification and analysis of multivariate observations. *Proceedings of the 5th Berkeley Symposium on Mathematical Statistics and Probability.*

Maindonald, J. (2004). Graphics = the human eye + data + theory. Talk given to the Sydney Summer Statistics Workshop, Sydney University, February 2004.

Man, M.Z., Wang, X., and Wang, Y. (2000). POWER SAGE: comparing statistical tests for SAGE experiments. *Bioinformatics* **16**, 953–959.

Manduchi, E. (2000). Generation of patterns from gene expression data by assigning confidence to differentially expressed genes. *Bioinformatics* **16**, 685–698.

Mar, J.C. and McLachlan, G.J. (2003). Model-based clustering in gene expression microarrays: an application to breast cancer data. *International Journal of Software Engineering and Knowledge Engineering* **13**, 579–592.

Marra, M.A., Hiller, L., and Waterston, R.H. (1998). Expressed sequence tags – establishing bridges between genomes. *Trends in Genetics* **14**, 4–7.

Marriott, F.H.C. (1974). *The Interpretation of Multiple Observations*. London: Academic Press.

Marron, J.S. and Todd, M. (2002). Distance weighted discrimination. *Technical report*. Chapel Hill: NC, Department of Statistics, University of North Carolina.

Mateos, A., Herrero, J., Tamames, J., and Dopazo, J. (2001). Supervised and hierarchical unsupervised neural networks for clustering both gene expression profiles and samples. In *Proceedings of CAMDA'01. Methods of Microarray Data Analysis, Vol. 2*. Boston: Kluwer.

McLachlan, G.J. (1977). A note on the choice of a weighting function to give an efficient method for estimating the probability of misclassification. *Pattern Recognition* **9**, 147–149.

McLachlan, G.J. (1982). The classification and mixture maximum likelihood approaches to cluster analysis. In *Handbook of Statistics*, Vol. 2, P.R. Krishnaiah and L. Kanal (Eds.). Amsterdam: North-Holland, pp. 199–208.

McLachlan, G.J. (1987). On bootstrapping the likelihood ratio test statistic for the number of components in a normal mixture. *Applied Statistics* **36**, 318–324.

McLachlan, G.J. (1992). *Discriminant Analysis and Statistical Pattern Recognition*. New York: Wiley.

McLachlan, G.J. (2004). Mixtures of linear mixed models for the clustering of correlated data. *Technical Report*. Brisbane: Department of Mathematics, University of Queensland.

McLachlan, G.J. and Basford, K.E. (1988). *Mixture Models: Inference and Applications to Clustering*. New York: Marcel Dekker.

McLachlan, G.J., Bean, R.W., and Peel, D. (2002). A mixture model-based approach to the clustering of microarray expression data. *Bioinformatics* **18**, 413–422.

McLachlan, G.J., Chang, S.U., Mar, J., and Ambroise, C. (2004). On the simultaneous use of clinical and microarray expression data in the cluster analysis of tissue samples. In *Conferences in Research and Practice in Information Technology*, Vol. 29, Y.-P. Chen (Ed.). Sydney: Australian Computer Society, pp. 161–166.

McLachlan, G.J. and Khan, N. (2004). On a resampling approach for tests on the number of clusters with mixture model-based clustering of tissue samples. *Journal of Multivariate Analysis* **90**, 90–105.

McLachlan, G.J. and Krishnan, T. (1997). *The EM Algorithm and Extensions*. New York: Wiley.

McLachlan, G.J. and Peel, D. (1997). On a resampling approach to choosing the number of components in normal mixture models. In *Computing Science and Statistics*, Vol. 28, L. Billard and N.I. Fisher (Eds.). Fairfax Station, VA: Interface Foundation of North America, pp. 260–266.

McLachlan, G.J. and Peel, D. (1998). Robust cluster analysis via mixtures of multivariate t-distributions, in: *Lecture Notes in Computer Science*, Vol. 1451, A. Amin, D. Dori, P. Pudil, and H. Freeman (Eds.). Berlin: Springer-Verlag.

McLachlan, G.J. and Peel, D. (2000a). Mixtures of factor analyzers. In *Proceedings of the Seventeenth International Conference on Machine Learning*, P. Langley (Ed.). San Francisco: Morgan Kaufmann.

McLachlan, G.J. and Peel, D. (2000b). *Finite Mixture Models*. New York: Wiley.

McLachlan, G.J., Peel, D., Basford, K.E., and Adams, P. (1999). Fitting of mixtures of normal and t-components, *Journal of Statistical Software* **4(2)**.

McLachlan, G.J., Peel, D., and Bean, R.W. (2003). Modelling high-dimensional data by mixtures of factor analyzers. *Computational Statistics and Data Analysis* **41**, 379–388.

McShane, L.M., Radmacher, M.D., Freidlin, B., Yu, R., Li, M.-C., and Simon, R. (2002). Methods for assessing reproducibility of clustering patterns observed in analyses of microarray data. *Bioinformatics* **18**, 1462–1469.

Medvedovic, M. and Sivaganesan, S. (2002). Bayesian infinite mixture model based clustering of gene expression profiles. *Bioinformatics* **18**, 1194–1206.

Meltzer, P.S. (2001). Large-scale genome analysis. In *Bioinformatics: A Practical Guide to the Analysis of Genes and Proteins*, A.D. Baxevanis and B.F.F. Ouellette (Eds.). New York: Wiley-Liss.

Meng, X.L. and van Dyk, D. (1997). The EM algorithm — an old folk song sung to a fast new tune (with discussion). *Journal of the Royal Statistical Society B* **59**, 511–567.

Mertens, B.J.A. (2003). Microarrays, pattern recognition and exploratory data analysis. *Statistics in Medicine* **22**, 1879–1899.

Miles, M.F. (2001). Microarrays: lost in a storm of data? *Nature Reviews Neurosciences* **2**, 441–443.

Miller, A.J. (1984). Selection of subsets of regression variables (with discussion). *Journal of the Royal Statistical Society A* **147**, 389–425.

Miller, A.J. (1990). *Subset Selection in Regression*. London: Chapman & Hall.

Model, F., Adorján, P., Olek, A., and Piepenbrock, C. (2001). Feature selection for DNA methylation based cancer classification. *Bioinformatics* **17**, S157–S164.

Moler, E.J. , Chow, M.L., and Mian, I.S. (2000). Analysis of molecular profile data using generative and discriminative methods. *Physiological Genomics* **4**, 109–126.

Molina, L. C., Belanche, L., and Nebot, A. (2002). Feature selection algorithms: A survey and experimental evaluation. In *IEEE International Conference on Data Mining (ICDM'02)*, Maebashi City, Japan, pp. 306–313.

Montgomery, D. (1999). Very large-scale arrays of biomolecules. *Nature Genetics* **23(S)**, 63.

Morgan, B.J.T. and Ray, A.P.G. (1995). Non-uniqueness and inversions in cluster analysis. *Applied Statistics* **44**, 117–134.

Mosteller, F. and Wallace, D.L. (1963). Inference in an authorship problem. *Journal of the American Statistical Association* **58**, 275–309.

Mukherjee, S., Tamayo, P., Rogers, S., Rifkin, R., Engle, A., Campbell, C., Golub, T.R., and Mesirov, J.P. (2003). Estimating dataset size requirements for classifying DNA microarray data. *Journal of Computational Biology* **10**, 119–142.

Mullis, K.B. (1990). The unusual origin of the polymerase chain reaction. *Scientific American* **256**, 56–65.

Munson, P. (2001). A 'consistency' test for determining the significance of gene expression changes on replicate samples and two convenient variance-stabilizing transformations. *GeneLogic Workshop on Low Level Analysis of Affymetrix GeneChip Data*, Bethesda, MD.

Muro, S., Takemasa, I., Oba, S., Matoba, R., Ueno, N., Maruyama, C., Yamashita, R., Sekimoto, M., Yamamoto, H., Nakamori, S., Monden, M., Ishii, S., and Kato, K. (2003). Identification of expressed genes linked to malignancy of human colorectal carcinoma by parametric clustering of quantitative expression data. *Genome Biology* **4(5)**, Article R21.

Murray, G.D. (1977). A cautionary note on selection of variables in discriminant analysis. *Applied Statistics* **26**, 246–250.

Myles, J.P. and Hand, D.J. (1990). The multi-class metric problem in nearest neighbour discrimination rules. *Pattern Recognition* **23**, 1291–1297.

Nallur, G., Luo, C., Fang, L., Cooley, S., Dave, V., Lambert, J., Kukanskis, K., Kingsmore, S., Lasken, R., and Schweitzer, B. (2001). Signal amplification by rolling circle amplification on DNA microarrays. *Nucleic Acids Research* **29**, e118.

National Institutes of Health (2001). National Center for Biotechnology Information, GenBank overview. `http://www.ncbi.nlm.nih.gov/Genbank/GenbankOverview.html` Bethesda, MD: NIH.

Newton, M.A. and Kendziorski, C. (2003). Parametric empirical Bayes methods for microarrays. In *The Analysis of Gene Expression Data: Methods and Software*, G. Parmigiani, E.S. Garrett, R.A. Irizarry, and S.L. Zeger (eds.), Springer: New York, pp. 254–271.

Newton, M.A., Kendziorski, C.M., Richmond, C.S., Blattner, F.R., and Tsui, K.W. (2001). On differential variability of expression ratios: improving statistical inference about gene expression changes from microarray data. *Journal of Computational Biology* **8**, 37–52.

Newton, M.A., Noueiry, A., Sarkar, D., and Ahlquist, P. (2004). Detecting differential gene expression with a semiparametric hierarchical mixture method. *Biostatistics* **5**, 155–176.

Ng, S.K. and McLachlan, G.J. (2004). Applying the EM algorithm in training neural networks: misconceptions and a new algorithm for multiclass classification. *IEEE Transactions on Neural Networks*. To appear.

Nguyen, D.V., Arpat, A.B., Wang, N., and Carroll, R.J. (2002). DNA microarray experiments: biological and technological aspects. *Biometrics* **58**, 701–717.

Nguyen, D.V. and Rocke, D.M. (2001). Classification of acute leukemia based on DNA microarray gene expressions using partial least squares. In *Methods of Microarray Data Analysis*, S.M. Lin and K.F. Johnson (Eds.). Dordrecht, The Netherlands: Kluwer, pp. 109–124.

Nguyen, D.V. and Rocke, D.M. (2002a). Tumor classification by partial least squares using microarray gene expression data. *Bioinformatics* **18**, 39–50.

Nguyen, D.V. and Rocke, D.M. (2002b). Multi-class cancer classification via partial least squares with gene expression profiles. *Bioinformatics* **18**, 1216–1226.

Nguyen, D.V. and Rocke, D.M. (2002c). Partial least squares proportional hazard regression for application to DNA microarray survival data. *Bioinformatics* **18**, 1625–1632.

Nikkila, J., Törönen, P., Kaski, S., Venna, J., Castrén, E. , and Wong, G. (2002). Analysis and visualization of gene expression data using self-organizing maps. *Neural Networks* **15**, 953–966.

Notterman, D.A., Alon, U., Sierk, A.J., and Levine, A.J. (2001). Transcriptional gene expression profiles of colorectal adenoma, adenocarcinoma, and normal tissue examined by oligonucleotide arrays. *Cancer Research* **61**, 3124–3130.

Novak, J.P., Sladek, R., and Hudson, T.J. (2003). Characterization of variability in large-scale gene expression data: implications for study design. *Genomics* **79**, 104–113.

Pan, W. (2002). A comparative review of statistical methods for discovering differentially expressed genes in replicated microarray experiments. *Bioinformatics* **18**, 546–554.

Pan, W. (2003). On the use of permutation in and the performance of a class of nonparametric methods to detect differential gene expression. *Bioinformatics* **19**, 1333–1340.

Pan, W., Lin, J., and Le, C.T. (2002a). Model-based cluster analysis of microarray gene-expression data. *Genome Biology* **3**, research0009.1–0009.8.

Pan, W., Lin, J., and Le, C. (2002b). How many replicates of arrays are required to detect gene expression changes in microarray experiments? A mixture model approach. *Genome Biology* **3**, research0022.1–0022.10.

Pan, W., Lin, J., and Le, C.T. (2003). A mixture model approach to detecting differentially expressed genes with microarray data. *Functional and Integrative Genomics* **3**, 117–124.

Park, P.J., Pagano, M., and Bonetti, M. (2001). A nonparametric scoring algorithm for identifying informative genes from microarray data. *Pacific Symposium on Biocomputing* **6**, 52–63.

Park, P.J., Tian, L., and Kohane, I.S. (2002). Linking gene expression data with patient survival times using partial least squares. *Bioinformatics* **18**, S120–S127.

Park, T., Yi, S.-G., Lee, S., Lee, S.Y., Yoo, D.-H., Ahn, J.-I., Lee, Y.-S. (2003). Statistical tests for identifying differentially expressed genes in time-course microarray experiments. *Bioinformatics* **19**, 694–703.

Parmigiani, G., Garrett, E.S., Irizarry, R.A., and Zeger, S.L. (Eds.) (2003). *The Analysis of Gene Expression Data.* New York: Springer-Verlag.

Pavlidis, P., Li, Q., and Noble, W.S. (2003) The effect of replication on gene expression microarray experiments. *Bioinformatics* **19**, 1620–1627.

Pavlidis, P., Weston, J., Cai, J., and Grundy, W.N. (2001). Gene functional classification from heterogeneous data. *RECOMB 2001: Proceedings of the Fifth Annual International Conference on Computational Biology.* New York: ACM Press, pp. 249–255.

Peel, D. and McLachlan, G.J. (2000). Robust mixture modelling using the t-distribution, *Statistics and Computing* **10**, 335–344.

Pepe, M.S., Longton, G., Anderson, G.L., and Schummer, M. (2003). Selecting differentially expressed genes from microarray experiments. *Biometrics* **59**, 133–142.

Perou, C.M., Jeffrey, S.S., van de Rijn, M., Rees, C.A., Eisen, M.B., Ross, D.T., Pergamenschikov, A., Williams, C.F., Zhu, S.X., Lee, J.C.F., Lashkari, D., Shalon, D., Brown, P.O., and Botstein, D. (1999). Distinctive gene expression patterns in human mammary epithelial cells and breast cancers. *Proceedings of the National Academy of Sciences USA* **96**, 9212–9217.

Phillips, D.B. and Smith, A.F.M. (1996). Bayesian model comparison via jump diffusions. In *Markov Chain Monte Carlo in Practice*, W.R. Gilks, S. Richardson, and D.J. Spiegelhalter (Eds.). London: Chapman & Hall, pp. 215–239.

Pittelkow, Y.E. and Wilson, S.R. (2003). Visualisation of gene expression data - the GE-biplot, the Chip-plot and the Gene-plot. *Statistical Applications in Genetics and Molecular Biology* **2** *No. 1*, Article 6.

Pittelkow, Y. and Wilson, S.R. (2004). Use of principal component analysis and of the GE-plot for the graphical exploration of gene expression data. Unpublished manuscript.

Platt, J.C., Cristianini, N., and Shawe-Taylor, J. (2000). Large Margin DAGs for multiclass classification. *Advances in Neural Information Processiong Systems* **12**, 547–553.

Pollack, J.R., Sørlie, T., Perou, C.M., Rees, C.A., Jeffrey, S.S., Lønning, P.E., Tibshirani, R., Botstein, D., Børresen-Dale, A.-L., and Brown, P.O. (2002). Microarray analysis reveals a major direct role of DNA copy number alteration in the transcriptional program of human breast tumors. *Proceedings of the National Academy of Sciences USA* **99**, 12963–12968.

Pollard, K.S. and van der Laan, M.J. (2002). Statistical inference for simultaneous clustering of gene expression data. *Mathematical Biosciences* **176**, 99–121.

Polsky-Cynkin, R., Parsons, G.H., Allerdt, L., Landes, G., Davis, G., and Rashtchian, A. (1985). Use of DNA immobilized on plastic and agarose supports to detect DNA by sandwich hybridization. *Clinical Chemistry* **31**, 1438–1443.

Pomeroy, S.L., Tamayo, P., Gaasenbeek, M., Sturla, L.M., Angelo, M., McLaughlin, M.E., Kim, J.Y.H., Goumnerova, L.C., Black, P.M., Lau, C., Allen, J.C., Zagzag, D., Olson, J.M., Curran, T., Wetmore, C., Biegel, J.A., Poggio, T., Mukherjee, S., Rifkin, R., Califano, A., Stolovitzky, G., Louis, D.N., Mesirov, J.P., Lander, E.S., and Golub, T.R. (2002). Prediction of central nervous system embryonal tumour outcome based on gene expression. *Nature* **415**, 436–442.

Pritchard, C.C., Hsu, L., Delrow, J., and Nelson, P.S. (2001). Project normal: defining normal variance in mouse gene expression. *Proceedings of the National Academy of Sciences USA* **98**, 13266–13271.

Pudil, P., Novovičová, J., and Kittler, J. (1994). Floating search methods in feature selection. *Pattern Recognition Letters* **15**, 1119–1125.

Quackenbush, J. (2001). Computational genetics: computational analysis of microarray data. *Nature Reviews Genetics* **2**, 418–427.

Quinlan, R. (1986). Induction of decision trees. *Machine Learning* **1**, 81–106.

Quinlan, R. (1993). C4.5: *Programs for Machine Learning*. San Mateo, CA: Morgan Kaufmann.

Radmacher, M.D., McShane, L.M., and Simon, R. (2002). A paradigm for class prediction using gene expression profiles. *Journal of Computational Biology* **9**, 505–511.

Ramaswamy, S., Ross, K.N., Lander, E.S., and Golub, T.R. (2003). A molecular signature of metastasis in primary solid tumors. *Nature Genetics* **33**, 49–54.

Ramaswamy, S., Tamayo, P., Rifkin, R., Mukherjee, S., Yeang, C.-H., Angelo, M., Ladd, C., Reich, M., Latulippe, E., Mesirov, J.P., Poggio, T., Gerald, W., Loda, M., Lander, E.S., and Golub, T.R. (2001). Multiclass cancer diagnosis using tumor gene expression signatures. *Proceedings of the National Academy of Sciences USA* **98**, 15149–15154.

Ramdas, L., Coombes, K.R., Baggerly, K., Abruzzo, L., Highsmith, W.E., Krogmann, T., Hamilton, S.R., and Zhang, W. (2001). Sources of nonlinearity in cDNA microarray expression measurements. *Genome Biology* **2**, research0047.1–0047.7.

Ramoni, M.F., Sebastiani, P., and Kohane, I.S. (2002). Cluster analysis of gene expression dynamics. *Proceedings of the National Academy of Sciences USA* **99**, 9121–9126.

Ramsey, J.M. (1999). The burgeoning power of the shrinking laboratory. *Nature Biotechnology* **17**, 1061–1962.

Rao, C.R. (1948). The utilization of multiple measurements in problems of biological classification. *Journal of the Royal Statistical Society B* **10**, 159–203.

Rao, C.R. (1954). A general theory of discrimination when the information about alternative population distributions is based on samples. *Annals of Mathematical Statistics* **25**, 651–670.

Rattray, M., Morrison, N., Hoyle, D., and Brass, A. (2001). DNA microarray normalisation, PCA and a related latent variable model, http://www.cs.man.ac.uk/ai/Papers/magnus/normalisation_draft_v1.1.ps.gz.

Raychaudhuri, S., Schutze, H., and Altman, R.B. (2002). Using text analysis to identify functionally coherent gene groups. *Genome Research* **12**, 1582-1590.

Reiner, A., Yekutieli, D., and Benjamini, Y. (2003). Identifying differentially expressed genes using false discovery rate controlling procedures. *Bioinformatics* **19**, 368–375.

Richardson, S. and Green, P.J. (1997). On Bayesian analysis of mixtures with an unknown number of components (with discussion). *Journal of the Royal Statistical Society B* **59**, 731–792. Correction (1998). *Journal of the Royal Statistical Society B* **60**, 661.

Ripley, B.D. (1996). *Pattern Recognition and Neural Networks.* Cambridge University Press.

Rocke, D.M. and Durbin, B. (2001). A model for measurement error for gene expression arrays. *Journal of Computational Biology* **8**, 557–569.

Rocke, D.M. and Durbin, B. (2003). Approximate variance-stabilizing transformations for a gene-expression microarray data. *Bioinformatics* **19**, 966–972.

Ross, D.T., Scherf, U., Eisen, M.B., Perou, C.M., Rees, C., Spellman, P., Iyer, V., Jeffrey, S.S., van de Rijn, M., Waltham, M., Pergamenschikov, A., Lee, J.C.F., Lashkari, D., Shalon, D., Myers, T.G., Weinstein, J.N., Botstein, D., and Brown, P.O. (2000). Systematic variation in gene expression patterns in human cancer cell lines. *Nature Genetics* **24** 227–235.

Roth, F.P. (2001). Bringing out the best features of expression data. *Genome Research* **11**, 1801–1802.

Sapir, M. and Churchill, G.A. (2000). Estimating the posterior probability of differential gene expression from microarray data. Poster, Bar Harbor, ME: The Jackson Laboratory.

Satterthwaite, F.E. (1946). An approximate distribution of estimates of variance components. *Biometrics* **2**, 110.

Schadt, E.E., Cheng, L., Cheng, S., and Wong, W.H. (2000). Analyzing high-density oligonucleotide gene expression array data. *Journal of Cellular Biochemistry* **80**, 192–202.

Schadt, E.E., Cheng, L., Ellis, B., and Wong, W.H. (2001). Feature extraction and normalization algorithms for high-density oligonucleotide gene expression array data. *Journal of Cellular Biochemistry* **84**, 120–125.

Schena, M. and Davis, R.W. (1999). Genes, genomes and chips. In *DNA Microarrays: A Practical Approach*, M. Schena (Ed.). New York: Oxford University Press.

Schena, M., Shalon, D., Davis, R.W., and Brown, P.O. (1995). Quantitative monitoring of gene expression patterns with a complementary DNA microarray. *Science* **270**, 467–470.

Schena, M., Shaon, D., Heller, R., Chai, A., Brown, P., and Davis, R.W. (1996). Parallel human genome analysis: microarray-based expression monitoring of 1000 genes. *Proceedings of the National Academy of Sciences USA* **93**, 10614–10619.

Scherf, U., Ross, D.T., Waltham, M., Smith, L.H., Lee, J.K., Tanabe, L., Kohn, K.W., Reinhold, W.C., Myers, T.G., Andrews, D.T., Scudiero, D.A., Eisen, M.B., Sausville, E.A., Pommier, Y., Botstein, D., Brown, P.O., and Weinstein, J.N. (2000). A gene expression database for the molecular pharmacology of cancer. *Nature Genetics* **24**, 236–244.

Schermer, M.J. (1999). Confocal scanning in microscopy in microarray detection. In *DNA Microarrays: A Practical Approach*, M. Schena (Ed.). New York: Oxford University Press.

Schiavo, R. and Hand, D.J. (2000). Ten more years of error rate research. *International Statistical Review* **68**, 295–310.

Schubert, C.M. (2003). Microarray to be used as routine clinical screen. *Nature Medicine* **9**, 9.

Schuchhardt, J., Beule, D., Malik, A., Wolski, E., Eickhoff, H., Lehrach, H., and Herzel, H. (2000). Normalization strategies for cDNA microarrays. *Nucleic Acids Research* **28**, e47.

Schwarz, G. (1978). Estimating the dimension of a model. *Annals of Statistics* **6**, 461–464.

Scott, A.J. and Symons, M.J. (1971). Clustering methods based on likelihood ratio criteria. *Biometrics* **27**, 387–397.

Schweder, T. and Spjøvtoll, E. (1982). Plots of p-values to evaluate many tests simultaneously. *Biometrika* **69**, 493–502.

Sebastiani, P., Gussoni, E., Kohane, I.S., and Ramoni, M.F. (2003). Statistical challenges in functional genomics. *Statistical Science* **18**, 33–71.

Seber, G.A.F. (1984). *Multivariate Observations*. New York: Wiley.

Segal, E., Taskar, B., Gasch, A., Friedman, N., and Koller, D. (2001). Rich probabilistic models for gene expression. *Bioinformatics* **17**, S243–S252.

Segal, E., Wang, H., and Koller, D. (2003a). Discovering molecular pathways from protein interaction and gene expression data. *Bioinformatics* **19**(Suppl. 1), i264–i272.

Segal, E.,, Yelensky, R., and Koller, D. (2003b). Genome-wide discovery of transcriptional modules from DNA sequence and gene expression. *Bioinformatics* **19**(Suppl. 1), i273–i282.

Shannon, W., Culverhouse, R., and Duncan, J.. (2002). Analyzing microarray data using cluster analysis. *Pharmacogenomics* **4**, 41–51.

Sherlock, G., Hernandex-Boussard, T., Kasarski, A., Binkley G., Matese, J.C., Dwight, S.S., Kaloper, M., Weng, S., Jin, H., Ball, C.A., Eisen, M.B., Spellman, P.T., Brown, P.O., Botstein, D., and Cherry, J.M. (2001). The Stanford microarray database. *Nucleic Acids Research* **29**, 152–155.

Shipp, M.A., Ross, K.N., Tamayo, P., Weng, A.P., Kutok, J.L., Aguiar, R.C.T., Gaasenbeek, M., Angelo, M., Reich, M., Pinkus, G.S., Ray, T.S., Koval, M.A., Last, K.W., Norton, A., Lister, T.A., Mesirov, J., Neuberg, D.S., Lander, E.S., Aster, J.C., and Golub, T.R. (2002). Diffuse large B-cell lymphoma outcome prediction by gene-expression profiling and supervised machine learning. *Nature Medicine* **8**, 68–74.

Sibson, R. (1978). Studies in the robustness of multidimensional scaling: Procrustes statistics. *Journal of the Royal Statistical Society B* **40**, 234–238.

Šidák, Z. (1967). Rectangular confidence regions for the means of multivariate normal distributions. *Journal of the American Statistical Association* **62**, 626–633.

Simes, R.J. (1986). An improved Bonferroni procedure for multiple tests of significance. *Biometrika* **73**, 751–754.

Simon, R., Radmacher, M.D., and Dobbin, K. (2002). Design of studies using DNA. *Genetic Epidemiology* **23**, 21–36.

Simon, R., Radmacher, M.D., Dobbin, K., and McShane, L.M. (2003). Pitfalls in the use of DNA microarray data for diagnostic and prognostic classification. *Journal of the National Cancer Institute* **95**, 14–18.

Sinclair, B. (1999). Everything's great when it sits on a chip. *The Scientist* **13**, 18.

Slonim, D. (2002). From patterns to pathways: gene expression data analysis comes of age. *Nature Genetics* **32(S)**, 502–508.

Smith, C.A.B. (1947). Some examples of discrimination. *Annals of Eugenics* **13**, 272–282.

Smyth, G.K. (2004). Linear models and empirical Bayes methods for assessing differential expression in microarray experiments. *Statistical Applications in Genetics and Molecular Biology* **3** *No. 1*, Article 3.

Smyth, G.K., Yang, Y.H., and Speed, T. (2002). Statistical issues in cDNA microarray data analysis. In *Functional Genomics*, M.J. Brownstein and A.B. Khodursky (Eds.), Methods in Molecular Biology Series. Totowa, NJ: Humana Press, pp. 111–136.

Smyth, P. (2000). Model selection for probabilistic clustering using cross-validated likelihood. *Statistics and Computing* **10**, 63–72.

Snapinn, S.M. and Knoke, J.D. (1989). Estimation of error rates in discriminant analysis with selection of variables. *Biometrics* **45**, 289–299.

Soille, P. (1999). *Morphological Image Analysis: Principles and Applications.* New York: Springer-Verlag.

Sørlie, T., Perou, C.M., Tibshirani, R., Turid, A., Geisler, S., Johnsen, H., Hastie, T., Eisen, M.B., van de Rijn, M., Jeffrey, S.S., Thorsen, T., Quist, H., Matese, J.C., Brown, P.O., Botstein, D., Lønning, P.E., and Børresen-Dale, A.-L. (2001). Gene expression patterns of breast carcinomas distinguish tumor subclasses with clinical implications. *Proceedings of the National Academy of Sciences USA* **98**, 10869–10874.

Southern, E.M. (1975). Detection of specific sequences among DNA fragments separated by gel electrophoresis. *Journal of Molecular Biology* **98**, 503–517.

Southern, E.M., Mir, K., and Shchepinov, M. (1999). Molecular interactions on microarrays. *Nature Genetics* **21(S)**, 5–9.

Spang, R. (2003). Diagnostic signatures from microarrays: a bioinformatics concept for personalized medicine. *Biosilico* **1**, 64–68.

Speed, T. (Ed.) (2003). *Statistical Analysis of Gene Expression Microarray Data.* Boca Raton, FL: Chapman & Hall/CRC.

Stears, R.L., Getts, R.C., and Gullans, S.R. (2000). A novel, sensitive detection system for high-density microarrays using dendrimer technology. *Physiological Genomics* **3**, 93–99.

Stone, C.J. (1977). Consistent nonparametric regression (with discussion). *Annals of Statistics* **5**, 595–645.

Stone, M. (1974). Cross-validatory choice and assessment of statistical predictions (with discussion). *Journal of the Royal Statistical Society B* **36**, 111–147.

Stone, M. and Brooks, R.J. (1990). Continuum regression: cross-validated sequentially constructed prediction embracing ordinary least squares, partial least squares, and principal components regression (with discussion). *Journal of the Royal Statistical Society B* **52**, 237–269.

Storey, J.D. (2002). A direct approach to false discovery rates. *Journal of the Royal Statistical Society B* **64**, 479–498.

Storey, J. (2004). The positive false discovery rate: a Bayesian interpretation and the *q*-value. *Annals of Statistics* **31**, 2013–2035.

Storey, J., Taylor, J.E., and Siegmund, D. (2004). Strong control, conservative point estimation and simultaneous conservative consistency of false discovery rates: a unified approach. *Journal of the Royal Statistical Society B* **66**, 187–205.

Storey, J.D. and Tibshirani, R. (2003a). SAM thresholding and false discovery rates for detecting differential gene expression in DNA microarrays. In *The Analysis of Gene Expression Data: Methods and Software*, G. Parmigiani, E.S. Garrett, R.A. Irizarry, and S.L. Zeger (Eds.), New York: Springer-Verlag.

Storey, J.D. and Tibshirani, R. (2003b). Statistical significance for genome-wide studies. *Proceedings of the National Academy of Sciences USA* **100**, 9440–9445.

Sturn, A., Quackenbush, J., and Trajanoski, Z. (2000). Genesis: cluster analysis of microarray data. *Bioinformatics* **18**, 207–208.

Tamayo, P., Slonim, D., Mesirov, J., Zhu, Q., Kitareewan, S., Dmitrovsky, E., Lander, E.S., and Golub, T.R. (1999). Interpreting patterns of gene expression with self-organizing maps: methods and application to hematopoietic differentiation. *Proceedings of the National Academy of Sciences USA* **96**, 2907–2912.

Tavazoie, S., Hughes, J.D., Campbell, M.J., Cho, R.J., and Church, G.M. (1999). Systematic determination of genetic network architecture. *Nature Genetics* **22**, 281–285.

Theilhaber, J., Bushnell, S., Jackson, A., and Fuchs, R. (2001). Bayesian estimation of fold-changes in the analysis of gene expression: the PFOLD algorithm. *Journal of Computational Biology* **8**, 585–614.

Theriault, T.P., Winder, S.C., and Gamble, R.C. (1999). Application of ink-jet printing technology to the manufacture of molecular arrays. In *DNA Microarrays: A Practical Approach*, M. Schena (Ed.). New York: Oxford University Press.

Thomas, J.G., Olson, J.M., Tapscott, S.J., and Zhao, L.P. (2001). An efficient and robust statistical modeling approach to discover differentially expressed genes using genomic expression profiles. *Genome Biology* **11**, 1227–1236.

Tibshirani, R.J. and Efron, B. (2002). Pre-validation and inference in microarrays. *Statistical Applications in Genetics and Molecular Biology* **1**: No. 1, Article 1.

Tibshirani, R., Hastie, T., Eisen, M., Ross, D., Botstein, D., and Brown, P. (1999). Clustering methods for the analysis of DNA microarray data. *Technical Report*. Stanford, CA: Department of Statistics, Stanford University.

Tibshirani, R.J., Hastie, T., Narasimhan, B., and Chu, G. (2002a). Diagnosis of multiple cancer types by shrunken centroids of gene expression. *Proceedings of the National Academy of Sciences USA* **99**, 6567–6572.

Tibshirani, R., Hastie, T., Narasimhan, B., and Chu, G. (2003). Class prediction by nearest shrunken centroids, with applications to DNA microarrays. *Statistical Science* **18**, 104–117.

Tibshirani, R., Hastie, T., Narasimhan, B., Eisen, M., Sherlock, G., Brown, P., and Botstein, D. (2002b). Exploratory screening of genes and clusters from microarray experiments. *Statistica Sinica* **12**, 47–59.

Tibshirani, R., Walther, G., and Hastie, T. (2001). Estimating the number of clusters in a data set via the gap statistic. *Journal of the Royal Statistical Society B* **63**, 411–423.

Tipping, M.E. and Bishop, C.M. (1999). Mixtures of probabilistic principal component analysers. *Neural Computation* **11**, 443–482.

Todeschini, R. (1989). k-Nearest neighbour method: the influence of data transformations and metrics. *Chemometrics and Intelligent Laboratory Systems* **6**, 213–220.

Toussaint, G.T. (1974). Bibliography on estimation of misclassification. *IEEE Transactions on Information Theory* **20**, 472–479.

Toussaint, G.T. and Sharpe, P.M. (1975). An efficient method for estimating the probability of misclassification applied to a problem in medical diagnosis. *Computers in Biology and Medicine* **4**, 269–278.

Troyanskaya, O.G., Cantor, M., Sherlock, G., Brown, P., Hastie, T., Tibshirani, R., Botstein, D., and Altman, R.B. (2001). Missing value estimation methods for DNA microarrays. *Bioinformatics* **17**, 520–525.

Troyanskaya, O.G., Dolinski, K., Owen, A.B., Altman, R.B., and Botstein, D. (2003). A Bayesian framework for combining heterogeneous data sources for gene function prediction (in *Saccharomyces cerevisiae*). *Proceedings of the National Academy of Sciences USA* **100**, 8348–8353.

Troyanskaya, O.G., Garber, M.E. Brown, P.O., Botstein, D., and Altman, R.B. (2002). Nonparametric methods for identifying differentially expressed genes in microarray data. *Bioinformatics* **18**, 1454–1461.

Tsai, C.-A., Hsueh, H.-M., and Chen, J.J. (2003). estimation of false discovery rates in multiple testing: application to gene microarray data. *Biometrics* **59**, 1071–1081.

Tusher, V.G., Tibshirani, R., and Chu, G. (2001). Significance analysis of microarrays applied to the ionizing radiation response. *Proceedings of the National Academy of Sciences USA* **98**, 5116–5121.

Valk, P.J.M., Verhaak, R.G.W., Beijen, M.A., Erpelinck, C.A.J., Barjesteh van Waalwijk van Doorn-Khosrovani, S., Boer, J.M., Beverloo, H.B., Moorhouse, M.J., van der Spek, P.J., Löwenberg, B., and Delwel, R. (2004). Prognostically useful gene-expression profiles in acute myeloid leukemia. *New England Journal of Medicine* **350**, 1617–1628.

van 't Veer, L.J., Dai, H., van de Vijver, M.J., He, Y.D., Hart, A.A.M., Mao, M., Peterse, H.L., van der Kooy, K., Marton, M.J., Witteveen, A.T., Schreiber, G.J., Kerkhoven, R.M., Roberts, C., Linsley, P.S., Bernards, R., and Friend, S.H. (2002). Gene expression profiling predicts clinical outcome of breast cancer. *Nature* **415**, 530–536.

van der Laan, M.J. and Bryan, J.F. (2000). Gene expression analysis with the parametric bootstrap. *Biostatistics* **2**, 445–461.

van de Vijver, M.J., He, Y.D., van 't Veer, L.J., Dai, H., Hart, A.A.M., Voskuil, D.W., Schreiber, G.J., Peterse, J.L., Roberts, C., Marton, M.J., Parrish, M., Atsma, D., Witteveen, A., Glas, A., Delahaye, L., van der Velde, T., Bartelink, H., Rodenhuis, S., Rutgers, E.T., Friend, S.H., and Bernards, R. (2002). A gene-expression signature as a predictor of survival in breast cancer. *New England Journal of Medicine* **347**, 1999–2009.

Vapnik, V. (1998). *Statistical Learning Theory*. New York: Wiley.

Velculescu, V.E., Zhang, L., Vogelstein, B., and Kinzler, K.W. (1995). Serial analysis of gene expression. *Science* **270**, 484–487.

Vilo, J., Brazma, A., Jonassen, I., Robinson, A., and Ukkonen, E. (2000). Mining for putative regulatory elements in the yeast genome using gene expression data. In *Proceedings of the Eighth Annual International Conference on Intelligent Systems for Molecular Biology*, R. Altman, T.L. Bailey, P. Bourne, M. Gribskov, T. Lengauer, and I. N. Shindyalov (Eds.). La Jolla, CA: AAAI Press, pp. 384–394.

Virtanen, C., Ishikawa, Y., Honjoh, D., Kimura, M., Shimane, M., Miyoshi, T., Nomura, H., and Jones, M.H. (2002). Integrated classification of lung tumors and cell lines by expression profiling. *Proceedings of the National Academy of Sciences USA* **99**, 12357–12362.

Wald, A. (1944). On a statistical problem arising in the classification of an individual into one of two groups. *Annals of Mathematical Statistics* **15**, 145–162.

Wall, M.E., Dyck, P.A., and Brettin, T.S. (2001). SVDMAN—singular value decomposition analysis of microarray data. *Bioinformatics* **17**, 566–568.

Ward, J.H. (1963). Hierarchical grouping to optimize an objective function. *Journal of the American Statistical Association* **58**, 236–244.

Warrington, J.A., Dee, S., and Trulson, M. (2000). Large-scale genomic analysis using Affymetrix GeneChip® probe arrays. In *Microarray Biochip Technology*, M. Schena (Ed.). Natick, MA: Eaton Publishing.

Weeraratna, A.T., Nagel, J.E., Mello-Coelho, V.V., and Taub, D.D. (2004). Gene expression profiling: from microarrays to medicine. *Journal of Clinical Immunology* **24**, 213–224.

Wei, P., Lin, J., and Le, C.T. (2002). Model-based cluster analysis of microarray gene-expression data. *Genome Biology* **3**, research0009.1–0009.8.

Wen, X., Fuhrman, S., Michaels, G.S., Carr, D.B., Smith, S., Barker, J.L., and Somogyi, R. (1997). Large-scale temporal gene expression mapping of central nervous system development. *Proceedings of the National Academy of Sciences USA* **95**, 334–339.

West, M., Blanchette, C., Dressman, H., Huang, E., Ishida, S., Sprang, R., Zuzan, H., Olson, J.A., Marks, J.R., and Nevins, J.R. (2001). Predicting the clinical status of human breast cancer by using gene expression profiles. *Proceedings of the National Academy of Sciences USA* **98**, 11462–11467.

Westfall, P.H. and Young, S.S. (1993). *Resampling Based Multiple Testing: Examples and Methods for p-Value Adjustment*. New York: Wiley.

Westfall, P.H., Zaykin, D.V., and Young, S.S. (2001). Multiple tests for genetic effects in association studies. In *Biostatistical Methods*, S. Looney (Ed.). Totowa, New Jersey: Humana, pp. 143–168.

Weston, J. and Watkins, C. (1999). Support vector machines for multi-class pattern recognition. In *Proceedings of ESANN99*, M. Verleysen (Ed.). Brussels: Facto Press, pp. 219–224.

Wigle, D.A., Jurisica, I., Radulovich, N., Pintilie, M., Rossant, J., Liu, N., Lu, C., Woodgett, J., Seiden, I., Johnston, M., Keshavjee, S., Darling, G., Winton, T., Breitkreutz, B.J., Jorgenson, P., Tyers, M., Shepherd, F.A., and Tsao, M.S. (2002). Molecular profiling of non-small cell lung cancer and correlation with disease-free survival. *Cancer Research* **62**, 3005–3008.

Wilcoxon, F. (1945). Individual comparisons by ranking methods. *Biometrics Bulletin* **1**, 80–83.

Wold, H. (1966). Estimation of principal components and related models by iterative least squares. In *Mulitvariate Analysis*, P. Krishnaiah (Ed.). New York: Academic Press, pp. 391–420.

Wolfinger, R.D., Gibson, G., Wolfinger, E.D., Bennett, L., Hamadeh, H., Bushel, P., Afshari, C., and Paules, R.S. (2001). Assessing gene significance from cDNA microarray expression data via mixed models. *Journal of Computational Biology* **8**, 625–637.

Wolfsberg, T.G. and Landsman, D. (2001). Expressed sequence tags (ESTs). In *Bioinformatics: A Practical Guide to the Analysis of Genes and Proteins*, A.D. Baxevanis, and B.F.F. Ouellette (Eds.). New York: Wiley-Liss.

Wouters, L., Göhlmann, H.W., Bijnens, L., Kass, S.U., Molenberghs, G., and Lewi, P.J. (2004). Graphical exploration of gene expression data: a comparative study of three multivariate methods. *Biometrics* **59**, 1131–1139.

Wu, H., Kerr, M.K., Cui, X., Churchill, G.A. (2003). MAANOVA: A software package for the analysis of spotted cDNA microarray experiments. In *The analysis of Gene Expression Data: Methods and Software*. New York: Springer-Verlag. To appear.

Wu, T.D. (2001). Analyzing gene expression data from DNA microarrays to identify candidate genes. *Journal of Pathology* **195**, 53–65.

Xiong, M., Li, W., Zhao, J., Jin, L., and Boerwinkle, E. (2001). Feature (gene) selection in gene expression-based tumor classification. *Molecular Genetics and Metabolism* **73**, 239–247.

Yang, Y.H., Buckley, M.J., Dudoit, S., and Speed, T. (2002a). Comparison of methods for image analysis on cDNA microarray data. *Journal of Computational and Graphical Statistics* **11**, 108–136.

Yang, Y.H., Dudoit, S., Luu, P., Lin, D.M., Peng, V., Ngai, J., and Speed, T.P. (2002b). Normalization for cDNA microarray data: a robust composite method addressing single and multiple slide systematic variation. *Nucleic Acids Research* **30**, e15.

Yang, Y.H., Dudoit, S., Luu, P., and Speed, T.P. (2001a). Normalization for cDNA microarray data. In *Microarrays: Optical Technologies and Informatics*, Vol. 4266 of Proceedings of SPIE.

Yang, Y.H., Dudoit, S., Luu, P., and Speed, T.P. (2001b). Normalization for cDNA microarray data. *Technical Report No. 589.* Berkeley: Department of Statistics, University of California.

Yeang, C.-H., Ramaswamy, S., Tamayo, P., Makherjee, S., Rifkin, R.M., Angelo, M., Reich, M., Lander, E., Mesirov, J., and Golub, T. (2001). Molecular classification of multiple tumor types. *Bioinformatics* **17**, S316–S322.

Yeung, K.Y., Fraley, C., Murua, A., Raftery, A.E., and Ruzzo, W.L. (2001a). Model-based clustering and data transformations for gene expression data. *Bioinformatics* **17**, 977–987.

Yeung, K.Y., Haynor, D.R., and Ruzzo, W.L. (2001b). Validating clustering for gene expression data. *Bioinformatics* **17**, 309–318.

Yeung, K.Y., Medvedovic, M., and Bumgarner, R.E. (2003). Clustering gene-expression data with repeated measurements. *Genome Biology* **4** *No.* 5, Article R34.

Yeung, K.Y. and Ruzzo, W.L. (2001). Principal component analysis for clustering gene expression data. *Bioinformatics* **17**, 763–774.

Yue, H., Eastman, P.S., Wang, B.B., Minor, J., Doctolero, M.H., Nuttall, R.L., Stack, R., Becker, J.W., Montgomery, J.R., Vainer, M., and Johnston, R. (2001). An evaluation of the performance of cDNA microarrays for detecting changes in global mRNA expression. *Nucleic Acids Research* **29**, e41.

Zarrinkar, P.P., Mainquist, J.K., Zamora, M., Stern, D., Welsh, J.B., Sapinoso, L.M., Hampton, G.M., and Lockhart, D.J. (2001). Arrays of arrays for high-throughput gene expression profiling. *Genome Research* **11**, 1256–1261.

Zhang, H., Yu, C.-Y., Singer, B., and Xiong, M. (2001). Recursive partitioning for tumor classification with gene expression microarray data. *Proceedings of the National Academy of Sciences USA* **98**, 6730–6735.

Zhang, K. and Zhao, H. (2000). Assessing reliability of gene clusters from gene expression data. *Functional and Integrative Genomics* **1**, 156–173.

Zhang, L., Miles, M.F., and Aldape, K.D. (2003). A model of molecular interactions on short oligonucleotide microarrays. *Nature Biotechnology* **21**, 818–821.

Zhang, W. and Shmulevich, I. (Eds.). (2003). *Computational and Statistical Approaches to Genomics.* Dordrecht: Kluwer.

Zhao, Y. and Pan, W. (2003). Modified nonparametric approaches to detecting differentially expressed genes in replicated microarray experiments. *Bioinformatics* **19**, 1046–1054.

Zhu, H., Cong, J.-P., Mamtora, G., Gingeras, T., and Shenk, T. (1998). Cellular gene expression altered by human cytomegalovirus: global monitoring with oligonucleotide arrays. *Proceedings of the National Academy of Sciences USA* **95**, 14470–14475.

Zhu, X., Ambroise, C., and McLachlan, G.J. (2004). Selection bias in working with the top genes in supervised classification of tissue samples. *Technical Report*, Brisbane, Queensland, Australia: University of Queensland, Department of Mathematics.

Author Index

Abruzzo, L.V., 268, 272, 288
Accelrys Inc., 267
Ach, R., 278
Adams, M.D., 267
Adams, P., 284
Adams, R., 7, 34, 267
Adorján, P., 284
Afshari, C.A., 280, 295
Agrawala, A.K., 275
Aguiar, R.C.T., 290
Aguilar, F., 272
Ahlquist, P., 286
Ahmed, A.A., 20, 267
Ahn, J.-I., 287
Aitkin, M., 72, 86, 87, 267
Aitman, T.J., 20, 267
Alberts, D., 270
Aldape, K.D., 44, 296
Alizadeh, A., 37, 61, 99, 126, 253, 267, 277
Allen, J.C., 288
Allerdt, L., 287
Allison, D.B., 136, 166, 168–171, 268, 275

Alon, U., 56, 100, 104, 108, 111, 112, 146, 147, 159, 179, 210, 226, 227, 239, 268, 286
Alter, O., 102, 268
Altman, R.B., 276, 288, 293
Ambroise, C., 219, 225–230, 232, 245, 268, 269, 284, 296
Andersen, C.A.F., 268
Anderson, D., 72, 267
Anderson, G.L., 287
Anderson, T.W., 192, 268
Andrews, D.T., 289
Angelo, M., 288, 290, 296
Antonellis, K.J., 279
Antonescu, C.R., 280
Antoniadis, A., 97, 223, 268
Ares, M., 271
Armitage, J.O., 267
Armour, C.D., 279
Arpat, A.B., 286
Ashburner, M., 270
Aster, J.C., 290
Åstrand, M., 42, 44, 268, 270
Atsma, D., 294

Audic, S., 23, 24, 268
Axon Instruments Inc., 34, 46, 268

Baggerly, K.A., 53–55, 268, 272, 278, 288
Bailey, T.L., 268
Bair, E., 271
Baldi, P., 139, 268
Ball, C.A., 290
Ballman, K.V., 44, 268
Banfield, J.D., 78, 268
Barash, Y., 174, 268
Bard, M., 279
Barjesteh van Waalwijk van Doorn-Khosrovani, S., 293
Barkai, N., 268
Barker, J.L., 294
Barnard, G.A., 87, 269
Barnhill, S., 277
Barron, A.R., 198, 282
Bartelink, H., 294
Basford, K.E., 69, 71, 283, 284
Bean, R.W., 283, 284
Beaudry, C., 270
Beazer-Barclay, Y.D., 279
Becker, J.W., 296
Bednarski, D.W., 275
Beer, D.J., 254, 269
Beheshti, J., 269
Beijen, M.A., 293
Belanche, L., 217, 285
Bellman, R.E., 192, 269
Ben-Dor, A., 111, 125, 179, 251, 269, 270, 278
Ben-Tovim Jones, L., 254–256, 259, 262, 266, 269
Benito, M., 207, 269
Benjamini, Y., 135, 141–143, 151, 269, 278, 289
Bennett, H.A., 279
Bennett, K.P., 205, 271
Bennett, L., 280, 295
Benson, D.A., 25, 269
Berens, M., 270
Bernards, R., 293, 294

Berthold, F., 280
Beucher, S., 34, 269
Beule, D., 290
Beverloo, H.B., 293
Bhattacharjee, A., 254, 269
Bhattacharyya, C., 207, 270
Bhaumik, D.K., 276
Bickel, D.R., 270
Biegel, J.A., 288
Biernacki, C., 85, 270
Bijnens, L., 295
Bild, A., 279
Binkley G., 290
BioDiscovery Inc., 35, 270
Bischof, L., 7, 34, 267
Bishop, C.M., 80, 207, 270, 293
Bissell, M.J., 270
Bittner, M.L., 65, 99, 127, 270, 272–274
Black, P.M., 288
Blanchette, C., 294
Blattner, F.R., 285
Blekas, K., 270
Bloomfield, C.D., 277
Bock, H.H., 35, 270
Boer, J.M., 293
Boerwinkle, E., 295
Boguski, M.S., 279
Boldrick, J.C., 135, 267, 274
Bolstad, B.M., 43, 44, 270, 276
Bonetti, M., 286
Borg, A., 278
Børresen-Dale, A.-L., 287, 291
Bortoluzzi, S., 270
Botstein, D., 10, 267, 268, 274, 276, 277, 279, 287, 289–293
Bozinov, D., 34–36, 270
Braga-Neto, U.M., 224, 225, 230, 270
Brass, A., 288
Brazma, A., 12, 15, 19, 20, 25, 270, 294
Bredensteiner, E.J., 205, 271
Breiman, L., 203, 210, 271
Breitkreutz, B.J., 295
Brenton, J.D., 267

Brettin, T.S., 294
Broberg, P., 139, 271
Brody, J.P., 271
Broët, P., 136, 138, 164, 165, 271
Brooks, R.J., 97, 292
Broom, B.M., 180, 273
Brown, C.S., 31, 36, 271
Brown, E.L., 278, 282
Brown, M.P.S., 271
Brown, P.O., 9, 224, 267, 268, 273, 274, 276, 277, 279, 287, 289–293
Bruhn, L., 269
Brunak, S., 268
Bruno, W.J., 174, 278
Bryan, J.F., 125, 134, 136, 271, 294
Buckley, M.J., 35, 271, 295
Bueno, R., 269
Bühlmann, P., 224, 250, 251, 273
Bullinger, L., 99, 271
Bulyk, M.L., 281
Bumgarner, R.E., 296
Bushel, P., 280, 295
Bushnell, S., 292
Byrd, J.C., 267
Byrne, M.C., 282

Cai, J., 287
Caldas, C., 267
Califano, A., 288
Caligiuri, M.A., 277
Caliński, T., 88, 271
Callow, M.J., 47, 271, 274
Cameron, G., 270
Campbell, C., 282, 285
Campbell, M.J., 292
Cantor, M., 293
Carpten, J., 270
Carr, D.B., 294
Carroll, R.J., 270, 286
Caskey, C.T., 7, 276
Castrén, E., 286
Celeux, G., 82, 85, 175, 176, 270, 271
Chadt, E.E., 271
Chai, A., 289

Chakraburtty, K., 279
Chan, W.C., 267, 277
Chang, C.-C., 206, 271
Chang, S.U., 284
Chang, T.K., 272
Chang, W.C., 92, 272
Chauvin, Y., 268
Chebil, G., 278
Chee, M.S., 282
Chen, C.-M., 279
Chen, G., 269
Chen, J.J., 136, 293
Chen, M.-H., 136, 279
Chen, Y., 34, 46, 53, 134, 270, 272–274, 278
Cheng, L., 289
Cheng, S., 289
Cheng, S.H., 279
Cherry, J.M., 290
Cheung, V.G., 10, 12, 20, 272
Childs, G., 272
Chilingaryan, A., 136, 272
Chinnaiyan, A.M., 72, 100, 172, 276
Cho, R.J., 292
Chow, M.L., 111, 272, 285
Chu, F.W., 9, 274
Chu, G., 58, 135, 292, 293
Chudin, E., 22, 23, 272
Church, G.M., 281, 292
Churchill, G.A., 47, 134, 136, 272, 280, 289, 295
Cimons, M., 272
Clarke, R., 282
Claverie, J.-M., 23, 24, 268
Cleveland, W.S., 43, 272
Coffey, E., 279
Colantuoni, C., 51, 52, 272
Coleman, D., 69, 75, 272
Coller, H., 277
Collin, F., 279
Cong, J.-P., 296
Conlon, E.M., 65, 276
Cooley, S., 285
Coombes, K.R., 268, 272, 278, 288
Cope, L.M., 44, 272, 276

Cox, D.R., 254, 259, 272, 276
Crammer, K., 272
Cristianini, N., 271, 275, 287
Crosby, L., 280
Cui, X., 134, 136, 272, 295
Culverhouse, R., 290
Curran, T., 288

Dai, H., 279, 293, 294
d'Alessi, F., 270
Dalton, R., 5, 6, 273
Danieli, G.A., 270
Darden, T.A., 282
Darling, G., 295
Datta, S., 67, 273
Dave, V., 285
Davis, G., 287
Davis, J.M., 276
Davis, R.E., 267
Davis, R.W., 11, 289
Dawid, A.P., 191, 273
Dayan, P., 79, 278
Dee, S., 294
De Francesco, L., 27, 273
Delahaye, L., 294
Delrow, J.J., 281, 288
Delwel, R., 293
Dempster, A.P., 70, 73, 74, 83, 273
DeRisi, J.L., 53, 134, 273
Dettling, M., 224, 250, 251, 273
Devlin, S.J., 43, 272
Diaconis, P., 166, 273
Díaz-Uriarte, R., 224, 251, 273
Dietrich, K., 270
Dmitrovsky, E., 292
Do, K.-A., 135, 145, 159, 160, 180, 273
Dobbin, K., 8, 273, 291
Doctolero, M.H., 296
Döhner, H., 271
Döhner, K., 271
Dolinski, K., 293
Domany, E., 276
Dong, H., 282
Dong, X., 272

Dopazo, J., 278, 283
Dougherty, E.R., 225, 270, 272
Downing, J.R., 277
Dressman, H., 279, 294
Du, Q., 269
Dubes, R.C., 64, 66, 273, 279
Dubnick, M., 267
Dudoit, S., 43, 87–89, 100, 125–127, 135, 141, 144, 172, 254, 271, 273, 274, 295
Duffy, N., 275
Dugas, M., 278
Duggan, D.J., 6, 10–12, 274
Duncan, J., 290
Durbin, B.P., 53, 54, 100, 274, 289
Dwight, S.S., 290
Dych, P.A., 294

Eastman, P.S., 296
Efron, B., 86, 123, 135, 139, 144, 145, 154, 158, 159, 213–215, 219, 225, 226, 274, 292
Eickhoff, H., 290
Eisen, M.B., 34, 61, 65, 124, 172, 267, 274, 277, 279, 287, 289–293
Ekins, R.P., 8, 9, 274
Elder, J.K., 21, 274
Elisseeff, A., 277
Elkan, C., 268
Ellis, B., 289
Engle, A., 285
Epstein, J.R., 24, 274
Erpelinck, C.A.J., 293
Evertsz, E., 14, 275

Fang, L., 285
Fayz, B., 282
Fazzio, T.G., 281
Fernández, J.R., 268
Fisher, R.A., 96, 191, 199, 275
Fix, E., 208, 275
Fiziev, P., 278
Fodor, S.P.A., 9, 275, 282

Follettie, M.T., 282
Fortina, P., 10, 11, 281
Foster, W.R., 22, 26, 27, 275
Fotiadis, D.I., 270
Fowlkes, E.B., 88, 275
Fraley, C., 62, 77, 124, 275, 296
Frank, I.E., 97, 275
Freidlin, B., 284
Fridlyand, J., 87–89, 125–127, 172, 273, 274
Friedman, H.P., 82, 275
Friedman, J.H., 97, 103, 206, 251, 271, 275, 277
Friedman, N., 174, 185, 268, 269, 290
Friend, S.H., 279, 293, 294
Fröhling, S., 271
Fuchs, R., 283, 292
Fuhrman, S., 294
Furey, T.S., 224, 271, 275
Furlanello, C., 205, 275

Gaasenbeek, M., 288, 290
Gabriel, K.R., 94, 275
Gachotte, D., 279
Gadbury, G.L., 144, 268, 275
Gallo, M.V., 282
Gamble, R.C., 292
Ganeshanandam, S., 218, 276
Garber, M.E., 254, 262, 263, 276, 293
Garrett, E.S., 136, 276, 287
Garthwaite, P.H., 97, 276
Gasch, A., 290
Gassenbeck, M., 277
Gautier, L., 44, 276
Gehan, E., 282
Geisler, S., 291
Geisser, S., 213, 276
Geller, S.C., 54, 276
Genovese, C.R., 135, 143, 276
Gerald, W., 288
Gerhold, D., 7, 276
Getts, R.C., 291
Getz, G., 109, 179, 276
Gevorgyan, N., 272
Ghahramani, Z., 79, 276

Gharib, T.G., 269
Ghosh, D., 65, 72, 100, 172, 276
Ghosh, S., 282
Gibbons, F.D., 276
Gibbons, R.D., 276
Gibson, G., 280, 295
Gillanders, E., 270
Gillette, M., 269
Gingeras, T.R., 282, 296
Giordano, T.J., 269
Gish, K., 268
Glas, A., 294
Glatfelter, A., 270
Gocayne, J.D., 267
Goh, L., 224, 251, 276
Göhlmann, H.W., 295
Goldstein, D.R., 65, 276
Golub, T.R., 99, 125, 226, 269, 277, 285, 288, 290, 292, 296
Gong, E.L., 271
Gong, G., 218, 277
Gooden, D., 270
Goodwin, P.C., 271
Gould, M.N., 280
Goumnerova, L.C., 288
Govaert, G., 82, 85, 270, 271
Grate, L.R., 270
Gray, R.J., 136, 279
Grayson, D.R., 276
Green, P.J., 85, 289
Green, S.C., 21, 277
Gregg, J.P., 276
Greiner, T.C., 267
Grever, M.R., 267
Grill, D., 268
Grundy, W.N., 271, 287
GSI Lumonics Inc., 34, 277
Guermeur, Y., 205, 277
Gui, J., 254, 282
Gullans, S.R., 291
Guo, X., 159, 277
Gupta, R., 275
Gussoni, E., 272, 290
Guyon, I., 205, 206, 224, 225, 227, 239, 277

Habbema, J.D.F., 97, 277
Hagerman, P., 276
Hamadeh, H., 295
Hamilton, S.R., 288
Hampton, G.M., 296
Hanash, S., 269
Hand, D.J., 83, 185, 210, 211, 277, 285, 290
Hansen, K.M., 69, 277
Harabasz, J., 88, 271
Hardin, J.S., 173, 174, 272, 274, 277
Hart, A.A.M., 293, 294
Hartigan, J.A., 62, 69, 88, 277
Hashimoto, R., 270
Hastie, T., 62, 68, 125, 173, 177–179, 185, 198, 224, 249, 253, 275, 277, 291–293
Hattori, M., 3, 277
Hatzigeorgiou, A., 278
Haussler, D., 271, 275, 280
Hawkins, D.M., 71, 274, 278, 282
Hayasaka, S., 269
Haynor, D.R., 296
Hayward, N., 270
He, Y.D., 279, 293, 294
Hedenfalk, I., 152, 278
Heller, M.J., 24, 278
Heller, R., 289
Hendrix, M., 270
Hennig, C., 78, 278
Henry, G., 272
Heo, M., 268, 275
Hermans, J., 97, 277
Hernandex-Boussard, T., 290
Herrero, J., 278, 283
Herzel, H., 290
Hess, K.R., 32, 268, 278
Highleyman, W.H., 213, 278
Highsmith, W.E., 272, 288
Hill, A.A., 40, 41, 278
Hiller, L., 283
Hinde, J., 72, 267
Hinton, G.E., 79, 276, 278
Hoaglin, D.C., 86, 278
Hobbs, B., 279

Hochberg, Y., 135, 141–143, 269, 278
Hodges, J.L., 208, 275
Hoffmann, R., 58, 278
Hollon, T., 16, 18, 26, 278
Holm, S., 140, 278
Holmes, I., 174, 278
Holmsen, A., 278
Honjoh, D., 294
Hood, L.E., 279
Hope, A.C.A., 87, 279
Horimoto, K., 279
Horng, Ch.-F., 279
Horton, H., 282
Hoyle, D., 288
Hsu, C.-W., 205, 206, 279
Hsu, L., 288
Hsueh, H.-M., 136, 293
Huang, A.T., 279
Huang, C.-C., 269
Huang, E., 106, 107, 249, 279, 294
Huang, X., 279
Huard, C., 277
Huber, R.M., 22, 26, 27, 275
Huber, W., 54, 100, 279
Hudson, J., 267, 279
Hudson, T.J., 175, 286
Hughes, J.D., 292
Hughes, T.R., 53, 279
Hunt, L.A., 83, 279
Hunter, C.P., 278

Iannettoni, M.D., 269
Ibrahim, J.G., 136, 279
Ideker, T., 136, 279
Imamura, M., 280
Inaki, Y., 280
Irizarry, R.A., 100, 270, 272, 276, 279
Ishida, S., 294
Ishii, S., 285
Ishikawa, Y., 294
Itami, A., 280
Iversen, E.S., 279
Iyer, N.G., 267
Iyer, V.R., 61, 273, 279, 289

Jackson, A., 292
Jaeger, J., 173, 279
Jaffee, H.A., 272
Jain, A.K., 64, 66, 273, 279
Jeffrey, S.S., 287, 289, 291
Jiang, Y., 270
Jin, H., 290
Jin, L., 295
Jin, W., 51, 280
John, G.H., 217, 281
Johnsen, H., 291
Johnson, B.E., 269
Johnson, M., 274
Johnson, N.L., 164, 280
Johnston, M., 295
Johnston, R., 296
Jonassen, I., 294
Jones, A.R., 279
Jones, D., 272
Jordan, B., 20, 280
Jordan, M.I., 270
Jorgensen, M.A., 83, 279
Jorgenson, P., 295
Jurisica, I., 295
Jurman, G., 275

Kaganoi, J., 280
Kallioniemi, A., 280
Kaloper, M., 290
Kamat, V., 272
Kan, T., 280
Karchin, R., 280
Kardia, S.L.R., 269
Karplus, K., 280
Karsch-Mizrachi, I., 269
Kasabov, N., 224, 251, 276
Kasarski, A., 290
Kaski, S., 286
Kass, S.U., 295
Kato, K., 285
Kaufman, L., 35, 68, 88, 125, 280
Kawabe, A., 280
Kellam, P., 16, 280
Kelley, J.M., 267

Kendziorski, C.M., 136, 160, 161, 280, 285
Kepler, T.B., 47, 280
Kerkhoven, R.M., 293
Kerlavage, A.R., 267
Kerr, M.K., 8, 49, 50, 56, 136, 138, 280, 295
Keshavjee, S., 295
Khan, J., 208, 280
Khan, N., 88–90, 125, 269, 284
Kidd, M.J., 279
Kim, J.Y.H., 288
Kimura, M., 294
Kimura, N., 21, 280
King, A.M., 279
King, H.C., 16, 280
Kingsmore, S., 285
Kinzler, K.W., 294
Kitareewan, S., 292
Kittler, J., 288
Knight, J., 5, 280
Knoke, J.D., 218, 291
Kobayashi, M., 282
Kohane, I.S., 272, 287, 288, 290
Kohavi, R., 217, 281
Kohn, K.W., 289
Koller, D., 290
Kooperberg, C., 36, 281
Kosaka, A., 272
Kotz, S., 164, 280
Koval, M.A., 290
Krawetz, S.A., 282
Kreder, D.E., 272
Kreßel, U., 206, 281
Kricka, L.J., 10, 11, 281
Krishnan, T., 70, 75, 83, 176, 284
Krogmann, T.A., 272, 288
Kroll, T.C., 56, 281
Krzanowski, W.J., 88, 218, 276, 281
Kucherlapati, R., 272
Kudo, M., 217, 281
Kuick, R., 269
Kukanskis, K., 285
Kuo, F.C., 281
Kuruvilla, F.A., 281

Kutok, J.L., 290

Lachenbruch, P.A., 213, 214, 281
Ladanyi, M., 280
Ladd, C., 269, 288
Lai, Y., 88, 281
Laird, N.M., 70, 273
Lambert, J., 285
Lambert-Lacroix, S., 97, 268
Lan, H., 280
Lance, G.N., 65, 281
Lander, E.S., 269, 277, 288, 290, 292, 296
Landes, G., 287
Landsman, D., 7, 24, 295
Lapointe, J., 99, 281
Lashkari, D., 279, 287, 289
Lasken, R., 285
Last, K.W., 290
Latulippe, E., 288
Lau, C., 288
Lavergne, C., 175, 271
Lawrence, C.E., 282
Lazzeroni, L., 125, 281
Le, C.T., 135, 136, 145, 159, 164, 286, 294
Leblanc, F., 97, 268
Lee, C.-K., 198, 268, 281
Lee, J.C.F., 279, 287, 289
Lee, J.K., 289
Lee, K.H., 274
Lee, M.-L.T., 52, 136, 138, 154, 169, 170, 175, 281
Lee, S., 287
Lee, S.Y., 287
Lee, Y., 198, 281
Lee, Y.-S., 287
Lehmann, E.L., 281
Lehrach, H., 290
Leja, D., 270
Leung, A.P., 274
Levin, A.M., 269
Levine, A.J., 268, 286
Levine, E., 276
Levy, R., 267, 277

Lewi, P.J., 295
Lewis, D.B., 267
Li, C., 42, 43, 269, 271, 281, 282
Li, F., 282
Li, H., 254, 282
Li, J.Q., 198, 282
Li, L., 282
Li, M.-C., 284
Li, Q., 175, 287
Li, W., 295
Li, Y., 282
Liao, M., 279
Likas, A., 270
Lin, C.-J., 145, 205, 206, 271, 279
Lin, D.M., 271, 295
Lin, J., 135, 136, 159, 164, 286, 294
Lin, L., 269
Linsley, P.S., 293
Lipman, D.J., 269
Lipshutz, R.J., 22, 282
Lister, T.A., 290
Liu, A., 224, 250, 282
Liu, D.K., 22, 282
Liu, J.S., 63, 72, 92, 100, 102, 103, 223, 282
Liu, L., 93, 282
Liu, N., 295
Lizyness, M.L., 269
Lockhart, D.J., 16, 21, 24, 282, 296
Loda, M., 269, 288
Loh, M.L., 277
Loh, W.-Y., 210, 282
Loman, N., 278
Long, A.D., 139, 268
Longton, G., 287
Lønning, P.E., 287, 291
Lönnstedt, I., 135, 136, 283
Lossos, I.S., 267
Louis, D.N., 288
Löwenberg, B., 293
Lu, A.R., 275
Lu, C., 295
Lu, L., 267
Lueders, J., 270
Lukashin, A.V., 283

Lum, P.Y., 279
Luo, C., 285
Luu, P., 295

Ma, C., 267
Mack, D., 268
MacQueen, J., 35, 283
Maeda, M., 280
Maindonald, J., 248, 283
Mainquist, J.K., 296
Makherjee, S., 296
Malik, A., 290
Mallows, C.L., 88, 275
Mamtora, G., 296
Man, M.Z., 283
Manduchi, E., 136, 283
Mannila, H., 185, 277
Mao, M., 293
Mar, J.C., 100, 112, 113, 121, 123, 131, 283, 284
Marincola, F., 270
Mark, E.J., 269
Marks, J.R., 294
Marra, M.A., 7, 283
Marriott, F.H.C., 62, 66, 72, 283
Marron, J.S., 207, 269, 283
Marti, G., 267
Martin, M., 136, 280
Martin, O., 175, 271
Martinez, N.W., 280
Marton, M.J., 279, 293, 294
Maruyama, C., 285
Massimi, A., 272
Mateos, A., 283
Matese, J.C., 290, 291
Matoba, R., 285
McCombie, W.R., 267
McLachlan, G.J., 63, 69–72, 75, 77–83, 85–90, 92, 97, 100, 108, 109, 112, 113, 121, 123, 125, 127, 131, 165, 173, 176, 185, 188, 189, 192, 208, 211, 215, 217, 219, 225–230, 232, 245, 249,

255, 268, 269, 283, 284, 286, 287, 296
McLaughlin, M.E., 288
McShane, L.M., 284, 288, 291
Medvedovic, M., 72, 100, 175, 284, 296
Mello-Coelho, V.V., 294
Meltzer, P.S., 16, 24, 270, 272–274, 278, 280, 284
Meng, X.L., 80, 284
Merler, S., 275
Merril, C.R., 267
Mertens, B.J.A., 284
Mesirov, J.P., 277, 285, 288, 290, 292, 296
Meulman, J.J., 103, 275
Meyer, F., 34, 269
Meyer, M.R., 279
Meyerson, M., 269
Mian, I.S., 270, 272, 285
Michaels, G.S., 294
Mickey, M.R., 213, 281
Miles, M.F., 44, 53, 284, 296
Miller, A.J., 218, 284
Milner, N., 274
Minor, J., 296
Mir, K.U., 274, 291
Misek, D.E., 269
Mittmann, M., 282
Miyoshi, T., 294
Model, F., 284
Molenberghs, G., 295
Moler, E.J., 227, 272, 285
Molina, F.J., 270
Molina, L.C., 217, 285
Monden, M., 285
Monico, K., 269
Montgomery, D., 27, 285
Montgomery, J.R., 296
Monti, S., 269
Moore, T., 267, 279
Moorhouse, M.J., 293
Moreno, R.F., 267
Morgan, B.J.T., 67, 285
Morgan, K.T., 280

Morley, M., 272
Morrison, N., 288
Mosteller, F., 214, 285
Mountz, J.D., 275
Müeller, P., 135, 273
Mukherjee, S., 285, 288
Muller, M.W., 71, 278
Mullis, K.B., 5, 285
Munson, P., 54, 285
Muro, S., 174, 285
Murray, G.D., 218, 285
Murua, A., 296
Myers, T.G., 289
Myles, J.P., 210, 285

Nachman, I., 269
Nagel, J.E., 294
Nakamori, S., 285
Nallur, G., 24, 285
Narasimhan, B., 292, 293
National Center for Biotechnology Information, National Institutes of Health, 285
Nebot, A., 217, 285
Nelson, P.S., 288
Neuberg, D.S., 290
Nevins, J.R., 279, 294
Newton, M.A., 53, 136, 155, 160, 162, 280, 285, 286
Ng, S.K., 208, 269, 286
Ngai, J., 295
Nguyen, D.V., 8, 223, 225, 254, 270, 286
Nielsen, H., 268
Nikkila, J., 286
Noble, W.S., 175, 287
Nomura, H., 294
Norton, A., 290
Notterman, D.A., 268, 286
Noueiry, A., 286
Novak, J.P., 175, 286
Novovičová, J., 288
Nuttall, R.L., 296

O'Connell, J., 278

Oba, S., 285
Oberg, A., 268
Oda, R., 280
Olde, B., 267
Olek, A., 284
Olshen, R.A., 271
Olson, J.A., 294
Olson, J.M., 288, 292
Olsson, H., 278
Orringer, M.B., 269
Ostell, J., 269
Owen, A.B., 125, 281, 293

Pacyna-Gengelbach, M., 276
Pagano, M., 286
Page, G.P., 275
Palumbo, M.J., 282
Pan, W., 53, 135, 136, 144–146, 158, 159, 164, 165, 277, 279, 286, 296
Park, P.J., 251, 254, 281, 286, 287
Park, T., 174, 287
Parker, J., 269
Parmigiani, G., 100, 136, 276, 287
Parrish, M., 294
Parsons, G.H., 287
Passador-Gurgel, G., 280
Paugam-Moisy, H., 277
Paules, R.S., 295
Pavlidis, P., 175, 287
Pedersen, L.G., 282
Peel, D., 69, 72, 75, 77–79, 82, 83, 85, 87, 283, 284, 287
Peng, V., 295
Penland, L., 273
Pepe, M.S., 134, 287
Pergamenschikov, A., 287, 289
Perou, C.M., 99, 269, 276, 287, 289, 291
Peterse, H.L., 293
Peterse, J.L., 294
Petersen, C., 280
Petersen, I., 276
Petersen, S., 276
Pevsner, J., 272

Phillips, D.B., 85, 287
Piepenbrock, C., 284
Pinkus, G.S., 290
Pintilie, M., 295
Pirrung, M.C., 275
Pittelkow, Y.E., 92, 94, 287
Pittman, J., 279
Platt, J.C., 205, 287
Poggio, T., 288
Pollack, J.R., 271, 287
Pollard, K.S., 271, 287
Pollock, P., 270
Polsky-Cynkin, R., 8, 287
Polymeropoulous, M.H., 267
Pomeroy, S.L., 288
Pommier, Y., 289
Popper Shaffer, J., 135, 274
Poustka, A., 279
Powell, J.I., 267
Pritchard, C., 277
Pritchard, C.C., 52, 59, 288
Prolla, T.A., 268
Pudil, P., 218, 288

Qi, H., 277
Quackenbush, J., 7, 11, 16, 288, 292
Quake, S.R., 271
Quinlan, R., 210, 288
Quist, H., 291

Rabert, D., 272
Radisky, D., 270
Radmacher, M.D., 270, 284, 288, 291
Radulovich, N., 295
Radvanyi, F., 136, 271
Raftery, A.E., 62, 77, 78, 124, 268, 275, 296
Rahnenführer, J., 34–36, 270
Ramaswamy, S., 224, 288, 296
Ramdas, L., 38, 288
Ramoni, M.F., 272, 288, 290
Ramsey, J.M., 24, 288
Rao, C.R., 192, 288
Rashtchian, A., 287
Rattray, M., 50, 56, 288

Ray, A.P.G., 67, 285
Ray, M., 273
Ray, T.S., 290
Raychaudhuri, S., 288
Read, J.L., 275
Reczko, M., 278
Rees, C.A., 287, 289
Reich, M., 288, 290, 296
Reiner, A., 135, 289
Reinhold, W.C., 289
Revow, M., 79, 278
Richards, W.G., 269
Richardson, S., 85, 136, 271, 289
Richmond, C.S., 285
Rifkin, R.M., 285, 288, 296
Riley, R.M., 280
Ringnér, M., 278, 280
Ripley, B.D., 62, 185, 289
Rizki, A., 270
Roberts, C.J., 279, 293, 294
Robinson, A., 270, 294
Rocke, D.M., 53, 54, 100, 173, 174, 223, 225, 254, 272, 274, 276, 277, 286, 289
Rodenhuis, S., 294
Rogers, S., 285
Romualdi, C., 270
Rosen, G.D., 276
Rosenwal, A., 267
Ross, D.T., 99, 279, 287, 289, 292
Ross, K.N., 288, 290
Rossant, J., 295
Roth, F.P., 276, 289
Rousseeuw, P.J., 35, 68, 88, 125, 280
Rubin, D.B., 70, 273
Rubin, E.M., 271
Rubin, J., 82, 275
Rutgers, E.T., 294
Ruzzo, W.L., 102, 173, 279, 296

Saal, L.H., 280
Sabet, H., 267
Sampas, N., 270
Sapinoso, L.M., 296
Sapir, M., 47, 289

Sarkar, D., 286
Sato, F., 280
Satterthwaite, F.E., 139, 289
Sausville, E.A., 289
Schadt, E.E., 22, 23, 42–44, 289
Schena, M., 10, 11, 53, 134, 289
Scherf, U., 279, 289
Schermer, M.J., 10, 12, 14, 15, 20, 290
Schiavo, R.A., 211, 290
Schlenk, R.F., 271
Schluens, K., 276
Schreiber, G.J., 293, 294
Schreiber, S.L., 281
Schubert, C.M., 221, 290
Schuchhardt, J., 45, 290
Schuler, G., 279
Schummer, M., 269, 275, 287
Schutze, H., 288
Schwab, M., 280
Schwarz, G., 79, 85, 290
Schweder, T., 143, 290
Schweitzer, B., 285
Scott, A.J., 81, 290
Scudiero, D.A., 289
Sebastiani, P., 272, 288, 290
Seber, G.A.F., 62, 94, 290
Seftor, E., 270
Segal, E., 171, 172, 290
Seiden, I., 295
Seidl, T., 278
Sekimoto, M., 285
Sengupta, R., 173, 279
Serafini, M., 275
Shalon, D., 279, 287, 289
Shamir, R., 125, 269
Shannon, W., 290
Sharma, R.P., 276
Sharpe, P.M., 215, 293
Shawe-Taylor, J., 287
Shchepinov, M., 291
Shenk, T., 296
Shepherd, F.A., 295
Sherlock, G., 15, 267, 290, 293
Shih, J.H., 273
Shimada, Y., 280

Shimane, M., 294
Shipp, M.A., 290
Shmulevich, I., 296
Shoemaker, D.D., 279
Sibson, R., 248, 290
Šidák, Z., 140, 291
Siegel, A.F., 279
Siegmund, D., 135, 142, 292
Sierk, A.J., 286
Simes, R.J., 142, 290
Simon, J., 279
Simon, R., 8, 270, 273, 284, 288, 291
Sinclair, B., 10, 11, 21, 291
Singer, B., 296
Singer, Y., 272
Sinha, A.A., 16, 280
Sivaganesan, S., 72, 100, 175, 284
Sklansky, J., 217, 281
Sklar, J., 281
Slade, D., 279
Sladek, R., 175, 286
Slonim, D.K., 222, 277, 278, 291, 292
Smith, A.F.M., 85, 287
Smith, C.A.B., 213, 291
Smith, L.H., 289
Smith, S., 294
Smyth, G.K., 34, 136, 291
Smyth, P., 86, 87, 185, 277, 291
Snapinn, S.M., 218, 291
Sohail, M., 274
Soille, P., 35, 291
Solas, D., 275
Somogyi, R., 294
Sondak, V., 270
Song, Q., 224, 251, 276
Sorger, P.K., 271
Sørlie, T., 287, 291
Sosnowski, R.D., 278
Southern, E.M., 9, 10, 21, 274, 277, 291
Spang, R., 107, 291
Speed, T.P., 100, 135, 136, 270–272, 274, 279, 283, 291, 295
Spellman, P.T., 274, 289, 290
Spjøtvoll, E., 143, 290

Sprang, R., 294
Stack, R., 296
Starink, P., 275
Staudt, L.M., 267, 277, 279
Staunton, J., 269
Stears, R.L., 12, 291
Stepaniants, S.B., 279
Stern, D., 296
Stivers, D.N., 268, 272, 278
Stolovitzky, G., 288
Stone, C.J., 209, 271, 291
Stone, M., 97, 213, 214, 292
Storey, J.D., 135, 136, 142–146, 148–153, 157, 274, 292
Stormo, D.G., 282
Stoughton, R., 279
Stryer, L., 275
Sturla, L.M., 288
Sturn, A., 292
Su, C., 271
Su, Y.A., 273
Sueltmann, H., 279
Sugarbaker, D.J., 269
Sugnet, C.W., 271
Suzuki, O., 280
Symons, M.J., 81, 290
Szaboi, A., 272

Takemasa, I., 285
Tamames, J., 283
Tamayo, P., 65, 124, 277, 285, 288, 290, 292, 296
Tanabe, L., 289
Tang, F., 135, 273
Tapscott, S.J., 292
Taskar, B., 290
Taub, D.D., 294
Tavazoie, S., 172, 292
Taylor, J.E., 135, 142, 292
Taylor, J.M.G., 269
Taylor, T.D., 3, 277
ten Krooden, J.A., 71, 278
Thaesler, Z., 276
Theilhaber, J., 36, 292
Theriault, T.P., 11, 22, 292

Therneau, T., 268
Thomas, D.G., 269
Thomas, J.G., 136, 292
Thorsen, T., 291
Thorsson, V., 279
Tian, L., 287
Tibshirani, R.J., 58, 62, 86–88, 123, 125, 135, 142, 144–146, 148, 150–153, 178, 185, 198, 202, 214, 219, 225, 226, 236, 238, 239, 267, 271, 274, 275, 277, 287, 291–293
Tipping, M.E., 80, 282, 293
Todd, M., 207, 283
Todeschini, R., 210, 293
Toh, H., 279
Törönen, P., 286
Toussaint, G.T., 214, 215, 293
Trajanoski, Z., 292
Tran, T., 267
Trent, J.M., 270, 272–274, 278, 279
Troyanskaya, O.G., 37, 136, 276, 293
Trulson, M., 294
Tsai, C.-A., 136, 138, 293
Tsao, M.S., 295
Tsou, M.-H., 279
Tsui, K.W., 285
Tsukiyama, T., 281
Tu, E., 278
Tucker-Kellogg, G., 278
Tukey, J.W., 69, 277
Turid, A., 291
Tusher, V.G., 58, 135, 139, 144, 146, 159, 274, 293
Tyers, M., 295

Ueno, N., 285
Ukkonen, E., 294

Vainer, M., 296
Valencia, A., 278
Valk, P.J.M., 99, 293
van 't Veer, L.J., 82, 100, 112, 113, 115, 124, 127, 128, 201,

203, 233, 234, 236, 245, 293, 294
van de Rijn, M., 276, 287, 289, 291
van de Vijver, M.J., 221, 233, 234, 236, 239, 293, 294
van der Kooy, K., 293
van der Laan, M.J., 125, 134, 136, 271, 287, 294
van der Spek, P.J., 293
van der Velde, T., 294
van Dyk, D.A., 80, 284
Vanichsetakul, N., 210, 282
Vapnik, V.N., 203, 205, 277, 294
Vardanyan, A., 272
Vasa, P., 269
Velculescu, V.E., 23, 294
Venna, J., 286
Venter, J.C., 267
Verfaillie, C.M., 277
Verhaak, R.G.W., 293
Vias, M., 267
Vilo, J., 19, 270, 294
Vingron, M., 279
Virtanen, C., 294
Vogelstein, B., 294
von Heydebreck, A., 279
Voskuil, D.W., 294

Wald, A., 192, 198, 294
Walker, R., 272
Wall, M.E., 294
Wallace, D.L., 214, 285
Walt, D.R., 274
Waltham, M., 289
Walther, G., 293
Wang, B.B., 296
Wang, C., 282
Wang, E., 270
Wang, H., 290
Wang, N., 286
Wang, X., 283
Wang, Y., 283
Ward, J.H., 65, 294
Warnke, R., 267
Warrington, J.A., 10, 11, 22, 23, 294

Wasserman, L., 135, 143, 276
Waterston, R.H., 283
Watkins, C., 205, 295
Watson, D., 275
Weber, G., 269
Weeraratna, A.T., 24, 294
Wei, J.S., 280
Wei, P., 294
Weinberg, C.R., 282
Weindruch, R., 268
Weinstein, J.N., 289
Weisenburger, D.D., 267
Welsh, J.B., 296
Wen, S., 180, 273
Wen, X., 294
Weng, A.P., 290
Weng, S., 290
West, M., 223, 253, 279, 294
Westermann, F., 280
Westfall, P.H., 135, 136, 141, 294
Weston, J., 205, 277, 287, 295
Wetmore, C., 288
Wheeler, D.L., 269
White, K.P., 280
Whitley, M.Z., 278
Whitmore, G.A., 281
Whyte, R.I., 276
Wigle, D.A., 254, 259, 295
Wilcoxon, F., 251, 295
Williams, B.A., 271
Williams, C.F., 287
Williams, W.T., 65, 281
Wilson, S.R., 92, 94, 287
Wilson, W., 267
Winder, S.C., 292
Winton, T., 295
Winzeler, E.A., 16, 24, 282
Witteveen, A.T., 293, 294
Wold, B.J., 271
Wold, H., 97, 295
Wolfinger, E.D., 295
Wolfinger, R.D., 50, 51, 136, 280, 295
Wölfl, S., 56, 281
Wolfsberg, T.G., 7, 24, 295
Wolski, E., 290

Womble, D.D., 282
Wong, G., 286
Wong, W.H., 42, 43, 269, 271, 282, 289
Woodgett, J., 295
Woodruff, D.L., 173, 174, 272, 277
Wouters, L., 92, 295
Wu, A., 267
Wu, H., 50, 295
Wu, J., 269
Wu, S., 272
Wu, T.D., 16, 18, 20, 51, 295
Wu, Z., 272

Xiang, D., 269
Xiao, H., 267
Xiong, M., 221, 229, 295, 296

Yakhini, Z., 125, 269, 270, 278
Yamamoto, H., 285
Yamasaki, S., 280
Yamashita, R., 285
Yang, L., 267
Yang, Y.H., 35, 43, 44, 46–48, 53, 274, 291, 295
Yao, B., 282
Ybarra, S., 268
Yeang, C.-H., 288, 296
Yekutieli, D., 135, 142, 151, 269, 289
Yelensky, R., 290
Yeung, K.Y., 72, 100, 102, 174, 175, 296
Yi, S.-G., 287
Ylvisaker, D., 166, 273
Yoo, D.-H., 287
Young, S.S., 135, 136, 141, 282, 294
Yu, C.-Y., 296
Yu, K., 83, 277
Yu, R., 284
Yu, X., 267
Yue, H., 296
Yukhananov, R.Y., 281

Zagzag, D., 288
Zamora, M., 296

Zarrinkar, P.P., 23, 24, 296
Zaykin, D.V., 136, 294
Zeger, S.L., 272
Zhang, H., 296
Zhang, J.L., 282
Zhang, K., 296
Zhang, L., 44, 294, 296
Zhang, W., 268, 278, 288, 296
Zhang, Y., 282
Zhao, H., 296
Zhao, J., 295
Zhao, L.P., 292
Zhao, Y., 158, 296
Zhu, H., 296
Zhu, Q., 292
Zhu, S.X., 287
Zhu, X., 245, 296
Zuzan, H., 294

Subject Index

AECM algorithm, 80
    mixtures of factor analyzers, for, 80
AIC, *see* Number of components, information criteria for, AIC
Akaike's criterion, *see* AIC
Alternating expectation–conditional maximization algorithm, *see* AECM algorithm

Bayesian estimation, 87, 159, 215. *See also* Markov chain Monte Carlo methods
    Dirichlet, 159
    gamma, 54, 161
    inverse gamma, 161
Bayesian information criterion, *see* BIC
Beta distribution, *see* Distributions, beta
BIC, 79, 85, 105, 165–166. *See also* Number of components, BIC

Bootstrap, 49, 85–88, 126–127, 144, 168, 213–216, 219–220, 226, 228–229, 233
    likelihood ratio test of number of components, for, *see* likelihood ratio test statistic, number of components, for

CART, *see* Discriminant analysis, classification trees, CART
Canonical variates, 94–97
Chi-squared distribution, *see* Distributions, chi-squared
Classification ML approach, 81–82
Classification of data
    Supervised, *See also* Discriminant analysis
    Unsupervised,
Clest method *see* Number of clusters, Clest method
Clustering. *See also* Cluster analysis
    genes, of, 171–184
    gene shaving, 173, 177–181, 183
    tissue samples, of, 99–132

EMMIX-GENE, 63, 83, 100, 103–109, 112–113, 115–124, 126, 128, 130–131, 171, 173, 198, 249, 255–256, 261–262

Cluster analysis
agglomerative methods, *see* hierarchical methods
decision-theoretic approach, 187–188
hierarchical methods, 61–68, 71, 99, 102, 112, 115–116, 125, 136, 180, 250, 253, 259, 262
agglomerative, 61–68, 102, 115
divisive, 64, 67, 71
finite mixture models, via, 63, 69–72, 74, 75
k-means, 35, 61, 68–69, 75–77, 81–82, 102, 105, 109, 127, 249
likelihood-based criteria, 71, 72, 81
mixture likelihood approach, 71, 72, 81
model-based approach, 65, 72, 84, 100, 102, 124, 130, 175, 249
advantages, 72, 84, 100, 102, 124, 130
spurious clusters, 125, 127, 178, 250

Cross-validation, 43, 111, 123, 125, 210, 213–216, 218–219, 224–241, 245–248, 260–261, 266.
half-sample validation, 215
ten-fold, 212, 226, 229, 231, 234–239, 241, 245–247, 260–261, 266

Data sets
Alizadeh data, 126–127
Alon data, 56–58, 100, 104, 108, 110–112, 130, 146–147, 159, 162–163, 179, 210–212, 226–227, 231, 239–242
Bittner data, 65, 99, 127
Golub data, 99, 125, 226
Harvard data, 254–255
Michigan data, 254–255
Ontario data, 254–256, 259–261, 256
Stanford data, 254–255, 261–266
van de Vijver data, 221, 233–237, 239–240, 243–245, 247, 254
van 't Veer data, 82, 100, 112–124, 127–131, 233–234, 236, 245

Differentially expressed genes, detection of, 58, 133–171
Fold change, 134
Multiplicity problem, 134–135. *See also* Multiple hypothesis testing.
SAM procedure, 146–148
t-test, 139
permutational methods, 144-145

Diagnostic tests, 187
error rates, assessment of, 187
sensitivity, 187
specificity, 187

Dirichlet distribution, *see* Distributions, Dirichlet

Discriminant analysis, 185–220
classification trees, 106, 210, 211
CART, 210, 212, 231–232.
C4.5, 210
diagnostic paradigm, 191
error rates. *See* Error rates
logistic discrimination, 201–202, 241
nearest-neighbor rules, 208–210, 251
neural networks, 207–208

normal mixture models, 53, 62, 65, 71, 74, 77–78, 80, 87–88, 102, 107, 112, 126, 130, 164–166, 175, 198, 248, 256

    Fisher's linear discriminant function, 96, 191, 199, 216, 217, 222, 226, 228–232

    heteroscedastic classes, 75–77, 193, 194, 197

    homoscedastic classes, 76, 77, 195, 196, 215, 217, 224

    plug-in sample rule, 191–192, 194, 197–198

    quadratic discriminant function, 194, 223, 225

    selection of feature variables, 204–205, 216–218

    suport vector machines, 111, 123, 198, 203–207, 216–217, 222, 224–237, 239–247, 260–261, 266

    multiple classes, 205–206

    recursive feature elimination, 205–206, 224–228, 230, 233–235, 237, 239, 241–243, 245–246, 260

    software, *see* Software, support vector machines

Distributions

    beta, 54, 166, 168, 169

    chi-squared, 79, 86, 127

    Dirichlet, 159

    F, 164–165, 224, 236, 249

    log normal, 161

    location model (conditional Gaussian distribution), 83

    multinomial, 186

    normal

        multivariate, 50, 69, 71, 75, 78, 81, 83, 90, 168, 192–193

        univariate, 164

    t, 78, 137, 139

    uniform, 88, 89, 166

    Weibull, 257

EM algorithm, mixtures, for 70, 72–75, 80–81, 83, 130, 176, 208

    convergence of

        local maximum, to a, 75, 81, 112

        saddle point, to a, 75

    factor analyzers, 81, 103, 106–109

    E-step, 73, 74, 81

    M-step, 74, 76, 81, 176

    single component, 131, 176

    normal components, 75–78, 80–81, 112, 126, 165, 173–174, 249

    root, choice of, 77

    starting points, 75, 81

    t components, 104–106

EMMIX, 77, 165

Error rates. 53, 96–97, 111, 123, 128, 131, 135, 140, 179, 186–191, 197, 207, 210–220, 222, 224–235, 239–240, 245–247. *See also* error-rate estimation

    apparent error, 210–213, 215–216, 218–219, 222, 224–226, 230–231

Error-rate estimation

    cross-validation. *See* Cross-validation

        external, 218–219

        leave-one-out (LOO), 213

        q-fold, 214

    bootstrap approach

        0.632 estimator, 214–215

        0.632+ estimator, 219–220

EST, 6, 7, 23, 25

Estimation, mixtures, for

    Bayesian methods, *see* Bayesian estimation

maximum likelihood, *see* ML estimation

Estimation, robust, 104
mixtures, for, *See also* Mixtures of *t* components
M-estimation of components, 42

Examples
spurious clusters, 125, 127, 196, 250
spurious local maximizers, 77, 80

Expectation–maximization algorithm, *see* EM algorithm, mixtures, for

Exponential distribution, *see* Distributions, exponential

Expressed sequence tag, *see* EST

Extensions of EM algorithm, *see* AECM algorithm

F-distribution, *see* Distributions, *F*

Factor analysis, *see* Factor analyzers

Factor analyzers, 78-81, 102–103, 106–109, 112, 114, 126, 127, 130–131, 174, 198, 245, 249, 256. *See also* Mixtures of factor analyzers
application of EM algorithm, 62–63, 78–81
probabilistic PCAs, link with, 80.

False discovery rate, *see* Multiple hypothesis testing, False discovery rate

False nondiscovery rate, *see* Multiple hypothesis testing, False nondiscovery rate

False positive rate, *see* Multiple hypothesis testing, single hypothesis, false positive rate

FDR, *see* Multiple hypothesis testing, False discovery rate

Finite mixtures, *see* Mixture models

Gamma distribution, *see* Distributions, gamma

Gap statistic, *see* Number of clusters, Gap statistic

Gene expression, 1–3

Genetics, 2

Genome, 2–3, 25–27

Genotype, 2

Hazard functions
proportional, 254, 259, 263, 265

ICL criterion, *see* Number of components, information criteria for, classification-based, integrated classification likelihood criterion

Identifiability, 50, 70

Image analysis. *See* Microarrays, preprocessing of, image analysis

Information criteria, *see* Number of components, information criteria for
Integrated classification likelihood criterion, 85. *See* Number of components, information criteria for, classification-based, integrated classification likelihood criterion

Likelihood ratio test statistic, 105

Likelihood ratio test statistic, number of components, for, 85, 86, 88
applications of, 104–105, 125–131, 159, 224
bootstrapping of, 85–87, 126
definition of, 79, 86, 159
effect on *P*-values, 87, 126
regularity conditions, breakdown of, 79, 85, 86, 127
simulation results, 87

Local maximizers, *see* ML estimation, local maximizers, multiplicity of

Log likelihood, *see* Mixture likelihood

Log normal distribution, *see* Distributions, log normal

LRTS, see Likelihood ratio test statistic; Likelihood ratio test statistic, number of components, for

Mahalanobis distance, 67, 69, 78, 92, 194, 197, 217

Markov chain Monte Carlo methods, 85. *See also* Bayesian estimation

Maximum likelihood estimation, *see* ML estimation

Maximum penalized likelihood estimation, *see* ML estimation, penalized,

MCMC or $(MC)^2$ methods, *see* Markov chain Monte Carlo methods

Metagenes, 99, 105–108, 112, 224, 249–251, 256, 258, 259, 262–265

Microarrays
  Affymetrix, 9, 10, 21–23, 26–28, 38–39, 44, 56, 125, 169–170, 254–255
  biology of, 2–4, 10–16
  definition of, 10
  cDNA, 6–7
  DNA, 1–12
  GeneChip, 1, 10, 21, 23, 26–27, 38–39
  history of, 8–10
  hybridization, 4, 9, 11–14, 16–17, 19–24, 27–28, 31–32, 38–41, 48–49, 51–52
  limitations of, 18–20
  missing values, 31–32, 36–37, 254
  mismatch, 22–23, 38–39, 42, 44

normalization, 38–52
  ANOVA, mixed model, 49–51, 58, 138, 172
  cyclic loess, 43, 44
  multiple-slide, 48
  nonlinear, 16, 32, 38, 42–44, 56, 63
  quantile, 43, 44, 56, 58, 147
  spiked-in, 40
  titration series, 46
  transformations, 53–55
  within-slide, 46–48
  oligonucleotide, 20–23, 38–44
  perfect match, 22–23, 38–39, 44
  preprocessing of
    cleaning, 32–38
    image processing, 32, 34
    normalization, *see* Microarrays, normalization
    standardization, 20
  RNA, 3–6, 9
  spiked, 39–41
  technology of, 7–24
  tools, 10–18

Mixed feature variables, *see* Mixture models, mixed variables, for

Mixing proportions
  definition of, 70
  estimation of, 74, 76

Mixture likelihood, *see also* ML estimation
  definition of, 81
  EM framework, in
    complete-data, 73, 74, 81, 85
    incomplete-data, 74
  local maxima of, *see* ML estimation, local maximizers, multiplicity of
  multiple maxima of, *see* ML estimation, local maximizers, multiplicity of
  unboundedness of, 77

Mixture models

definition of, 62, 69

estimation of, 69–90. *See also*
Estimation, mixtures, for

incomplete structure of, 73, 74

nonparametric estimation of, *see*
Mixing distribution, non-
parametric estimation of,

parametric formulation of, 160–
166

Mixtures of factor analyzers, 78–80,
102–103, 106–107

application of AECM algorithm,
80

examples, 108–109, 112, 114,
121, 123, 126–127, 130–
131, 245

likelihood ratio statistic, number
of factors, for, 125–128

mixtures of probabilistic PCAs,
link with, 80

Mixtures of normals, 62, 81, 173, 249,
256

application of EM algorithm, 81,
112, 126, 164–165, 173,
174, 249

M-step, sufficient statistics,
74–75

choice of covariance matrices,
76–82

choice of root, 77

global maximizer, 74, 75, 77

heteroscedastic components, 75–
76

homoscedastic components, 76

local maximizers, spurious, 77,
80

singularities, 98

spherical components, 76

Mixtures of probabilistic PCAs, *see*
Mixtures of factor analyz-
ers, mixtures of probabilistic PCAs, link with

Mixtures of survival functions

censored data, with, 254, 256–
257, 259, 263, 266

example, 257

long-term survivor model, 257,
259, 263, 266

Weibull components, 257

Mixtures of t components, 104–106

Mixtures of uniform components,
168, 169

Mixtures of Weibulls, 257

MLE, *see* ML estimation

ML estimation, 48, 50, 70, 72, 75, 77–
78, 83, 86, 194–195, 197

choice of root of likelihood equation, 77

EM algorithm for, *see* EM algorithm, mixtures, for

global maximizer, 74–75, 77

local maximizers, multiplicity
of, 77. *See also* Mixtures of
normals, local maximizers,
spurious

penalized, 85

properties of MLE, 69–72. *See
also* Mixtures of normals,
properties of MLE

Monte Carlo methods, *see also*
Markov chain Monte Carlo
methods

bootstrap, for, 68–70

Multinomial distribution, *see* Distributions, multinomial

Multiple hypothesis testing

Approaches to, 137–143

Bayesian, 159

Empirical Bayes, 158–160

Nonparametric, 159

Parametric, 136

Mixture model, 154–158

Bayes risk, 156–158

False discovery rate, 135, 136,
141–160

Positive false discovery rate,
143

False nondiscovery rate, 143

Positive false nondiscovery
rate, 143

Familywise error rate, 135, 140, 142
q-value, 148–154
single hypothesis, 137–138
false positive rate, 134–136, 138–141, 148–149, 151–156
P-value, 140–144, 148–154, 162, 165–170, 173
Multivariate normal distribution, *see* Distributions, normal, multivariate
Multivariate normal mixtures, *see* Mixtures of normals
Multivariate t distribution, 78. *See also* Distributions, t

Neural networks, 207–208
Nonparametric MLE, *see* Mixing distribution, nonparametric estimation of
Normal distribution, *see* Distributions, normal
Normalization of microarrays. *See* Microarrays, preprocessing of, normalization
Notation, 100–101
Number of clusters. *See also* Number of components
Clest method, 87–89, 91, 127.
Gap statistic, 87, 125, 178, 180.
Number of components, 71, 84–85
bias-correction of log likelihood, 213
cross-validation, 125
information criteria for
AIC, 85, 165
BIC, 79, 85, 105, 165, 166
information criteria for, classification-based,
classification likelihood criterion, 81–82, 85
integrated classification likelihood criterion, 85–86

likelihood ratio, *see* Likelihood ratio test statistic, number of components, for
order of a mixture model, 84–85
simulation results on, 88–91
testing for, 84, 104

Outlier detection from a mixture, 78, 104
Order of a mixture model, 84, 85 *See also* Number of components approaches for assessing

PAM, *see* Partitioning around medoids
Partial least squares, 97–98, 223–225, 254
Partitioning around medoids, 35, 68
PCR, 5, 6, 16–17, 24, 28
pFDR, *see* Multiple hypothesis testing, positive false discovery rate
pFNR, *see* Multiple hypothesis testing, positive false nondiscovery rate
Polymerase chain reaction, *see* PCR
Posterior probabilities of component membership, 70–71, 73–74, 77, 80–81, 86, 136, 164, 169, 177, 186, 191, 194–195, 198, 201, 207, 209–210, 228, 238, 259
Positive false discovery rate, *see* Multiple hypothesis testing, positive false discovery rate
Principal component analysis, 252–253. *See also* Mixtures of probabilistic PCAs
Prior densities, *see* Bayesian estimation, prior densities
Probabilistic principal component analyzers, *see* Factor analyzers, probabilistic PCAs, link with
P-value, *see* Multiple hypothesis testing, single hypothesis, P-value

q-value, *see* Multiple hypothesis testing, hypothesis, q-value

Resampling approach, *see* Likelihood ratio test statistic, bootstrapping, of

Robust estimation, *see* Estimation, robust; Mixtures of t components, robust estimation via

SAGE, 23–24

Selection bias, 123, 128, 218–219, 225–233, 245–248

Self-organizing maps, 61, 68

Simulation,
  posterior distribution, of, *see* Markov chain Monte Carlo methods, posterior simulation

Singular value decomposition, 37, 93, 96, 128, 178, 223, 259. *See also* Principal component analysis

SOM, *see* Self-organizing maps

Software
  Bioconductor, 44
  DNA-chip Analyzer, 43
  EMMIX-GENE, 63, 83, 100, 103–109, 112–113, 115–124, 126–128, 130–131, 171, 173, 198, 249, 255–256, 261–262.
  GeneClust, 180–183
  GenePix, 34, 46
  GeneShaving, 180–183
  QuantArray, 34
  ScanAlyze, 34
  SNOMAD, 51, 52
  Spot, 35
  rfe, 206
  support vector machines, 206

Spurious clusters, *see* Clustering, spurious clusters

Spurious local maximizers, *see* ML estimation, local maximizers, multiplicity of

Standard error estimation, 216

Starting values, *see* EM algorithm, starting points

Sufficient statistic, *see* Mixtures of normals, application of EM algorithm, M-step, sufficient statistics

Survival analysis, *see* Mixtures of survival functions

SVD, *see* Singular value decomposition

t distribution, *see* Distributions, t; Mixtures of t components

Transformations, *see* Microarrays, normalization, transformations

Tree-structured classifiers, *see* Discriminant analysis, classification trees

Uniform distribution, *see* Distributions, uniform

Univariate normal mixtures, *see* Mixtures of normals

WILEY SERIES IN PROBABILITY AND STATISTICS
ESTABLISHED BY WALTER A. SHEWHART AND SAMUEL S. WILKS

Editors: *David J. Balding, Noel A. C. Cressie, Nicholas I. Fisher,*
Iain M. Johnstone, J. B. Kadane, Geert Molenberghs. Louise M. Ryan,
David W. Scott, Adrian F. M. Smith, Jozef L. Teugels
Editors Emeriti: *Vic Barnett, J. Stuart Hunter, David G. Kendall*

The *Wiley Series in Probability and Statistics* is well established and authoritative. It covers many topics of current research interest in both pure and applied statistics and probability theory. Written by leading statisticians and institutions, the titles span both state-of-the-art developments in the field and classical methods.

Reflecting the wide range of current research in statistics, the series encompasses applied, methodological and theoretical statistics, ranging from applications and new techniques made possible by advances in computerized practice to rigorous treatment of theoretical approaches.

This series provides essential and invaluable reading for all statisticians, whether in academia, industry, government, or research.

ABRAHAM and LEDOLTER · Statistical Methods for Forecasting
AGRESTI · Analysis of Ordinal Categorical Data
AGRESTI · An Introduction to Categorical Data Analysis
AGRESTI · Categorical Data Analysis, *Second Edition*
ALTMAN, GILL, and McDONALD · Numerical Issues in Statistical Computing for the
    Social Scientist
AMARATUNGA and CABRERA · Exploration and Analysis of DNA Microarray and
    Protein Array Data
ANDĚL · Mathematics of Chance
ANDERSON · An Introduction to Multivariate Statistical Analysis, *Third Edition*
*ANDERSON · The Statistical Analysis of Time Series
ANDERSON, AUQUIER, HAUCK, OAKES, VANDAELE, and WEISBERG ·
    Statistical Methods for Comparative Studies
ANDERSON and LOYNES · The Teaching of Practical Statistics
ARMITAGE and DAVID (editors) · Advances in Biometry
ARNOLD, BALAKRISHNAN, and NAGARAJA · Records
*ARTHANARI and DODGE · Mathematical Programming in Statistics
*BAILEY · The Elements of Stochastic Processes with Applications to the Natural
    Sciences
BALAKRISHNAN and KOUTRAS · Runs and Scans with Applications
BARNETT · Comparative Statistical Inference, *Third Edition*
BARNETT and LEWIS · Outliers in Statistical Data, *Third Edition*
BARTOSZYNSKI and NIEWIADOMSKA-BUGAJ · Probability and Statistical Inference
BASILEVSKY · Statistical Factor Analysis and Related Methods: Theory and
    Applications
BASU and RIGDON · Statistical Methods for the Reliability of Repairable Systems
BATES and WATTS · Nonlinear Regression Analysis and Its Applications
BECHHOFER, SANTNER, and GOLDSMAN · Design and Analysis of Experiments for
    Statistical Selection, Screening, and Multiple Comparisons
BELSLEY · Conditioning Diagnostics: Collinearity and Weak Data in Regression

*Now available in a lower priced paperback edition in the Wiley Classics Library.

* BELSLEY, KUH, and WELSCH · Regression Diagnostics: Identifying Influential Data and Sources of Collinearity

BENDAT and PIERSOL · Random Data: Analysis and Measurement Procedures, *Third Edition*

BERRY, CHALONER, and GEWEKE · Bayesian Analysis in Statistics and Econometrics: Essays in Honor of Arnold Zellner

BERNARDO and SMITH · Bayesian Theory

BHAT and MILLER · Elements of Applied Stochastic Processes, *Third Edition*

BHATTACHARYA and WAYMIRE · Stochastic Processes with Applications

BILLINGSLEY · Convergence of Probability Measures, *Second Edition*

BILLINGSLEY · Probability and Measure, *Third Edition*

BIRKES and DODGE · Alternative Methods of Regression

BLISCHKE AND MURTHY (editors) · Case Studies in Reliability and Maintenance

BLISCHKE AND MURTHY · Reliability: Modeling, Prediction, and Optimization

BLOOMFIELD · Fourier Analysis of Time Series: An Introduction, *Second Edition*

BOLLEN · Structural Equations with Latent Variables

BOROVKOV · Ergodicity and Stability of Stochastic Processes

BOULEAU · Numerical Methods for Stochastic Processes

BOX · Bayesian Inference in Statistical Analysis

BOX · R. A. Fisher, the Life of a Scientist

BOX and DRAPER · Empirical Model-Building and Response Surfaces

*BOX and DRAPER · Evolutionary Operation: A Statistical Method for Process Improvement

BOX, HUNTER, and HUNTER · Statistics for Experimenters: An Introduction to Design, Data Analysis, and Model Building

BOX and LUCEÑO · Statistical Control by Monitoring and Feedback Adjustment

BRANDIMARTE · Numerical Methods in Finance: A MATLAB-Based Introduction

BROWN and HOLLANDER · Statistics: A Biomedical Introduction

BRUNNER, DOMHOF, and LANGER · Nonparametric Analysis of Longitudinal Data in Factorial Experiments

BUCKLEW · Large Deviation Techniques in Decision, Simulation, and Estimation

CAIROLI and DALANG · Sequential Stochastic Optimization

CHAN · Time Series: Applications to Finance

CHATTERJEE and HADI · Sensitivity Analysis in Linear Regression

CHATTERJEE and PRICE · Regression Analysis by Example, *Third Edition*

CHERNICK · Bootstrap Methods: A Practitioner's Guide

CHERNICK and FRIIS · Introductory Biostatistics for the Health Sciences

CHILÈS and DELFINER · Geostatistics: Modeling Spatial Uncertainty

CHOW and LIU · Design and Analysis of Clinical Trials: Concepts and Methodologies, *Second Edition*

CLARKE and DISNEY · Probability and Random Processes: A First Course with Applications, *Second Edition*

*COCHRAN and COX · Experimental Designs, *Second Edition*

CONGDON · Applied Bayesian Modelling

CONGDON · Bayesian Statistical Modelling

CONOVER · Practical Nonparametric Statistics, *Third Edition*

COOK · Regression Graphics

COOK and WEISBERG · Applied Regression Including Computing and Graphics

COOK and WEISBERG · An Introduction to Regression Graphics

CORNELL · Experiments with Mixtures, Designs, Models, and the Analysis of Mixture Data, *Third Edition*

COVER and THOMAS · Elements of Information Theory

COX · A Handbook of Introductory Statistical Methods

*Now available in a lower priced paperback edition in the Wiley Classics Library.

*COX · Planning of Experiments

CRESSIE · Statistics for Spatial Data, *Revised Edition*

CSÖRGŐ and HORVÁTH · Limit Theorems in Change Point Analysis

DANIEL · Applications of Statistics to Industrial Experimentation

DANIEL · Biostatistics: A Foundation for Analysis in the Health Sciences, *Eighth Edition*

*DANIEL · Fitting Equations to Data: Computer Analysis of Multifactor Data, *Second Edition*

DASU and JOHNSON · Exploratory Data Mining and Data Cleaning

DAVID and NAGARAJA · Order Statistics, *Third Edition*

*DEGROOT, FIENBERG, and KADANE · Statistics and the Law

DEL CASTILLO · Statistical Process Adjustment for Quality Control

DeMARIS · Regression with Social Data: Modeling Continuous and Limited Response Variables

DEMIDENKO · Mixed Models: Theory and Applications

DENISON, HOLMES, MALLICK and SMITH · Bayesian Methods for Nonlinear Classification and Regression

DETTE and STUDDEN · The Theory of Canonical Moments with Applications in Statistics, Probability, and Analysis

DEY and MUKERJEE · Fractional Factorial Plans

DILLON and GOLDSTEIN · Multivariate Analysis: Methods and Applications

DODGE · Alternative Methods of Regression

*DODGE and ROMIG · Sampling Inspection Tables, *Second Edition*

*DOOB · Stochastic Processes

DOWDY, WEARDEN, and CHILKO · Statistics for Research, *Third Edition*

DRAPER and SMITH · Applied Regression Analysis, *Third Edition*

DRYDEN and MARDIA · Statistical Shape Analysis

DUDEWICZ and MISHRA · Modern Mathematical Statistics

DUNN and CLARK · Basic Statistics: A Primer for the Biomedical Sciences, *Third Edition*

DUPUIS and ELLIS · A Weak Convergence Approach to the Theory of Large Deviations

*ELANDT-JOHNSON and JOHNSON · Survival Models and Data Analysis

ENDERS · Applied Econometric Time Series

ETHIER and KURTZ · Markov Processes: Characterization and Convergence

EVANS, HASTINGS, and PEACOCK · Statistical Distributions, *Third Edition*

FELLER · An Introduction to Probability Theory and Its Applications, Volume I, *Third Edition,* Revised; Volume II, *Second Edition*

FISHER and VAN BELLE · Biostatistics: A Methodology for the Health Sciences

FITZMAURICE, LAIRD, and WARE · Applied Longitudinal Analysis

*FLEISS · The Design and Analysis of Clinical Experiments

FLEISS · Statistical Methods for Rates and Proportions, *Third Edition*

FLEMING and HARRINGTON · Counting Processes and Survival Analysis

FULLER · Introduction to Statistical Time Series, *Second Edition*

FULLER · Measurement Error Models

GALLANT · Nonlinear Statistical Models

GHOSH, MUKHOPADHYAY, and SEN · Sequential Estimation

GIESBRECHT and GUMPERTZ · Planning, Construction, and Statistical Analysis of Comparative Experiments

GIFI · Nonlinear Multivariate Analysis

GLASSERMAN and YAO · Monotone Structure in Discrete-Event Systems

GNANADESIKAN · Methods for Statistical Data Analysis of Multivariate Observations, *Second Edition*

GOLDSTEIN and LEWIS · Assessment: Problems, Development, and Statistical Issues

GREENWOOD and NIKULIN · A Guide to Chi-Squared Testing

*Now available in a lower priced paperback edition in the Wiley Classics Library.

GROSS and HARRIS · Fundamentals of Queueing Theory, *Third Edition*
*HAHN and SHAPIRO · Statistical Models in Engineering
HAHN and MEEKER · Statistical Intervals: A Guide for Practitioners
HALD · A History of Probability and Statistics and their Applications Before 1750
HALD · A History of Mathematical Statistics from 1750 to 1930
HAMPEL · Robust Statistics: The Approach Based on Influence Functions
HANNAN and DEISTLER · The Statistical Theory of Linear Systems
HEIBERGER · Computation for the Analysis of Designed Experiments
HEDAYAT and SINHA · Design and Inference in Finite Population Sampling
HELLER · MACSYMA for Statisticians
HINKELMAN and KEMPTHORNE: · Design and Analysis of Experiments, Volume 1:
    Introduction to Experimental Design
HOAGLIN, MOSTELLER, and TUKEY · Exploratory Approach to Analysis
    of Variance
HOAGLIN, MOSTELLER, and TUKEY · Exploring Data Tables, Trends and Shapes
*HOAGLIN, MOSTELLER, and TUKEY · Understanding Robust and Exploratory
    Data Analysis
HOCHBERG and TAMHANE · Multiple Comparison Procedures
HOCKING · Methods and Applications of Linear Models: Regression and the Analysis
    of Variance, *Second Edition*
HOEL · Introduction to Mathematical Statistics, *Fifth Edition*
HOGG and KLUGMAN · Loss Distributions
HOLLANDER and WOLFE · Nonparametric Statistical Methods, *Second Edition*
HOSMER and LEMESHOW · Applied Logistic Regression, *Second Edition*
HOSMER and LEMESHOW · Applied Survival Analysis: Regression Modeling of
    Time to Event Data
HUBER · Robust Statistics
HUBERTY · Applied Discriminant Analysis
HUNT and KENNEDY · Financial Derivatives in Theory and Practice
HUSKOVA, BERAN, and DUPAC · Collected Works of Jaroslav Hajek—
    with Commentary
HUZURBAZAR · Flowgraph Models for Multistate Time-to-Event Data
IMAN and CONOVER · A Modern Approach to Statistics
JACKSON · A User's Guide to Principle Components
JOHN · Statistical Methods in Engineering and Quality Assurance
JOHNSON · Multivariate Statistical Simulation
JOHNSON and BALAKRISHNAN · Advances in the Theory and Practice of Statistics: A
    Volume in Honor of Samuel Kotz
JOHNSON and BHATTACHARYYA · Statistics: Principles and Methods, *Fifth Edition*
JOHNSON and KOTZ · Distributions in Statistics
JOHNSON and KOTZ (editors) · Leading Personalities in Statistical Sciences: From the
    Seventeenth Century to the Present
JOHNSON, KOTZ, and BALAKRISHNAN · Continuous Univariate Distributions,
    Volume 1, *Second Edition*
JOHNSON, KOTZ, and BALAKRISHNAN · Continuous Univariate Distributions,
    Volume 2, *Second Edition*
JOHNSON, KOTZ, and BALAKRISHNAN · Discrete Multivariate Distributions
JOHNSON, KOTZ, and KEMP · Univariate Discrete Distributions, *Second Edition*
JUDGE, GRIFFITHS, HILL, LÜTKEPOHL, and LEE · The Theory and Practice of
    Econometrics, *Second Edition*
JUREČKOVÁ and SEN · Robust Statistical Procedures: Aymptotics and Interrelations
JUREK and MASON · Operator-Limit Distributions in Probability Theory
KADANE · Bayesian Methods and Ethics in a Clinical Trial Design

*Now available in a lower priced paperback edition in the Wiley Classics Library.

KADANE AND SCHUM · A Probabilistic Analysis of the Sacco and Vanzetti Evidence
KALBFLEISCH and PRENTICE · The Statistical Analysis of Failure Time Data, *Second Edition*
KASS and VOS · Geometrical Foundations of Asymptotic Inference
KAUFMAN and ROUSSEEUW · Finding Groups in Data: An Introduction to Cluster Analysis
KEDEM and FOKIANOS · Regression Models for Time Series Analysis
KENDALL, BARDEN, CARNE, and LE · Shape and Shape Theory
KHURI · Advanced Calculus with Applications in Statistics, *Second Edition*
KHURI, MATHEW, and SINHA · Statistical Tests for Mixed Linear Models
*KISH · Statistical Design for Research
KLEIBER and KOTZ · Statistical Size Distributions in Economics and Actuarial Sciences
KLUGMAN, PANJER, and WILLMOT · Loss Models: From Data to Decisions
KLUGMAN, PANJER, and WILLMOT · Solutions Manual to Accompany Loss Models: From Data to Decisions
KOTZ, BALAKRISHNAN, and JOHNSON · Continuous Multivariate Distributions, Volume 1, *Second Edition*
KOTZ and JOHNSON (editors) · Encyclopedia of Statistical Sciences: Volumes 1 to 9 with Index
KOTZ and JOHNSON (editors) · Encyclopedia of Statistical Sciences: Supplement Volume
KOTZ, READ, and BANKS (editors) · Encyclopedia of Statistical Sciences: Update Volume 1
KOTZ, READ, and BANKS (editors) · Encyclopedia of Statistical Sciences: Update Volume 2
KOVALENKO, KUZNETZOV, and PEGG · Mathematical Theory of Reliability of Time-Dependent Systems with Practical Applications
LACHIN · Biostatistical Methods: The Assessment of Relative Risks
LAD · Operational Subjective Statistical Methods: A Mathematical, Philosophical, and Historical Introduction
LAMPERTI · Probability: A Survey of the Mathematical Theory, *Second Edition*
LANGE, RYAN, BILLARD, BRILLINGER, CONQUEST, and GREENHOUSE · Case Studies in Biometry
LARSON · Introduction to Probability Theory and Statistical Inference, *Third Edition*
LAWLESS · Statistical Models and Methods for Lifetime Data, *Second Edition*
LAWSON · Statistical Methods in Spatial Epidemiology
LE · Applied Categorical Data Analysis
LE · Applied Survival Analysis
LEE and WANG · Statistical Methods for Survival Data Analysis, *Third Edition*
LePAGE and BILLARD · Exploring the Limits of Bootstrap
LEYLAND and GOLDSTEIN (editors) · Multilevel Modelling of Health Statistics
LIAO · Statistical Group Comparison
LINDVALL · Lectures on the Coupling Method
LINHART and ZUCCHINI · Model Selection
LITTLE and RUBIN · Statistical Analysis with Missing Data, *Second Edition*
LLOYD · The Statistical Analysis of Categorical Data
MAGNUS and NEUDECKER · Matrix Differential Calculus with Applications in Statistics and Econometrics, *Revised Edition*
MALLER and ZHOU · Survival Analysis with Long Term Survivors
MALLOWS · Design, Data, and Analysis by Some Friends of Cuthbert Daniel
MANN, SCHAFER, and SINGPURWALLA · Methods for Statistical Analysis of Reliability and Life Data
MANTON, WOODBURY, and TOLLEY · Statistical Applications Using Fuzzy Sets
MARCHETTE · Random Graphs for Statistical Pattern Recognition

*Now available in a lower priced paperback edition in the Wiley Classics Library.

MARDIA and JUPP · Directional Statistics

MASON, GUNST, and HESS · Statistical Design and Analysis of Experiments with Applications to Engineering and Science, *Second Edition*

McCULLOCH and SEARLE · Generalized, Linear, and Mixed Models

McFADDEN · Management of Data in Clinical Trials

* McLACHLAN · Discriminant Analysis and Statistical Pattern Recognition

McLACHLAN, DO, and AMBROISE · Analyzing Microarray Gene Expression Data

McLACHLAN and KRISHNAN · The EM Algorithm and Extensions

McLACHLAN and PEEL · Finite Mixture Models

McNEIL · Epidemiological Research Methods

MEEKER and ESCOBAR · Statistical Methods for Reliability Data

MEERSCHAERT and SCHEFFLER · Limit Distributions for Sums of Independent Random Vectors: Heavy Tails in Theory and Practice

MICKEY, DUNN, and CLARK · Applied Statistics: Analysis of Variance and Regression, *Third Edition*

*MILLER · Survival Analysis, *Second Edition*

MONTGOMERY, PECK, and VINING · Introduction to Linear Regression Analysis, *Third Edition*

MORGENTHALER and TUKEY · Configural Polysampling: A Route to Practical Robustness

MUIRHEAD · Aspects of Multivariate Statistical Theory

MULLER and STOYAN · Comparison Methods for Stochastic Models and Risks

MURRAY · X-STAT 2.0 Statistical Experimentation, Design Data Analysis, and Nonlinear Optimization

MURTHY, XIE, and JIANG · Weibull Models

MYERS and MONTGOMERY · Response Surface Methodology: Process and Product Optimization Using Designed Experiments, *Second Edition*

MYERS, MONTGOMERY, and VINING · Generalized Linear Models. With Applications in Engineering and the Sciences

NELSON · Accelerated Testing, Statistical Models, Test Plans, and Data Analyses

NELSON · Applied Life Data Analysis

NEWMAN · Biostatistical Methods in Epidemiology

OCHI · Applied Probability and Stochastic Processes in Engineering and Physical Sciences

OKABE, BOOTS, SUGIHARA, and CHIU · Spatial Tesselations: Concepts and Applications of Voronoi Diagrams, *Second Edition*

OLIVER and SMITH · Influence Diagrams, Belief Nets and Decision Analysis

PALTA · Quantitative Methods in Population Health: Extensions of Ordinary Regressions

PANKRATZ · Forecasting with Dynamic Regression Models

PANKRATZ · Forecasting with Univariate Box-Jenkins Models: Concepts and Cases

*PARZEN · Modern Probability Theory and Its Applications

PEÑA, TIAO, and TSAY · A Course in Time Series Analysis

PIANTADOSI · Clinical Trials: A Methodologic Perspective

PORT · Theoretical Probability for Applications

POURAHMADI · Foundations of Time Series Analysis and Prediction Theory

PRESS · Bayesian Statistics: Principles, Models, and Applications

PRESS · Subjective and Objective Bayesian Statistics, *Second Edition*

PRESS and TANUR · The Subjectivity of Scientists and the Bayesian Approach

PUKELSHEIM · Optimal Experimental Design

PURI, VILAPLANA, and WERTZ · New Perspectives in Theoretical and Applied Statistics

PUTERMAN · Markov Decision Processes: Discrete Stochastic Dynamic Programming

*RAO · Linear Statistical Inference and Its Applications, *Second Edition*

RAUSAND and HØYLAND · System Reliability Theory: Models, Statistical Methods, and Applications, *Second Edition*

*Now available in a lower priced paperback edition in the Wiley Classics Library.

RENCHER · Linear Models in Statistics

RENCHER · Methods of Multivariate Analysis, *Second Edition*

RENCHER · Multivariate Statistical Inference with Applications

* RIPLEY · Spatial Statistics

RIPLEY · Stochastic Simulation

ROBINSON · Practical Strategies for Experimenting

ROHATGI and SALEH · An Introduction to Probability and Statistics, *Second Edition*

ROLSKI, SCHMIDLI, SCHMIDT, and TEUGELS · Stochastic Processes for Insurance and Finance

ROSENBERGER and LACHIN · Randomization in Clinical Trials: Theory and Practice

ROSS · Introduction to Probability and Statistics for Engineers and Scientists

ROUSSEEUW and LEROY · Robust Regression and Outlier Detection

RUBIN · Multiple Imputation for Nonresponse in Surveys

RUBINSTEIN · Simulation and the Monte Carlo Method

RUBINSTEIN and MELAMED · Modern Simulation and Modeling

RYAN · Modern Regression Methods

RYAN · Statistical Methods for Quality Improvement, *Second Edition*

SALTELLI, CHAN, and SCOTT (editors) · Sensitivity Analysis

*SCHEFFE · The Analysis of Variance

SCHIMEK · Smoothing and Regression: Approaches, Computation, and Application

SCHOTT · Matrix Analysis for Statistics

SCHOUTENS · Levy Processes in Finance: Pricing Financial Derivatives

SCHUSS · Theory and Applications of Stochastic Differential Equations

SCOTT · Multivariate Density Estimation: Theory, Practice, and Visualization

*SEARLE · Linear Models

SEARLE · Linear Models for Unbalanced Data

SEARLE · Matrix Algebra Useful for Statistics

SEARLE, CASELLA, and McCULLOCH · Variance Components

SEARLE and WILLETT · Matrix Algebra for Applied Economics

SEBER and LEE · Linear Regression Analysis, *Second Edition*

*SEBER · Multivariate Observations

SEBER and WILD · Nonlinear Regression

SENNOTT · Stochastic Dynamic Programming and the Control of Queueing Systems

*SERFLING · Approximation Theorems of Mathematical Statistics

SHAFER and VOVK · Probability and Finance: It's Only a Game!

SILVAPULLE and SEN · Constrained Statistical Inference: Order, Inequality and Shape Constraints

SMALL and McLEISH · Hilbert Space Methods in Probability and Statistical Inference

SRIVASTAVA · Methods of Multivariate Statistics

STAPLETON · Linear Statistical Models

STAUDTE and SHEATHER · Robust Estimation and Testing

STOYAN, KENDALL, and MECKE · Stochastic Geometry and Its Applications, *Second Edition*

STOYAN and STOYAN · Fractals, Random Shapes and Point Fields: Methods of Geometrical Statistics

STYAN · The Collected Papers of T. W. Anderson: 1943–1985

SUTTON, ABRAMS, JONES, SHELDON, and SONG · Methods for Meta-Analysis in Medical Research

TANAKA · Time Series Analysis: Nonstationary and Noninvertible Distribution Theory

THOMPSON · Empirical Model Building

THOMPSON · Sampling, *Second Edition*

THOMPSON · Simulation: A Modeler's Approach

THOMPSON and SEBER · Adaptive Sampling

THOMPSON, WILLIAMS, and FINDLAY · Models for Investors in Real World Markets

*Now available in a lower priced paperback edition in the Wiley Classics Library.

TIAO, BISGAARD, HILL, PEÑA, and STIGLER (editors) · Box on Quality and Discovery: with Design, Control, and Robustness

TIERNEY · LISP-STAT: An Object-Oriented Environment for Statistical Computing and Dynamic Graphics

TSAY · Analysis of Financial Time Series

UPTON and FINGLETON · Spatial Data Analysis by Example, Volume II: Categorical and Directional Data

VAN BELLE · Statistical Rules of Thumb

VAN BELLE, FISHER, HEAGERTY, and LUMLEY · Biostatistics: A Methodology for the Health Sciences, *Second Edition*

VESTRUP · The Theory of Measures and Integration

VIDAKOVIC · Statistical Modeling by Wavelets

VINOD and REAGLE · Preparing for the Worst: Incorporating Downside Risk in Stock Market Investments

WALLER and GOTWAY · Applied Spatial Statistics for Public Health Data

WEERAHANDI · Generalized Inference in Repeated Measures: Exact MANOVA and Mixed Models

WEISBERG · Applied Linear Regression, *Third Edition*

WELSH · Aspects of Statistical Inference

WESTFALL and YOUNG · Resampling-Based Multiple Testing: Examples and Methods for *p*-Value Adjustment

WHITTAKER · Graphical Models in Applied Multivariate Statistics

WINKER · Optimization Heuristics in Economics: Applications of Threshold Accepting

WONNACOTT and WONNACOTT · Econometrics, *Second Edition*

WOODING · Planning Pharmaceutical Clinical Trials: Basic Statistical Principles

WOOLSON and CLARKE · Statistical Methods for the Analysis of Biomedical Data, *Second Edition*

WU and HAMADA · Experiments: Planning, Analysis, and Parameter Design Optimization

YANG · The Construction Theory of Denumerable Markov Processes

*ZELLNER · An Introduction to Bayesian Inference in Econometrics

ZHOU, OBUCHOWSKI, and McCLISH · Statistical Methods in Diagnostic Medicine